AF698049

Mathematical Principles of Fuzzy Logic

THE KLUWER INTERNATIONAL SERIES IN ENGINEERING AND COMPUTER SCIENCE

MATHEMATICAL PRINCIPLES OF FUZZY LOGIC

VILÉM NOVÁK
University of Ostrava
Institute for Research and Applications of Fuzzy Modeling
Brafova 7, 701 03 Ostrava 1, Czech Republic

IRINA PERFILIEVA
Moscow State Academy of Instrument Making
Stromynka 20, 107846 Moscow, Russia

and

University of Ostrava
Institute for Research and Applications of Fuzzy Modeling
Brafova 7, 701 03 Ostrava 1, Czech Republic

JIŘÍ MOČKOŘ
University of Ostrava
Institute for Research and Applications of Fuzzy Modeling
Brafova 7, 701 03 Ostrava 1, Czech Republic

Kluwer Academic Publishers
Boston/Dordrecht/London

Distributors for North, Central and South America:
Kluwer Academic Publishers
101 Philip Drive
Assinippi Park
Norwell, Massachusetts 02061 USA
Telephone (781) 871-6600
Fax (781) 871-6528
E-Mail <kluwer@wkap.com>

Distributors for all other countries:
Kluwer Academic Publishers Group
Distribution Centre
Post Office Box 322
3300 AH Dordrecht, THE NETHERLANDS
Telephone 31 78 6392 392
Fax 31 78 6546 474
E-Mail <services@wkap.nl>

Electronic Services <http://www.wkap.nl>

Library of Congress Cataloging-in-Publication Data

Novák, Vilém.
Mathematical principles of fuzzy logic / Vilem Novak, Irina Perfilieva and Jiri Mockor.
p. cm. -- (The Kluwer international series in engineering and computer science ; secs 517)
Includes bibliographical references and index.
ISBN 0-7923-8595-0 (alk. paper)
1. Fuzzy logic. I. Perfilieva, Irina. 1953- . II. Močkoř, Jiří. III. Title. IV. Series.
QA9.64.N68 1999
511.3--dc21 99-37210
CIP

Copyright © 1999 by Kluwer Academic Publishers

All rights reserved. No part of this publication may be reproduced, stored in a retrieval system or transmitted in any form or by any means, mechanical, photo-copying, recording, or otherwise, without the prior written permission of the publisher, Kluwer Academic Publishers, 101 Philip Drive, Assinippi Park, Norwell, Massachusetts 02061

Printed on acid-free paper.

Printed in the United States of America

To David and Martin,
to Vitalij,
to Soňa, Marek and Kateřina,
with love

Contents

Preface

This book is an attempt to provide a systematic course of the formal theory of fuzzy logic. We made a lot of effort to be precise, but at the same time to explain the motivation and interpretation of all the results and, if possible, to accompany the theory by examples.

There are a lot of other books on various aspects of fuzzy logic. Our book is more specific from the point of view of several aspects. First, it is based on logical formalism demonstrating that fuzzy logic is a well developed logical theory. Second, it includes the theory of functional systems in fuzzy logic, which provide explanation of what, and how can be represented by formulas of fuzzy logic calculi. Third, except for the generalization of the classical way of interpretation within the environment of fuzzy sets constructed over classical sets, it also presents much more general interpretation of fuzzy logic within the environment of other proper categories of fuzzy sets stemming either from the topos theory, or even generalizing the latter. Last but not least, the leading philosophical point of view is presentation of fuzzy logic especially as the theory of vagueness as well as the theory of the common-sense human reasoning, which is based on the use of natural language, the distinguished feature of which is vagueness of its semantics.

We expect the book to be read by people interested in fuzzy logic and related areas, and also by logicians, mathematicians and computer scientists interested in mathematical aspects of fuzzy logic. It can be used in special courses of fuzzy logic, artificial and computational intelligence, in master and post-gradual university studies, in advanced courses on various applications including fuzzy control, decision-making and others.

The book is divided into eight chapters. The first chapter is introductory and it provides motivation for the development of fuzzy logic, describes its structure and outlines its potential for applications. Fuzzy logic is there divided into that in narrow sense (FLn), which is a special many-valued logic aiming at description of the vagueness phenomenon and that in broader sense (FLb), whose aim is to provide a formal theory for modeling of natural human deduction based on the use of natural language. It is argued that characterization of the vagueness phenomenon is fundamental for further development of fuzzy logic as well as its applications.

The second chapter is an overview of the basic algebraic concepts necessary for characterization of the structure of truth values and for understanding to the subsequent chapters.

The third chapter briefly reminds basic concepts of classical logic, and also the notation which is employed further.

The main chapters of the book are fourth to seventh. The fourth chapter contains explanation of fuzzy logic in narrow sense. We confine ourselves mainly to fuzzy logic based on the Łukasiewicz algebra of the truth values. Many reasons have been given arguing that this requirement is a necessary consequence of the assumptions, which seem to be natural and convincing. Among them, essential are continuity of the connectives (this follows from the conviction that vagueness phenomenon requires continuity) and completeness (balance between syntax and semantics). We prove lemmas and theorems characterizing behaviour of FLn, including deduction, contradiction, and others. The main result is the completeness theorem. Besides this, we also outline fundamentals of the model theory and discuss questions concerning computability.

The fifth chapter describes functional systems related to propositional and predicate calculi of fuzzy logic. A notable role here is played by the well known McNaughton theorem. We provide a constructive proof of it and then, deduce a canonical representation of formulas of propositional fuzzy logic as well as some special representation of formulas of predicate fuzzy logic. A special attention has been paid to the generalization of disjunctive and conjunctive normal forms into fuzzy logic. Among the main properties of them, the ability to represent approximately continuous functions including evaluation of the quality of the approximation has been investigated.

The sixth chapter extends the previous theory to obtain a theory of parts of natural language to constitute fuzzy logic in broader sense. We present a formalization of the concepts of intension and extension, evaluating linguistic syntagms and linguistic description (a set of IF-THEN rules), and demonstrate some theorems characterizing their behaviour.

The seventh chapter is more abstract and deals with fuzzy sets and fuzzy logic within category theory. First, we present properties of three possible categories of fuzzy sets. The subsequent sections provide the reader with a general picture of fuzzy logic interpreted in topos and its place in topos logic. We tried to show two possible approaches. The first one is based on topos theory and Heyting algebra structures defined on the set of subobjects. The second one is based on the MV-algebra structures defined on these sets. In this book, we present some of the results and consequences, which could be useful from the point of view of fuzzy set theory. In particular, we tried to show explicitly similarities between the obtained categories and topoi, and to show how the internal logic in these categories could be directly developed.

The book is concluded by a short, eighth chapter giving a brief overview of the history of fuzzy logic and outlining some of its actual problems to be solved in the future.

Sections 2.4, 3.4 and Chapter 7 have been written by J. Močkoř. Chapter 5 has been written by I. Perfilieva and Chapters 4 and 6 by V. Novák. The rest has has been written by by the latter two authors together.

This book has been prepared mostly in the Institute for Research and Applications of Fuzzy Modeling of the University of Ostrava, Czech Republic, which has been established on the basis of the project VS 96037 of the Ministry of Education, Youth and Physical Training of the Czech Republic. Some parts have been prepared also in the Moscow State Academy of Instrument-Making and Informatics, Russia, and Institute of the Theory of Information and Automation of the Academy of Sciences of the Czech Republic.

We wish to express our thanks to all who helped us in the preparation of this book. On the first place, we thank to people whose work has been for us the main source of knowledge and inspiration, namely P. Hájek from UI AS ČR, U. Höhle from the University of Wuppertal, D. Mundici form the University of Milano and A. diNola from the University of Salerno.

Further, we thank to the people who have read the manuscript and helped us to improve it on many places, namely to R. Bělohlávek and A. Dvořák from IRAFM, University of Ostrava, and to S. Gottwald from the University of Leipzig.

We want also to thank to our Chinese friends, M. Ying and G. Chen from Beijing Tschingua University, and G. Wang from ShaanXi Normal University for their hospitality, which helped to write some parts of this book during the stay of the first two authors in China. Last but not least, we want to thank to Prof. T. Kreck, director of the Institut für Mathematik in Oberwolfach, and also to other workers of it for providing us unexceptionable working conditions in which significant portions of the book have been written.

Ostrava, May 1999

Vilém Novák, Irina Perfilieva, Jiří Močkoř

1 FUZZY LOGIC: WHAT, WHY, FOR WHICH?

The discussion about the philosophical background and the role of fuzzy logic in science has not been finished till now. Since this book focusses on the mathematical principles of fuzzy logic, it is necessary to explain our point of view on the questions and concepts which led to it. Our aim is to demonstrate the reader that the key role in the roots of fuzzy logic is played by the vagueness phenomenon, which gives rise to problems not addressed by classical logic, but which are important to solve. For this purpose, we divided the chapter as follows.

In Section 1.1, we analyze two complementary phenomena, namely uncertainty and vagueness. A detailed discussion of both of them should reveal the reader that fuzzy logic deals with the latter while probability and few other theories[1] deal with the former. Section 1.2 continues discussion on vagueness and demonstrates the way how it is captured by fuzzy set theory, which is unseparably joined with fuzzy logic. Sections 1.3 and 1.4 give answer to the questions, what is fuzzy logic, what is its agenda and what is it useful for.

Fuzzy set theory as well as fuzzy logic originated from ideas of Lotfi A. Zadeh. His seminal paper [136] published in 1965 contains the concept of fuzzy set and establishes basic principles of its theory. The most elaborated form of the basic idea of fuzzy logic can be found in [140, 141].

[1] In connection with fuzzy logic, one may meet, for example, the possibility theory. As will be seen, this addresses uncertainty rather than vagueness.

1.1 Vagueness and Uncertainty

The discussion in this section focuses on two phenomena whose importance in science raised especially in this century, namely *uncertainty* and *vagueness*. Both these concepts characterize situations in which we regard[2] phenomena surrounding us; they are concerned with the amount of knowledge we have at disposal (or can have at disposal), which, however, is limited (mostly in principle). Our main goal is to show that these phenomena are two, rather complementary facets of a more general phenomenon which we will call *indeterminacy*[3].

Uncertainty. This phenomenon[4] emerges due to the *lack of knowledge* about the *occurrence* of some *event*. It is encountered when an experiment (process, test, etc.) is to proceed, the result of which is not known to us. Let us stress that there is no uncertainty after the experiment was realized and the result is known to us.

Note that the word "occurrence" inherently involves time, i.e. uncertainty is always connected with the question whether the given event may be regarded within some time period, or not. This becomes apparent on the typical example with tossing a player's cube. The phenomenon to occur is *the number of dots on the cube* and it occurs after the experiment (i.e. tossing the cube one times) has been realized.

A specific form of uncertainty is *randomness*. This has first been touched in science by the probability theory which has been founded by Jakob Bernoulli (1654-1705). He himself has not abandoned the deterministic point of view and took his theory as an attempt to characterize our expectation of the process of occurring phenomena in the world. Randomness was considered to raise mainly due to extensive number of outer influences causing ambiguity in the result, thus preventing us to describe it as deterministic. As late as in the 20th century, mainly in connection with important discoveries in atomic physics and quantum mechanics, randomness had been recognized to be unavoidable and a deep feature of the nature. In quantum mechanics , the concept of randomness is fundamental since the theory provides only the probability of results.

As already outlined, the mathematical model (i.e. quantified characterization) of the uncertainty phenomenon is provided especially by the *probability theory*. In everyday terminology, probability can be thought of as a numerical measure of the likelihood that a particular event will occur. Recall that the

[2] By "regarding" we mean learning the world in the most general way, i.e. by senses as well as by reason.

[3] We have chosen the term "indeterminacy" to encompass both phenomena discussed below, namely uncertainty as well as vagueness. This term, in the meaning very close to ours, can be found also in other sources, for example in Merriam Webster or Encyclopedia Brittanica. Note that in the literature, for example on fuzzy set theory, one may meet the term "uncertainty" in our meaning of "indeterminacy".

[4] The concept of "phenomenon" is left unexplained and taken as primitive. We suppose that the reader has some idea of what phenomenon is and we will better give some examples than attempt at providing a kind of exhausting philosophical explanation which, all in all, still might turn out to be unsatisfactory.

probability values are taken from the scale $[0, 1]$ where values near 1 indicate that an event is *likely to occur* while those near 0 indicate the opposite. The probability of 0.5 means that an event is equally likely to occur as well as not to occur. There are also other mathematical theories addressing the uncertainty phenomenon, for example the possibility theory, belief measures and others.

Vagueness. This phenomenon raises during the process of grouping together objects[5] having some property φ (of objects). Its result will be called a *grouping* of objects. In a slightly formal way, the grouping can be written as

$$X = \{x \mid x \text{ has the property } \varphi\} \tag{1.1}$$

where x varies over objects.

It is important to stress that, in general, the grouping X *cannot be taken as a set* since the property φ may not make us possible to characterize the grouping X precisely and unambiguously; there can exist *borderline* elements x for which it is unclear whether they have the property φ (and thus belong to X), or not.

For example, is it possible to imagine "all tall people", "all beautiful flowers on the meadow", "all witty novels", etc? These are typical examples of vaguely formed groupings of objects put together using, sometimes extremely complicated, property (try to define "beautiful flower"!). On the other hand, it is always possible to characterize, at least some *typical objects*, i.e. objects having typically the property in concern. For example, everybody can show a "blue sweater" or "huge building" but it is impossible to show "all huge buildings". The objects, which can be unambiguously decided to have φ, will be called *prototypes*. In general, we say that X, which is delineated using a property enabling borderline cases, has *unsharp boundaries*.

The following example will often be brought back in the sequel.

EXAMPLE 1.1

A typical vague property is *to be a small natural number*. Can we imagine all the small natural numbers? Clearly, 0 is small, 1 is small as well, etc. But where does this sequence finish? The only sure fact is that there exists a number, say $1,000,000,000$, which is not small.

An attempt at sharp (exact) explanation would mean to be able to find a small natural number n, $0 < n < 1,000,000,000$ such that $n + 1$ is not small. However, such a conclusion can hardly be defended. If n is small then $n+1$ must

---

[5]The concepts of object and property are also taken as primitive. P. Vopěnka in [130] provides the following characterization.

An *object* is a phenomenon which we separate in the world, i.e. a phenomenon to which we concede its individuality making it distinct from the other phenomena. Objects are usually accompanied by other *kind* of phenomena called *properties*. The same property may be applicable to more than one object. Thus, for example, a "pen" is an object accompanied by the properties "to be blue, narrow, fine", etc., but we may also have "blue flowers", "blue ink", "blue eyes", etc. It is natural for the human mind to construe as objects so much phenomena as possible. Thus, properties may be also often viewed as objects (cf., for example, "roundness, colour, length", etc.).

be small, as well. Hence, there is *no last small* number before $1,000,000,000$ and *no first number* $n < 1,000,000,000$ which is *not small.* We can distinguish small numbers from big ones but we are not able to say unambiguously about *each* number whether it is small or not. The property of "small" is vague and small numbers disappear inside the sequence of numbers ranging from 0 to 1,000,000,000. Consequently, they form a vague grouping of numbers. □

Vagueness has not been investigated so long as uncertainty. The interest in it appeared in science as late as in this century. One of the first philosophical papers about it has been published by B. Russell in 1923 (see [117]). Another important one [10] was written by M. Black 14 years later. However, the real interest came only after the fuzzy set theory had been founded. Let us stress that vagueness stems from the way how people regard the world and phenomena in it. Our mind forms concepts via some kind of "idealization" process. However, the nature itself is only in a partial coincidence with such idealization and namely, there is no vagueness in the nature.

Vagueness is an opposite to exactness and we argue that it cannot be avoided in the human way of regarding the world. Any attempt to explain an extensive detailed description necessarily leads to using vague concepts since precise description contains abundant number of details. To understand it, we must group them together — and this can hardly be done precisely. It is likely that the explanation would significantly rely on the use of natural language since vagueness is often connected with its use. However, the problem lays deeper, in the way how people regard the phenomena around them.

We may thus conclude that the increase of exactness leads to an increase in the amount of information whose relevancy then decreases until a point is reached after which the preciseness and relevancy are mutually excluding characteristics. This is the famous *incompatibility principle* described by L. A. Zadeh in [139]. It demonstrates that vagueness is necessary to convey relevant information.

Vagueness should be distinguished from generality and from ambiguity. In our terms, to be more general means to take into account more (various) groupings of objects, while ambiguity occurs in the language when more alternative meanings are assigned to the same word expression. For example, "colour" is more general than "red", since the former concerns various groupings of objects being red, green, etc. and each concrete colour represents a vague property.

Continuity of vagueness. A typical feature of vagueness is its continuity. This means that if an object has a vague property and another one differs very little from it, then it must have the same property. In other words, a small difference between objects cannot lead to abrupt change in the decision of whether either of them has, or has not a vague property. The transition from having a (vague) property to not having it is smooth.

As an example, let us recall from [10] the "museum of applied logic" consisting of a series of "chairs". It begins with a typical chair and ends up with a lump of wood apparently not being a chair. Each two neighbouring chairs in the series differ by some small piece cut from one of them. But since this piece can be arbitrarily small, there are no two neighbouring members in this series such that one is a chair and the second one is not. Consequently, the meaning of "chair" is vague and the transition from chair to non-chair is continuous. Let us stress that the continuity feature of the vagueness phenomenon will play an important role throughout this book.

A mathematical theory of the vagueness phenomenon is most successfully provided by fuzzy logic. Another deep mathematical theory addressing, among others, also vagueness is the Alternative Set Theory (AST) (see [129, 130] and possibly also [89]).

Comparison of uncertainty and vagueness. Let us now compare both discussed phenomena, which, as noted above, are two facets of the indeterminacy. It is principal for the uncertainty that occurrence of some phenomenon (one of many ones) is a result of an experiment. The substance of the former is irrelevant. More phenomena may occur, but we have not enough knowledge to determine, which one. The situation may thus be characterized as a classification of one result to fall into some of many classes.

Vagueness concerns the way how the *phenomenon itself* is delineated and disregards whether it occurs or not. The phenomenon is represented by "one" unsharp grouping of "many" objects. Hence, we may briefly say that vagueness raises as *many-to-one* relation while uncertainty raises as *one-to-many.* It has to be stressed that in the reality, we encounter indeterminacy *with both its facets* present.

The difference between vagueness and uncertainty may also be characterized from another side on the basis of the conflict between *actuality* and *potentiality* as discussed in the set theory. In classical set theory, every set is understood to be *actualized*, i.e. we imagine all its elements to be already existing and at disposal to us in one moment. This concerns both finite as well as infinite sets. Though we can always see only part of the infinite set, our reasoning about any set stems from the assumption that it is at our disposal as a whole. On the other hand, most events around us are only *potential*, i.e. they may, but need not, to occur or happen.

Thence, the difference between actuality and potentiality corresponds to the difference between vagueness and uncertainty. Vagueness stems from the *actualized* non-sharply delineated groupings while uncertainty is encountered when dealing with still non-actualized ranges of objects. The latter means that we may speculate about the range X, but only part of it indeed exists. Furthermore, once an actualized (i.e. already existing) grouping of objects from the range X is at our disposal, it has sense to speak only about the *truth* of the fact that some element belongs to it. Indeed, let an object y we are thinking of be created. If we learn that it has a property φ (cf. (1.1)), we know that it falls into X[6], i.e. we know that it is true that $y \in X$. However, in general it is *uncertain* whether y *will be* created (will exist) or not and thus, we cannot speak about the truth of $y \in X$.

[6]More precisely, it falls into the existing part of X.

1.2 Vagueness and Fuzzy Sets

In the previous section, we have widely discussed the vagueness phenomenon, which for a long time has been neglected by the European science. "Black-and-white" reasoning is apparent especially in mathematics: either the fact holds, or it does not, and nothing else. In science, we cannot avoid this way of thinking as any kind of releasing from it led to serious mistakes in the past. On the other hand, strictly two-valued thinking brings us nearer to the world of Platon than to the common-day-to-day life. The latter is not so simple. There are no purely good or purely bad people; if few drops fall down then it is difficult to say whether it rains; and even white snow is often a little black (especially in industrial towns). Vagueness is hidden in the way how people regard phenomena surrounding them.

Mathematics faces the problem, how vague groupings of objects can be characterized. Classical set theory does not carry any vagueness since there was no interest in it. *Fuzzy set* theory is an attempt to find such an approximation of vague groupings which in situations where, for example, natural language plays a significant role, would be more convenient than classical set theory.

Graded approach. Throughout this book, we will follow a very general principle, which we will call the *graded approach*, or as a kind of jargon, the *fuzzy approach*. This means the use of a scale when characterizing a relation between object and its property. The graded approach seems to be a general principle of the human mind, which uses it when trying to specify whether the object possesses the property fully or only partly, though the given property is vague. For example, we often say "almost white dress", "very strong engine", "too unpleasant situation", etc. In all these examples, we may encounter hidden degrees of intensity of the property in concern.

The graded approach will be mathematized by means of a special scale being an ordered set. In order to be sufficiently general and to be able to capture the continuity feature of vagueness, we will suppose that this set is uncountable. Furthermore, to be able to represent various manipulations with the properties partly possessed by objects, we will endow the scale by additional operations. The result is a specific algebra. There are several classes of algebras used as scales in the graded approach. They will be studied in Chapter 2.

The result of the graded approach to vague groupings are fuzzy sets. We present some basic ideas of the fuzzy set theory in the next subsection.

Fuzzy set theory. The decision whether the concrete object x has the property φ (e.g. "small", "tall", "light", "blue", etc.) is tantamount to the question whether it is *true* that x has it (in positive case, we will usually write $\varphi(x)$). However, such a question cannot be unambiguously answered. A reasonable solution consists in using some kind of a scale whose elements would express various *degrees of truth* of $\varphi(x)$.

Let U be a sufficiently large set from which we take the objects x. This set is called the *universe*. Note that this assumption is not restrictive since such a set always exists. For example, when talking about height of people (in cm), we may put $U = \{x \in \mathbb{R} \mid 0 \leq x \leq 300\}$. Furthermore, let L be a scale of truth values having the smallest $\mathbf{0}$ and greatest $\mathbf{1}$ elements, respectively. We usually put $L = [0,1]$ but this is only an unnecessary commonsense caused by the fact that this interval is very natural and transparent. Thus, $\mathbf{1}$ expresses that $\varphi(x)$ holds (x has the property φ) without any doubt while $\mathbf{0}$ means that $\varphi(x)$ does not hold at all. The other values mean that $\varphi(x)$ holds only partly.

DEFINITION 1.1
The fuzzy set *A is identified with a function*

$$A : U \longrightarrow L \tag{1.2}$$

assigning a value $A(x) \in L$ to each element $x \in U$. The value $A(x) \in L$ is the membership degree *of x in A.*

The function (1.2) is also called the *membership function* of the fuzzy set A. We will therefore use the same symbol for both and write $A \underset{\sim}{\subset} U$ if A is a fuzzy set in the universe U. Explicitly, we will write down the fuzzy set as

$$\{\, {}^{A(x)}/x \mid x \in U\} \tag{1.3}$$

where the couple ${}^{A(x)}/x$ means "the element x belongs to A with the membership degree $A(x)$", $A(x) \in L$. The value $A(x)$ expresses the *degree of truth* that the element x belongs to A.

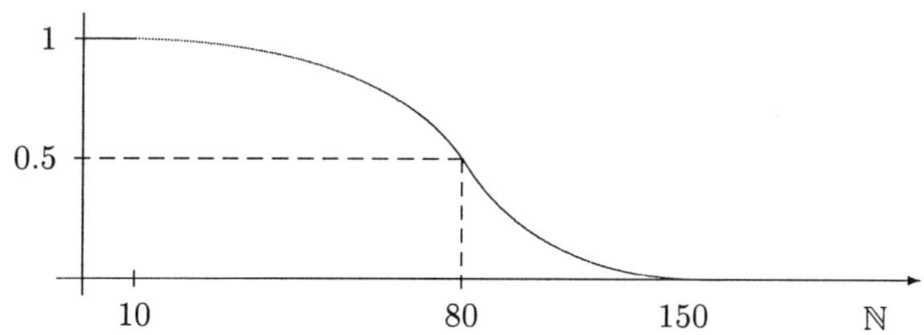

Figure 1.1. Membership function of the fuzzy set of "small numbers".

Recall Example 1.1 of the vague grouping of *small numbers*. In the fuzzy set theory, it can be approximated by the fuzzy set given by the (continuous) membership function depicted on Figure 1.1. In this picture, the numbers smaller than or equal to 10 are "surely small", i.e. they are prototypes of the property "small". Then the *truth* of the fact that the given number x is *small* diminishes until $x = 150$ beyond which it completely vanishes. Thus, the fuzzy set of small numbers can be written as

$$\text{Small}(x) = \{\, {}^{1}/0, \ldots, {}^{1}/10, \ldots, {}^{0.5}/80, \ldots, {}^{0.1}/149\}.$$

Of course, the concrete numbers depend on the context in which we consider the property of "being a small number". However, the course of the membership function is the same in all contexts.

Note that the membership function is a generalization of the characteristic function of the ordinary set. The crucial difference between the ordinary set and fuzzy set is thus in using of the scale L.

Few basic concepts of fuzzy set theory. Let us now very briefly overview some basic notions of the fuzzy set theory. More details can be found in the extensive literature (see e.g.[22, 58, 85, 87], or from newer ones [57]).

Let the fuzzy set $A \underset{\sim}{\subset} U$ be given by the membership function (1.2) where $L = [0,1]$. Let $A, B \underset{\sim}{\subset} U$. Then $A \subseteq B$ if $A(x) \leq B(x)$ holds for all $x \in U$. The set of all fuzzy sets on U is

$$\mathcal{F}(U) = \{A \mid A \underset{\sim}{\subset} U\} = L^U.$$

Note that $\mathcal{F}(U)$ contains also all ordinary subsets of U.

Given the fuzzy set $A \underset{\sim}{\subset} U$, it can be characterized in several ways. The following are important ordinary sets defined on the basis of A.

(i) *Support*

$$\text{Supp}(A) = \{x \mid A(x) > 0\}, \tag{1.4}$$

(ii) *a-cut*

$$A_a = \{x \mid A(x) \geq a\}, \qquad a \in (0,1], \tag{1.5}$$

(iii) *Kernel*

$$\text{Ker}(A) = \{x \mid A(x) = 1\}. \tag{1.6}$$

The fuzzy set A is *normal* if $\text{Ker}(A) \neq \emptyset$. The *empty fuzzy set* is defined by

$$\emptyset = \{\,{}^0/x \mid x \in U\}.$$

Obviously, this is the ordinary empty set.

The basic operations with fuzzy sets are defined using their membership functions. Given fuzzy sets $A, B \underset{\sim}{\subset} U$.

(i) *Union* of A and B is a fuzzy set $A \cup B \underset{\sim}{\subset} U$ with the membership function

$$C = A \cup B, \quad \text{iff} \quad C(x) = A(x) \vee B(x), \qquad x \in U \tag{1.7}$$

where $\vee$ denotes 'supremum'.

(ii) *Intersection* of A and B is a fuzzy set $A \cap B \underset{\sim}{\subset} U$ with the membership function

$$C = A \cap B, \quad \text{iff} \quad C(x) = A(x) \wedge B(x), \qquad x \in U \tag{1.8}$$

where $\wedge$ denotes 'infimum'. Both union as well as intersection of two fuzzy sets may be extended to arbitrary classes of fuzzy sets provided that suprema and infima of arbitrary sets exist in L.

(iii) *Complement* $\bar{A}$ is the fuzzy set

$$\bar{A}(x) = \neg A(x) = 1 - A(x), \qquad x \in U. \tag{1.9}$$

Note that the above definition of the basic operations with fuzzy sets is not the only possibility. The operations of infimum, supremum and negation used in equations (1.7)–(1.9) belong to more general classes of operations, which are analyzed in Section 2.3.

We finish with the definition of the concept of a fuzzy relation which generalizes the classical concept of relation and which is quite often used in the sequel.

DEFINITION 1.2
Let $U_1, \ldots, U_n$ be sets. An n-ary fuzzy relation R is a fuzzy set

$$R \underset{\sim}{\subseteq} U_1 \times \cdots \times U_n.$$

1.3 What Is Fuzzy Logic

The term "fuzzy logic" has been used since the late sixties. At first, it had the meaning of any logic possessing more than two truth values. Later on, after the famous paper of L. A. Zadeh [141] it received two other meanings, namely the *theory of approximate reasoning* and the *theory of linguistic logic.* The latter, somewhat marginal theory, is one of logics whose truth values are expressions of natural language (for example, *true, more or less true*, etc.). The former is the main most often used meaning[7].

In general, fuzzy logic can be characterized as the *many-valued logic with special properties aiming at modeling of the vagueness phenomenon and some parts of the meaning of natural language via graded approach.* L. A. Zadeh formulates paradigm of fuzzy logic in the preface of [71] as follows: "In a narrow sense, fuzzy logic, FLn, is a logical system which aims at formalization of approximate reasoning. In this sense FLn is an extension of multivalued logic. However, the agenda of FLn is quite different from that of traditional multivalued logics. In particular, such key concepts as the concept of a linguistic variable, canonical form, fuzzy if-then rule, fuzzy quantification and defuzzification, predicate modification, truth qualification, the extension principle, the compositional rule of inference and interpolative reasoning, among others, are not addressed in traditional systems."

In this book, we will differentiate fuzzy logic more subtly and distinguish fuzzy logic in narrow and broader sense. *Fuzzy logic in narrow sense*, FLn, is a special many-valued logic which aims at providing formal background for the graded approach to vagueness. *Fuzzy logic in broader sense*, FLb, is an extension of FLn and it aims at developing mathematical model of natural human reasoning, in which principal role is played by the natural language.

[7]Unfortunately, nowadays the meaning of the term "fuzzy logic" is often blurred by interpreting it as any kind of formalism based on the fuzzy set theory, which is used in the applications. We will avoid such understanding in this book.

It must be stressed that, in the same way as classical logic, fuzzy logic presented in this book is mostly *truth functional.* This means that the truth value of a compound formula is a function of the truth values of its constituents. We have no significant reason to abandon this assumption.

In general, fuzzy logic is the result of the graded approach to the formal logical systems. It is important to stress that this is not a futile result of striving for endless generalization. Due to the graded approach, fuzzy logic provides solution of some, classically non-solvable problems. For example, there are well known ancient paradoxes of *sorites* [8] (heap) and *falakros* (bald men). Let us remind them briefly.

> *One grain of wheat does not make a heap. Neither make it two grains, three, etc. Hence, there are no heaps.*

The essence of the falakros paradox is the same.

> *A man having no hair, or only one hair is apparently bald. The same holds for a man with two hair, etc. Hence, all men are bald.*

These paradoxes rise if we understand the properties "to be a heap" and "to be bald" sharply, i.e. if we neglect their vagueness. Note that both these properties have the same character as the property *to be small* applied to natural numbers, which has already been discussed in Example 1.1.

How vagueness of the these properties acts in the logical reasoning? Apparently, classical two-valued logic is not able to cope with it. The situation is as follows.

Let $\mathbb{FN}(x)$ denote the proposition "the number x is small" (x be natural). The problem lays in the truth verification of implication $\mathbb{FN}(x) \Rightarrow \mathbb{FN}(x+1)$, i.e. classical induction cannot be applied to vague property. If the latter is taken classically (i.e. absolute true) then starting from $\mathbb{FN}(0)$ we necessarily arrive to the above paradoxes. However, for different subsequents of x, the verification of the proposition $\mathbb{FN}(x) \Rightarrow \mathbb{FN}(x+1)$ is different. In general, verification that x is small does not imply that we will be able to verify that also $x+1$ is small with the same effort. For example, if we verify that 1000 is small by counting one thousand lines then verification that 1001 is also small means that we must count one line more, i.e. our effort to verify that 1001 is small is a little (imperceptibly) greater than that for 1000. Consequently, the above implication is not fully convincing. A solution offered by fuzzy logic is to assume that the implication $\mathbb{FN}(x) \Rightarrow \mathbb{FN}(x+1)$ is true only in some degree close to 1, say $1-\varepsilon$ where $\varepsilon > 0$. Then sorites (as well as falakros) paradox disappears. We will give precise formulation of this reasoning in Chapter 4. Other example, where FLn offers an interesting solution, is *the liar paradox* (see [43]).

---

[8]The name "sorites" derives form the Greek word "soros" which means "heap" and originally refers to a puzzle attributed to Megarian logician Eubulides of Miletus, who formulated also other known puzzles such as *The Falakros* or *The Liar.* In the up-to-date logic, the name *sorites paradox* refers to a sequence of syllogisms starting with a true statement and ending with a false one.

Due to the discussion about vagueness phenomenon and the graded (fuzzy) approach to it, we will consistently take the latter as the core of generalization of as much classical concepts as possible. Most formal logical systems consist of two levels: syntax and semantics. The consistent graded approach will lead us to introduce grades in both levels. Hence, we will deal with the graded consequence operation, and furthermore, with evaluated formulas, fuzzy sets of axioms, degree of provability, graded interpretation of formulas in the models, etc.

The formal apparatus developed in this book provides us the following outcome. *First*, because graded approach serves us for modeling of the vagueness phenomenon, we are able to deal with *any possible truth value* on both levels, namely syntactical (make graded syntactical derivations) as well as semantical (evaluate truth degrees of formulas in the interpretation). Note that unlike other multivalued logics, all truth values in FLn are *equal in their importance*, i.e. there are no designated truth values.

Second, FLn and especially FLb as the extension of the latter seem to be working theories, which may be used as formal apparatus explicating the approximate reasoning schemes. We will deal with this question in Chapters 5 and 6.

Third, FLn is a generalization of classical logic. It opens different, more general way for explanation of, at least some of the classical problems. We may also expect new problems of classical mathematics influenced directly by FLn.

Our way of introducing the graded approach both in syntax as well as semantics is not the only possibility. For example, P. Hájek in his book [41] uses only classical syntax, because he has shown that the graded syntax of FLn can be modeled within classical Łukasiewicz logic. However, to be able to realize this, he uses specific formulas and modified understanding of the syntactical consequence.

The duality in generalization is faced here: either we take the graded approach and show that all classical concepts are special case of it, or take classical concepts and show that the graded approach can be embedded within them. In this book, we took the former view while P. Hájek took in his book the latter one.

The development of fuzzy logic is far from being complete. Since it has a specific agenda, which, as will be seen later, is closely connected with modeling of the semantics of some parts of natural language, various its extensions are possible. Let us mention some of them.

- *Extension by new connectives* (unary, binary, or other kinds of them). This is motivated by the necessity to interpret the so called *close coordination* in natural language (roughly speaking, the connection of adjectives with nouns) locally, i.e. with respect to the linguistic context. For example, the linguistic connective 'and' cannot be interpreted by only one kind of conjunction in all situations. In fuzzy logic, generalization of the conjunction is represented by the, so called, t-norms (see Chapter 2), which are certain binary operations on the interval $[0, 1]$. The concrete connective depends on the mutual relation between the conjuncts, thus leading to the use various t-norms (minimum, product, Łukasiewicz product, or other kind).

Furthermore, this reveals also the possibility to model the meaning of the *linguistic modifiers* (words such as "very, roughly", etc.) and the *linguistic quantifiers* (words such as "most, a lot of", etc.). In both cases, significant role may be played by specific unary connectives. It is interesting to study conditions under which it is possible to extend FLn by additional connectives without harming its basic properties.

- *Extension by modalities* (possible, necessary, usual, etc.), i.e. pushing fuzzy logic also towards modeling of some acts of the uncertainty phenomenon. The goal is to include them in such a way that modalities would naturally become part of the whole system of FLn.
- *Extension by nonstandard inference rules.* Such rules may provide more sophisticated inference schemes of human reasoning. For example, a problem related to nonstandard rules is realization of reasoning schemes not well elaborated so far, namely *abduction* (deriving assumptions from conclusions), default reasoning, and others. These schemes seem to fit well the program of the development of fuzzy logic.

This list is by no means finished and we challenge the reader to take it rather as an outline of possible ways of extension of fuzzy logic (not all of them are encompassed in this book), and to search his own ones.

1.4 Outline of the Agenda of Fuzzy Logic

This section is devoted to some items of the L. A. Zadeh's fuzzy logic agenda mentioned in Section 1.3. Our aim is to prepare some notions used later and to give the reader a clearer understanding to what we are speaking about. The section may be omitted in first reading.

Linguistic variable. One of the fundamental concepts introduced by L. A. Zadeh in [141] is that of *linguistic variable*. It is the quintuple

$$\langle \mathcal{X}, T(\mathcal{X}), U, G, M \rangle,$$

where $\mathcal{X}$ is the *name of the variable*, $T(\mathcal{X})$ is the set of its values (term set) which are linguistic expressions (*syntagms*[9]), U is the universe, G syntactical rule using which we can form syntagms $\mathcal{A}, \mathcal{B}, \ldots \in \mathcal{T}(\mathcal{X})$, and M is semantical rule, using which every syntagm $\mathcal{A} \in T(\mathcal{X})$ is assigned its meaning being a fuzzy set A in the universe U, $A \underset{\sim}{\subset} U$.

A typical example of the linguistic variable is $\mathcal{X} := Age$. Its term set $T(Age)$ consists of the syntagms such as *young, very young, medium age, quite old, more or less young, not old but not young*, etc. The universe $U \subseteq \mathbb{R}$ is some

[9]A syntagm is any part of the sentence having a special meaning and formed according to the grammatical rules.

set of real numbers (note that we may speak about age of various things). The syntactic rule G may be a context-free grammar. The semantic rule M assigns meaning to the terms from $T(\text{Age})$ being various modifications of fuzzy sets depicted on Fig. 1.2. Clearly, there is a lot of other linguistic variables, such as "height, size, temperature, press", etc.

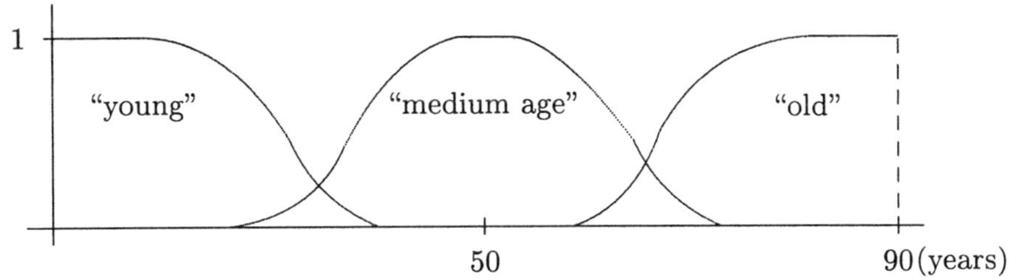

Figure 1.2. Possible forms of the fuzzy sets assigned as the meaning to the basic syntagms "small", "medium age", "old" from $T(\text{Age})$.

Approximate reasoning. Linguistic variables have a quite wide scope of applications. The most important is their use in the *approximate reasoning scheme*, such as the behaviour of the car driver below:

Condition: IF the obstacle is *near* AND the car speed is *big*
THEN break *very much*
IF the obstacle is *far* AND the car speed is *rather small*
THEN break *a little*
Observation: the obstacle is *quite near* AND the car speed is *big*

Conclusion: break *quite much*

This scheme contains vague expressions both in the condition consisting of the so called *fuzzy IF-THEN rules*, as well as in the observation. Note that such scheme is quite natural for the human mind. As a matter of fact, when driving a car the outer conditions vary so much that we could hardly be able to drive without ability to cope with vaguely stated rules. This is a very strong feature of our mind possible mainly due to its ability to cope with the vagueness phenomenon. Any attempt to give precise solution of tasks like this (think, e.g. about solution of the parking a car) necessarily fails. And it is a great challenge to find a formal system enabling to mimic human mind (at least in approximate reasoning schemes like that above).

Fuzzy logic offers a model of the above approximate reasoning scheme. The original proposal of L. A. Zadeh is the so called *generalized modus ponens*. In

Chapter 5, we present the concept of *fuzzy logic in broader sense* (FLb) which includes the approximate reasoning scheme, but its goal is more general — to model natural human reasoning, in which principal role is played by natural language.

Fuzzy quantifiers. Even more complicated situation is faced when using fuzzy quantifiers, i.e. words or syntagms such as "many, most, often, almost, quite much", etc. For example, we might extend the above scheme by the following rule:

> *In danger, most obstacles should be overcome left, provided that no car is approaching.*

Other example with fuzzy quantifiers can be:

> *Many young people like modern music but few of them play an instrument. How many of all young people like modern music and play an instrument?*

Remark to the linguistic semantics and fuzzy logic in broader sense. As already stressed, the agenda of fuzzy logic is connected with the meaning of some expressions of natural language. Hence, it is desirable to be sufficiently precise, especially with respect to the results of the classical linguistic theory. However, this is not always the case till now. For example, one may see that the concept of linguistic variable presented above is not fully satisfactory. The meaning of syntagms depends also on the context, yet there are some general laws inside. One of the reasons for the criticism is neglecting of the concepts of intension and extension.

Intension of the meaning, roughly speaking, is the property contained in it. It does not depend on the context and expresses the very content of the meaning.

Extension is the reflection of the intension in the universe and it leads to groupings of objects. Thus, for example, the intension of "young" presents itself in various worlds equally but leading to various groupings of "young objects". Apparently, "young dog" has the meaning different from "young man" and even the latter may be different in Africa, Europe, or Japan.

One immediately sees that most of the examples given above have been demonstrated on extensions, but with intensions on mind. The concept of the linguistic variable apparently deals only with extensions. To make it to fit better the linguistic theory, it is necessary to encompass also the concept of intension.

The necessity to include linguistic semantics in the fuzzy logic led to the concept of the fuzzy logic in broader sense (FLb) where the intension is interpreted as a special fuzzy set of formulas. The extension is then its interpretation in various models (cf. Chapter 6). On the basis of it, a model of the linguistic semantics of a selected part of natural language consisting of the so called evaluating syntagms is elaborated. Let us stress that FLb is based on FLn, which provides necessary tools for making the inference schemes to work, and for proving some reasonable properties of them. FLb has, in our opinion, a potential for the development of the Zadeh's fuzzy logic agenda and for proving interesting properties of it.

2 ALGEBRAIC STRUCTURES FOR LOGICAL CALCULI

As stated in the previous chapter, fuzzy logic presented in this book is mostly truth functional. The truth values are taken from the support of some algebraic structure and its operations are assigned to the logical connectives. Hence, the truth value of a compound formula is obtained using algebraic operations of the given algebra from its constituents.

In this chapter, we present an overview of basic algebraic structures, which are used for truth values of various logical calculi. The first two sections contain presentation of Boolean algebras, residuated lattices and MV-algebras. All these structures play an important role in the development of logic. The third section is an overview of specific operations in the interval $[0, 1]$ called t-norms, which are natural interpretation of the connective AND in fuzzy logic. The chapter is closed with overview of topos theory and lattice of subobjects in topos.

Notational convention. The following notational convention will be employed throughout this book. If there is some correspondence among various kinds of mathematical objects then they will be denoted by the same letter from different font families. For example, if $\mathcal{L}$ is an algebra then its support is denoted by the corresponding roman letter L. In general, $\mathcal{A}, \mathbf{A}, A, a$ denote objects being somehow related to each other.

The symbol $:=$ means "is defined by". By $\mathbb{N}$ we denote the set of natural numbers. The sets of rational and real numbers are denoted by $\mathbb{Q}$ and $\mathbb{R}$, respectively. The expression "if and only if" is usually written as "iff". Similarly, the expression "with respect to" is often shortened as "w.r.t". The symbols I, J, K usually serve as auxiliary to denote some (arbitrary) sets of indices (subscripts, superscripts, etc.). Besides this we assume that the reader is acquainted with the common mathematical notation. However, if we consider a symbol to be more specific then we introduce in the text.

2.1 Algebras for Logics

2.1.1 Boolean algebras

We start this section with the study of abstract Boolean algebras. Their definition will be given using the full list of generating identities. We choose this approach among others, especially to be able to show the existing close relation between the generating identities and logical axioms of classical logical calculus. Moreover, the way for generalization leading to non-classical logical algebras such as MV-algebras, etc. is thus opened. Having on mind the second purpose, we will suggest two different definitions of a Boolean algebra. The first, most familiar one, uses lattice operations, and the second one uses ring operations. It is worth to be mentioned that in the case of a Boolean algebra, both approaches are interchangeable while in other, non-classical logical algebras, both of them are present. We will conclude the subsection with a variant of the representation theorem given by Tarski and show some consequences of it leading to normal forms.

Boolean algebra as a special lattice. A Boolean algebra is introduced here as a lattice with special properties. This is the way well established in the literature (see e.g. [16]). For better readability, we recall that a *lattice* is an ordered set containing with each pair of elements their least upper bound (supremum) as well as the greatest lower bound (infimum). Since sup and inf are uniquely determined, they can be considered as two binary operations, namely $\vee$ (*join*) and $\wedge$ (*meet*) respectively, so that the lattice becomes at the same time an algebra endowed with these two operations. In symbols, we write

$$\sup(a, b) = a \vee b, \qquad \inf(a, b) = a \wedge b.$$

The following lemma exposes the basic lattice properties. These are normally included in its definition though they can easily be deduced from the definition given above.

LEMMA 2.1
Let L be a lattice. Then the following identities (also called laws) are true:

$$a \vee (b \vee c) = (a \vee b) \vee c, \qquad a \wedge (b \wedge c) = (a \wedge b) \wedge c, \qquad \text{(associativity)} \tag{2.1}$$

$$a \vee b = b \vee a, \qquad a \wedge b = b \wedge a, \qquad \text{(commutativity)} \tag{2.2}$$

$$a \vee (a \wedge b) = a, \qquad a \wedge (a \vee b) = a, \qquad \text{(absorption)} \tag{2.3}$$

$$a \vee a = a, \qquad a \wedge a = a. \qquad \text{(idempotency)} \tag{2.4}$$

Conversely, if $\mathcal{L} = \langle L, \vee, \wedge \rangle$ is an algebra with two binary operations fulfilling (2.1)– (2.3) then an ordering $\leq$ *could be defined on L as follows:*

$$a \leq b \quad \textit{iff} \quad a \vee b = b, \quad (\textit{equivalently} \quad a \wedge b = a),$$

so that the ordered set $(L, \leq)$ forms a lattice where

$$\sup(a, b) = a \vee b \quad \textit{and} \quad \inf(a, b) = a \wedge b.$$

The proof of this lemma is not complicated and uses only the definition and the uniqueness of sup and inf.

Important instances of lattices are *distributive lattices* and *lattices with complements*.

DEFINITION 2.1
A lattice L is distributive if the following identities (distributivity laws) are true:

$$a \vee (b \wedge c) = (a \vee b) \wedge (a \vee c), \tag{2.5}$$
$$a \wedge (b \vee c) = (a \wedge b) \vee (a \wedge c). \tag{2.6}$$

DEFINITION 2.2
A lattice L is a lattice with complements if it has the least element $\mathbf{0}$ and a unary operation of complement "$'$", so that the following identities are true:

$$a'' = a, \qquad \textit{(law of double negation)} \tag{2.7}$$
$$(a \vee b)' = a' \wedge b', \qquad \textit{(the first de Morgan law)} \tag{2.8}$$
$$a \wedge a' = \mathbf{0}. \qquad \textit{(law of contradiction)} \tag{2.9}$$

Let us put $\mathbf{1} = \mathbf{0}'$. It is easy to show that $\mathbf{1}$ is the greatest element of L. The interested reader can find more information about lattices, for example, in [9, 37].

DEFINITION 2.3
A Boolean algebra is a distributive lattice with complements.

With respect to this definition, a Boolean algebra is also called a *Boolean lattice*. The full list of the generating identities is given by (2.1)–(2.9). In symbols, a Boolean algebra is denoted by

$$\mathcal{L} = \langle L, \vee, \wedge, ', \mathbf{0}, \mathbf{1} \rangle.$$

In the theory of Boolean algebras, the complement is usually denoted by '$'$'. However, one may also meet the symbol '$-$' and in the papers on logic, the symbol '$\neg$'. The latter one is most often used in this book.

A Boolean algebra $\mathcal{L}$ is *finite* if its support L is finite. In the sequel, we will refer to the following examples of Boolean algebras.

EXAMPLE 2.1 (BOOLEAN ALGEBRA FOR CLASSICAL LOGIC)
This algebra serves for the truth evaluation of formulas of classical logic. Two values, namely "false" and "true", which are usually denoted by $\mathbf{0}$ and $\mathbf{1}$,

respectively, are sufficient for this purpose. Thus, the Boolean algebra in this case is based on the set consisting of two elements only, i.e.

$$\mathcal{L}_B = \langle \{\mathbf{0}, \mathbf{1}\}, \vee, \wedge, \neg \rangle$$

where the operations "$\vee, \wedge, \neg$" are defined using the following tables:

$\vee$	0	1
0	0	1
1	1	1

$\wedge$	0	1
0	0	0
1	0	1

$\neg$	
0	1
1	0

Since these operations are heavily used in classical logic, they are known under the names corresponding to the respective logical connectives "disjunction", "conjunction" and "negation".

The other two known logical connectives of "implication" and "equivalence" enrich the set of the basic operations above by $\rightarrow$ and $\leftrightarrow$, respectively defined by the following tables:

$\rightarrow$	0	1
0	1	1
1	0	1

$\leftrightarrow$	0	1
0	1	0
1	0	1

Note that the operations $\rightarrow, \leftrightarrow$ can be derived from the basic ones using the following expressions:

$$a \rightarrow b = a' \vee b \qquad a \leftrightarrow b = (a \rightarrow b) \wedge (b \rightarrow a).$$

□

EXAMPLE 2.2 (BOOLEAN ALGEBRA OF SUBSETS)
Let A be an arbitrary nonempty set and $P(A) = \{\mathbf{0}, \mathbf{1}\}^A$ be a set of all its subsets. The Boolean algebra based on this set is

$$\mathcal{B}(A) = \langle \{\mathbf{0}, \mathbf{1}\}^A, \cup, \cap, ^-, \emptyset, A \rangle$$

where "$\cup$","$\cap$" and "$^-$" are the ordinary operations of union, intersection and complement, respectively. □

EXAMPLE 2.3 ((ALGEBRA OF BOOLEAN FUNCTIONS))
Let P_2^n, $n \geq 1$, be a set of all functions $f(x_1, \ldots, x_n)$ defined on $\{\mathbf{0}, \mathbf{1}\}$ and taking values from this set. Such functions are called *Boolean*. The respective Boolean algebra based on P_2^n is

$$\mathcal{L}_P^n = \langle P_2^n, \vee, \wedge, \neg, \mathbf{0}, \mathbf{1} \rangle$$

where the operations "$\vee$", "$\wedge$" and "$\neg$" are defined pointwise, and $\mathbf{0}, \mathbf{1}$ denote the respective constant functions. □

EXAMPLE 2.4 (BOOLEAN ALGEBRA OF PROPOSITIONS)
This algebra is based on the set of all propositions, that is, the sentences which can be evaluated either by "true" or by "false". Every two propositions P, Q

are assigned the following three compound ones, namely "P or Q", "P and Q", "not P" being the result of the basic operations "$\vee$", "$\wedge$" and "$\neg$", respectively. Two propositions are equal if their evaluations coincide. The least element $\mathbf{0}$ of this algebra is a proposition evaluated by "false". □

Among all Boolean algebras, special attention will be paid to those having generating identities true for any (infinite) families of their elements.

DEFINITION 2.4
A Boolean algebra $\mathcal{L}$ is complete if the underlying lattice is complete that is, each subset has the least upper bound as well as the greatest lower bound.

In this case $\mathbf{0} = \inf L = \sup \emptyset$ and $\mathbf{1} = \sup L = \inf \emptyset$.

DEFINITION 2.5
A Boolean algebra is completely distributive if for any family of its elements $(a_{ij})_{i\in I, j\in J}$ the equality

$$\bigwedge_{i\in I}\left(\bigvee_{j\in J} a_{ij}\right) = \bigvee_{\alpha\in J^I}\left(\bigwedge_{i\in I} a_{i\alpha(i)}\right)$$

holds, provided that the elements defined on both its sides exist.

The example of complete and completely distributive Boolean algebra is given by $\mathcal{B}(A)$ (see Example 2.2) for any set A.

Boolean algebra as a special ring. In comparison with the previous way of introducing of the Boolean algebra, the algebra we suggest now is not so popular. However, it is important for us due to necessity for further generalizations.

DEFINITION 2.6
A Boolean ring is a commutative ring $\mathcal{R} = \langle L, +, \cdot, ', \mathbf{0}, \mathbf{1}\rangle$ with the unary operation of complement $a' = \mathbf{1} - a$, unit element, and such that every element is idempotent, i.e., $a \cdot a = a$.

The set of generating identities for a Boolean ring is the following:

$$\begin{aligned} a + (b + c) &= (a + b) + c, & a \cdot (b \cdot c) &= (a \cdot b) \cdot c, \\ a + b &= b + a, & a \cdot b &= b \cdot a, \\ a + (-a) &= \mathbf{0}, & a \cdot a &= a, \\ a + \mathbf{0} &= a, & a \cdot \mathbf{1} &= a, \\ a \cdot (b + c) &= a \cdot b + a \cdot c. \end{aligned} \tag{2.10}$$

LEMMA 2.2
In a Boolean ring, each element is equal to its opposite, i.e. $a = -a$.

PROOF: Using (2.10), we obtain a sequence of equalities, true for any two elements a, b, namely $a+b = (a+b)\cdot(a+b) = a\cdot a+a\cdot b+b\cdot a+b\cdot b = a+a\cdot b+b\cdot a+b$.

This gives $a \cdot b + b \cdot a = \mathbf{0}$. Setting $a = b$, the equations $a + a = \mathbf{0}$ and $a = -a$ follow. □

The following proposition establishes the equivalence between Boolean algebra and ring.

LEMMA 2.3
Let $\mathcal{R} = \langle L, +, \cdot, ', \mathbf{0}, \mathbf{1} \rangle$ be a Boolean ring and let

$$\begin{aligned} a \vee b &= a + b + a \cdot b, \\ a \wedge b &= a \cdot b. \end{aligned}$$

Then the algebra $\mathcal{L} = \langle L, \vee, \wedge, ', \mathbf{0}, \mathbf{1} \rangle$ is Boolean. Conversely, if in a Boolean algebra $\mathcal{L} = \langle L, \vee, \wedge, ', \mathbf{0}, \mathbf{1} \rangle$ the ring operations are defined using the formulas

$$\begin{aligned} a + b &= (a \wedge b') \vee (a' \wedge b), \\ a \cdot b &= a \wedge b \end{aligned}$$

then the algebra $\mathcal{R} = \langle L, +, \cdot, ', \mathbf{0}, \mathbf{1} \rangle$ is a Boolean ring.

PROOF: To prove both parts of this lemma, it is sufficient to derive the first group of generating identities from the second one and vice-versa. As an example, we will demonstrate it by choosing one identity from each side.

(a) The distributivity of $\vee$ with respect to $\wedge$.

For the proof, we will transform each side of the identity to the same expression.

$$\begin{aligned} a \vee (b \wedge c) &= a + b \cdot c + a \cdot b \cdot c, \\ (a \vee b) \wedge (a \vee c) &= (a + b + a \cdot b) \cdot (a + c + a \cdot c) = \\ &= a \cdot a + a \cdot c + a \cdot a \cdot c + b \cdot a + b \cdot c + b \cdot a \cdot c + \\ &\quad + a \cdot b \cdot a + a \cdot b \cdot c + a \cdot b \cdot a \cdot c = \\ &= a + b \cdot c + a \cdot b \cdot c. \end{aligned}$$

(b) The distributivity of $\cdot$ with respect to $+$.

The method of the proof will be the same as above.

$$\begin{aligned} a \cdot (b + c) &= a \wedge ((b \wedge c') \vee (b' \wedge c)), \\ (a \cdot b) + (a \cdot c) &= ((a \wedge b) \wedge (a \wedge c)') \vee ((a \wedge b)' \wedge (a \wedge c)) = \\ &= (a \wedge b \wedge a') \vee (a \wedge b \wedge c') \vee (a' \wedge a \wedge c) \vee (b' \wedge a \wedge c) = \\ &= a \wedge ((b \wedge c') \vee (b' \wedge c)). \end{aligned}$$

□

LEMMA 2.4
In a Boolean ring with lattice operations $\vee$ and $\wedge$, the following identities are true:

(a) $a + a' = \mathbf{1}$,

(b) $a \wedge b = (a + b') \cdot b$,

(c) $a \vee b = a \cdot b' + b$,

(d) $(a + b') \cdot b = (b + a') \cdot a$,

(e) $a \cdot b' + b = b \cdot a' + a$.

PROOF: (a) $a + a' = a + (\mathbf{1} - a) = \mathbf{1} + (a - a) = \mathbf{1}$,
(b) $(a + b') \cdot b = (a + (\mathbf{1} - b)) \cdot b = a \cdot b + b - b \cdot b = a \cdot b + (b - b) = a \cdot b = a \wedge b$,
(c) $a \cdot b' + b = a \cdot (\mathbf{1} - b) + b = a - a \cdot b + b = a + b + a \cdot b = a \vee b$,
(d) follows from (b),
(e) follows from (c). □

Duality. Note that the generating identities (2.1)–(2.9) of a Boolean algebra are such that if we at the same time replace $\mathbf{0}$ by $\mathbf{1}$, $\mathbf{1}$ by $\mathbf{0}$, $\vee$ by $\wedge$ and $\wedge$ by $\vee$, then we again obtain generating identities. This rule is known as the *duality principle*. Equalities obtained one from the other by applying the duality principle are called *dual*.

Validity of this principle can easily be checked for (2.1)– (2.6) since both, formula as well as its dual are there present. We show that (2.8) and (2.9) have also their duals and that they are identities. Indeed, from the identity $(a \vee b)' = a' \wedge b'$ (the first de Morgan law) we can easily deduce the identity

$$a' \vee b' = (a' \vee b')'' = (a'' \wedge b'')' = (a \wedge b)',$$

which is called the *second de Morgan law*.

Similarly, from the identity $a \wedge a' = \mathbf{0}$ (the law of contradiction) we deduce the dual identity

$$\mathbf{1} = (a \wedge a')' = (a'' \wedge a')' = (a' \vee a)'' = a' \vee a,$$

which is called the *law of the excluded middle*.

Important application of the duality principle gives us the representation of each element of a Boolean algebra in two normal forms. We are going to show it below.

Representation theorem and normal forms. Among the examples of a Boolean algebra, the most important is Example 2.2 because it serves as a canonical representation of any Boolean algebra. The following theorem known as the representation theorem justifies this assertion. Usually, this theorem is connected with the name of M. H. Stone as he suggested the most general formulation. Here we present a simpler version first proved by A. Tarski.

THEOREM 2.1
Every complete and completely distributive Boolean algebra is isomorphic to $\mathcal{B}(A)$ for some set A.

PROOF: Let $\mathcal{L}$ be the given algebra with the set of elements L. The following evident equality

$$\mathbf{1} = \bigwedge_{a \in L} (a \vee a')$$

could be rewritten using the complete distributivity into

$$\mathbf{1} = \bigvee_{C \in 2^L} \left(\left(\bigwedge_{a \in C} a \right) \wedge \left(\bigwedge_{b \notin C} b' \right) \right). \tag{2.11}$$

Put $t(C) = (\bigwedge_{a \in C} a) \wedge (\bigwedge_{b \notin C} b')$. We show that $t(C)$ is either $\mathbf{0}$ or an atom of the algebra $\mathcal{L}$ (i.e. minimal element different from $\mathbf{0}$). Indeed, suppose that $t(C) \neq \mathbf{0}$ but there exists $c \in L, c \neq \mathbf{0}$ and $c \leq t(C)$. In general, two cases are possible:

(a) $c \in C$, whence $t(C) \leq c$ and consequently, $c = t(C)$;

(b) $c \notin C$, whence $t(C) \leq c'$ which together with $c \leq t(C)$ gives $c \leq c'$. Consequently, $c = c \wedge c' = \mathbf{0}$, which contradicts with the choice of c.

Let A denote the set of all atoms $t(C) \neq \mathbf{0}$. We show that $\mathcal{L}$ is isomorphic to $\mathcal{B}(A)$. To do it, we assign to every subset of the set A its least upper bound in L (which exists due to the completeness of $\mathcal{L}$) and conversely, to each element $c \in L$ we assign the subset $\{t(C) \mid t(C) \leq c\}$. By (2.11), $c = c \wedge (\bigvee_{C \in 2^L} t(C))$ and, consequently, $c = \bigvee\{t(C) \mid t(C) \leq c\}$ or c is the least upper bound of the subset corresponding to it.

Hence, we have obtained a one-to-one correspondence between $\{\mathbf{0}, \mathbf{1}\}^A$ and L, which, as may easily be seen, preserves the ordering and thus, it is the isomorphism. □

COROLLARY 2.1
Every finite Boolean algebra is isomorphic to $\mathcal{B}(A)$ for some finite set A.

PROOF: The proof easily follows from the fact that each finite Boolean algebra is complete and completely distributive. In this case, A is evidently the set of all atoms of the given Boolean algebra. □

REMARK 2.1
A generalization of Corollary 2.1 stating that every Boolean algebra is isomorphic to some algebra of sets is known as Stone representation theorem.

COROLLARY 2.2
Let $\mathcal{L}$ be a finite Boolean algebra and A be its set of atoms. Then each element x of $\mathcal{L}$ can uniquely be represented in two forms, namely

$$x = \bigvee_{\substack{a \in A \\ a \leq x}} (a \wedge \bigwedge_{b \in A \setminus \{a\}} b'), \tag{2.12}$$

$$x = \bigwedge_{\substack{a \in A \\ a \leq x'}} (a' \vee \bigvee_{b \in A \setminus \{a\}} b) \tag{2.13}$$

which are called the perfect disjunctive normal form and the perfect conjunctive normal form, respectively.

PROOF: First, we prove (2.12). Let x be an arbitrary element from $\mathcal{L}$. Then the equality $x = \bigvee_{\substack{a \in A \\ a \leq x}} a$ directly follows from the proof of Theorem 2.1. Hence, it is sufficient to prove that if $a, b \in A$ and $b \in A \setminus \{a\}$ then $a \wedge b' = a$. Indeed, since a and b are atoms and $b \in A \setminus \{a\}$ then $a \wedge b = \mathbf{0}$ which finally gives $a = a \wedge (b \vee b') = (a \wedge b) \vee (a \wedge b') = \mathbf{0} \vee (a \wedge b') = (a \wedge b')$.

To prove (2.13), it is sufficient to represent the complement x' in the perfect disjunctive normal form and then apply the first de Morgan law (2.8):

$$x' = \bigvee_{\substack{a \in A \\ a \leq x'}} (a \wedge \bigwedge_{b \in A \setminus \{a\}} b'),$$

$$x'' = \left(\bigvee_{\substack{a \in A \\ a \leq x'}} (a \wedge \bigwedge_{b \in A \setminus \{a\}} b') \right)', \quad \text{and finally}$$

$$x = \bigwedge_{\substack{a \in A \\ a \leq x'}} \left(a' \vee \bigvee_{b \in A \setminus \{a\}} b \right).$$

□

2.1.2 *Residuated lattices and MV-algebras*

In order to have more than two values for evaluation of formulas, many attempts have been done to generalize the classical algebra for logic (see Example 2.1). On this way, two special structures took a distinguished role, namely residuated lattices and MV-algebras. In this subsection, we will briefly overview the main properties of them.

Suppose that the set L of truth values forms a lattice with respect to inf ($\wedge$) and sup ($\vee$) and it contains the least ($\mathbf{0}$) and the greatest ($\mathbf{1}$) elements. It has been demonstrated (see, e.g. [57]) that there is no possibility to define a unary operation of complement on L so that all the laws (2.1)–(2.9) of the Boolean algebra hold true. Therefore, on the way of generalization we may only increase the number of operations and distribute various laws among them. We will see that in the case of MV-algebras, this requires to add two specific operations of multiplication ($\otimes$) and addition ($\oplus$) and then to share the full list of Boolean laws over five operations $\vee$, $\wedge$, $\oplus$, $\otimes$, $\neg$ and two constants $\mathbf{0}, \mathbf{1}$. Moreover, the operations $\oplus$ and $\otimes$ have at the same time some of the properties of the ring operations taken from the Boolean rings. As may be expected, if L degenerates to two-element set then the operations $\oplus$ and $\otimes$ coincide with $\vee$ and $\wedge$, respectively.

A residuated lattice can be placed between Boolean and MV- algebras. The list of its operations contains, besides the lattice ones, also two additional operations called multiplication and residuation. In the case that $L = \{0, 1\}$, these coincide with $\wedge$ and $\rightarrow$, respectively.

The results in this subsection are due to many authors. Among them, let us name L. P. Belluce, A. DiNola, P. Hájek, U. Höhle, S. Gottwald, D. Mundici, J. Pavelka, and others. A lot of results, especially those concerning residuated lattices, are taken from the book [51].

Residuated lattices.

DEFINITION 2.7
A residuated lattice is an algebra

$$\mathcal{L} = \langle L, \vee, \wedge, \otimes, \rightarrow, \mathbf{0}, \mathbf{1} \rangle. \tag{2.14}$$

with four binary operations and two constants such that

(i) $\langle L, \vee, \wedge, \mathbf{0}, \mathbf{1} \rangle$ *is a lattice with the ordering* $\leq$ *defined using the operations* $\vee, \wedge$ *as usual, and* $\mathbf{0}, \mathbf{1}$ *are its least and the greatest elements, respectively;*

(ii) $\langle L, \otimes, \mathbf{1} \rangle$ *is a commutative monoid, that is,* $\otimes$ *is a commutative and associative operation with the identity* $a \otimes \mathbf{1} = a$*;*

(iii) the operation $\otimes$ *is isotone in both arguments;*

(iv) the operation $\rightarrow$ *is a residuation operation with respect to* $\otimes$*, i.e.*

$$a \otimes b \leq c \quad \text{iff} \quad a \leq b \rightarrow c. \tag{2.15}$$

We see that in comparison with the ordinary lattice, this structure is enriched by a special couple $\langle \otimes, \rightarrow \rangle$ of operations called the *adjoint couple.* The defining relation (2.15) is called the *adjunction property.* We will call $\otimes$ the *multiplication* and $\rightarrow$ the *residuation*.

Let us remark that there are several other names for residuated lattice, namely integral commutative, residuated *l*-monoid (or *cl*-monoid, if it is complete); residuated commutative Abelian semigroup; complete lattice ordered semigroup. The term "residuated lattice" has been introduced by Dilworth and Ward in [18, 19].

The following examples of residuated lattices will be quite often referred to.

EXAMPLE 2.5 (BOOLEAN ALGEBRA FOR CLASSICAL LOGIC, CF. EXAMPLE 2.1)

$$\mathcal{L}_B = \langle \{\mathbf{0}, \mathbf{1}\}, \vee, \wedge, \rightarrow, \mathbf{0}, \mathbf{1} \rangle$$

where $\rightarrow$ is the classical implication, is the simplest residuated lattice (the multiplication $\otimes = \wedge$). In general, every Boolean algebra is a residuated lattice when putting $a \rightarrow b = a' \vee b$. □

EXAMPLE 2.6 (GÖDEL ALGEBRA)

$$\mathcal{L}_G = \langle [0,1], \vee, \wedge, \rightarrow_G, 0, 1 \rangle$$

where the multiplication $\otimes = \wedge$ and

$$a \rightarrow_G b = \begin{cases} 1 & \text{if} \quad a \leq b, \\ b & \text{if} \quad b < a. \end{cases} \tag{2.16}$$

Gödel algebra is a special case of the *Heyting algebra* used in intuitionistic logic. We will return to Heyting algebras in Subsection 2.4.3. □

EXAMPLE 2.7 (GOGUEN ALGEBRA)

$$\mathcal{L}_P = \langle [0,1], \vee, \wedge, \odot, \rightarrow_P, 0, 1\rangle$$

where the multiplication $\odot = \cdot$ is the ordinary product of reals and

$$a \rightarrow_P b = \begin{cases} 1 & \text{if} \quad a \leq b, \\ \frac{b}{a} & \text{if} \quad b < a. \end{cases} \tag{2.17}$$

□

EXAMPLE 2.8 (ŁUKASIEWICZ ALGEBRA)

$$\mathcal{L}_Ł = \langle [0,1], \vee, \wedge, \otimes, \rightarrow_Ł, 0, 1\rangle \tag{2.18}$$

where

$$a \otimes b = 0 \vee (a + b - 1), \qquad \text{(Łukasiewicz conjunction)} \tag{2.19}$$

$$a \rightarrow_Ł b = 1 \wedge (1 - a + b). \qquad \text{(Łukasiewicz implication)} \tag{2.20}$$

□

EXAMPLE 2.9 (RESIDUATED LATTICE OF $[0,1]$-VALUED FUNCTIONS)
Let X be a nonempty set. For each two functions $f, g \in [0,1]^X$ we put

$$(f \boldsymbol{\vee} g)(x) = f(x) \vee g(x), \qquad (f \boldsymbol{\otimes} g)(x) = f(x) \otimes g(x), \tag{2.21}$$

$$(f \boldsymbol{\wedge} g)(x) = f(x) \wedge g(x), \qquad (f \boldsymbol{\rightarrow} g)(x) = f(x) \rightarrow g(x) \tag{2.22}$$

for all $x \in X$ where $\vee, \wedge, \otimes, \rightarrow$ are the corresponding operations of the Łukasiewicz algebra (Example 2.8). Furthermore, let $\mathbf{1}$ and $\mathbf{0}$ be constant functions taking the values 1 and 0, respectively. Then

$$\mathcal{L}^{func}(X) = \langle [0,1]^X, \boldsymbol{\vee}, \boldsymbol{\wedge}, \boldsymbol{\otimes}, \boldsymbol{\rightarrow}, \mathbf{0}, \mathbf{1}\rangle$$

is a residuated lattice. □

The subsequent lemma shows some commonly used properties of residuated lattices.

LEMMA 2.5
Let $\mathcal{L}$ be a residuated lattice given by (2.14). Then for every $a, b, c \in L$ the following holds true.

(a) $a \otimes b \leq a, \qquad a \otimes b \leq b, \qquad a \otimes b \leq a \wedge b,$

(b) $b \leq a \rightarrow b,$

(c) $a \otimes (a \to b) \leq b, \qquad b \leq a \to (a \otimes b)$,

(d) *if* $a \leq b$ *then*

$$c \to a \leq c \to b \qquad \textit{(isotonicity in the second argument)},$$
$$a \to c \geq b \to c \qquad \textit{(antitonicity in the first argument)},$$

(e) $a \otimes (a \to \mathbf{0}) = \mathbf{0}$,

(f) $a \to (b \to c) = (a \otimes b) \to c$,

(g) $a \leq b$ *iff* $a \to b = \mathbf{1}, \qquad a = \mathbf{1} \to a$,

(h) $(a \vee b) \otimes c = (a \otimes c) \vee (b \otimes c)$,

(i) $a \vee b \leq ((a \to b) \to b) \wedge ((b \to a) \to a)$.

PROOF: (a) is an easy consequence of isotonicity, and (b)–(c) of the adjunction property.

(d) Using (c), we obtain $c \otimes (c \to a) \leq a \leq b$, which follows the first inequality. The second inequality immediately follows from $a \otimes (b \to c) \leq b \otimes (b \to c) \leq c$.

(e)–(f) follow from the adjunction and (c).

(g) $a \otimes \mathbf{1} \leq b$, i.e. $\mathbf{1} \leq a \to b$. The second part is obtained from $\mathbf{1} \to a = \mathbf{1} \otimes (\mathbf{1} \to a) \leq a \leq \mathbf{1} \to a$ due to (b).

(h) $a \otimes c \leq (a \vee b) \otimes c$ and $b \otimes c \leq (a \vee b) \otimes c$. Thus, $(a \otimes c) \vee (b \otimes c) \leq (a \vee b) \otimes c$.

On the other hand, $a \otimes c \leq (a \otimes c) \vee (b \otimes c)$ and $b \otimes c \leq (a \otimes c) \vee (b \otimes c)$. Then $(a \vee b) \leq c \to (a \otimes c) \vee (b \otimes c)$ by the adjunction and properties of supremum. Again by the adjunction, we obtain $(a \vee b) \otimes c \leq (a \otimes c) \vee (b \otimes c)$.

(i) Using (a), (c) and (h), we obtain the inequality $(a \vee b) \otimes (a \to b) = (a \otimes (a \to b)) \vee (b \otimes (a \to b)) \leq b$ and similarly for for a. Hence, $a \vee b \leq (a \to b) \to b$ as well as $a \vee b \leq (b \to a) \to a$ which gives the required inequality. □

To be convinced that a residuated lattice is a generalization of the Boolean one, let us prove the following lemma.

LEMMA 2.6
Let $\mathcal{L}$ *be a residuated lattice given by (2.14). If*

$$\mathbf{1} = a \vee (a \to \mathbf{0})$$

holds for every $a \in L$ *then* $\mathcal{L}$ *is a Boolean lattice and* $\otimes = \wedge$.

PROOF: It is necessary to prove that the distributive law as well as the laws of double negation, de Morgan and contradiction are valid. We will prove idempotency and the first two mentioned properties.

Put $a' = a \to \mathbf{0}$. Then using Lemma 2.5(e) and (h) we obtain $a = a \otimes \mathbf{1} = a \otimes (a \vee a') = (a \otimes a) \vee \mathbf{0} = a \otimes a$ and thus, $\otimes$ is idempotent. Then $a \wedge b = (a \wedge b) \otimes (a \wedge b) \leq a \otimes b$, i.e. $\otimes = \wedge$. From this and Lemma 2.5(h), the distributivity easily follows.

The law of double negation is proved by the following sequence of equalities:

$$a = a \wedge \mathbf{1} = a \wedge (a' \vee a'') = (a \wedge a') \vee (a \wedge a'') = a \wedge a'' =$$
$$= (a \wedge a'') \vee (a' \wedge a'') = a'' \wedge (a \vee a') = a'' \wedge \mathbf{1} = a''.$$

□

Additional operations and their properties. To be used as an algebra for many-valued logic, residuated lattices should be completed by additional operations corresponding (among others) to the logical connectives. Analogously as in classical logic, they can be obtained as derived operations from the basic ones.

$$\neg a = a \to \mathbf{0} \qquad \text{(negation)}, \tag{2.23}$$

$$a \leftrightarrow b = (a \to b) \wedge (b \to a) \qquad \text{(biresiduation)}, \tag{2.24}$$

$$a \oplus b = \neg(\neg a \otimes \neg b) \qquad \text{(addition)}, \tag{2.25}$$

$$a^n = \underbrace{a \otimes \cdots \otimes a}_{n-\text{times}} \qquad (n\text{-fold multiplication}), \tag{2.26}$$

$$na = \underbrace{a \oplus \cdots \oplus a}_{n-\text{times}} \qquad (n\text{-fold addition}). \tag{2.27}$$

It follows from the antitonicity of $\to$ in the first variable (Lemma 2.5(d)) that the negation is order reversing, i.e.

$$a \leq b \quad \text{iff} \quad \neg b \leq \neg a. \tag{2.28}$$

LEMMA 2.7
Let $\mathcal{L}$ be a residuated lattice given by (2.14). Then the following holds true for every $a, b, c, d \in L$.

(a) $a \leftrightarrow \mathbf{1} = a, \qquad a = b \;\; \textit{iff} \;\; a \leftrightarrow b = \mathbf{1}$,
(b) $(a \leftrightarrow b) \wedge (c \leftrightarrow d) \leq (a \wedge c) \leftrightarrow (b \wedge d)$,
(c) $(a \leftrightarrow b) \vee (c \leftrightarrow d) \leq (a \vee c) \leftrightarrow (b \vee d)$,
(d) $(a \leftrightarrow b) \otimes (c \leftrightarrow d) \leq (a \otimes c) \leftrightarrow (b \otimes d)$,
(e) $(a \leftrightarrow b) \otimes (c \leftrightarrow d) \leq (a \to c) \leftrightarrow (b \to d)$,
(f) $(a \leftrightarrow b) \otimes (b \leftrightarrow c) \leq (a \leftrightarrow c)$.

PROOF: We will prove only one of the inequalities. The other ones are proved using analogous technique and thus, left to the reader.

(b) For any $a, b, c, d \in L$ we have

$$\begin{aligned} ((a \leftrightarrow b) \wedge (c \leftrightarrow d)) \otimes (a \wedge c) &\leq ((a \to b) \wedge (c \to d)) \otimes (a \wedge c) \\ &\leq ((a \to b) \otimes a) \wedge ((c \to d) \otimes c) \\ &\leq b \wedge d. \end{aligned}$$

Hence, $(a \leftrightarrow b) \wedge (c \leftrightarrow d) \leq (a \wedge c) \to (b \wedge d)$. By symmetry of biresiduation, we obtain also $(a \leftrightarrow b) \wedge (c \leftrightarrow d) \leq (b \wedge d) \to (a \wedge c)$ which yields the required inequality. □

The following operation is trivial in Boolean lattices and thus, it makes sense only for the residuated ones. It partially compensates the absence of the law of idempotency and is called the *square root*. This concept has been introduced by U. Höhle.

DEFINITION 2.8
Let $\mathcal{L}$ be a residuated lattice with the support L. If there exists a unary operation $\surd : L \longrightarrow L$ fulfilling the following properties

(S1) $\sqrt{a} \otimes \sqrt{a} = a$,

(S2) $b \otimes b \leq a$ *implies* $b \leq \sqrt{a}$

for every $a, b \in L$, then we say that $\mathcal{L}$ has square roots.

Obviously, $\surd$ is uniquely determined.

EXAMPLE 2.10
Łukasiewicz algebra $\mathcal{L}_{\text{Ł}}$ has square roots given by

$$\sqrt{a} = \frac{a+1}{2}$$

where $a \in [0,1]$. In general,

$$\underbrace{\surd \cdots \sqrt{a}}_{k-\text{times}} = \sqrt[2^k]{a} = \frac{a + 2^k - 1}{2^k}.$$

Therefore,

$$\sqrt[2^k]{0} = \frac{2^k - 1}{2^k} \quad \text{and} \quad \neg \sqrt[2^k]{0} = \frac{1}{2^k}.$$

Goguen algebra $\mathcal{L}_P$ has square roots coinciding with the ordinary square roots of real numbers. □

The following lemma summarizes some of the properties of square roots.

LEMMA 2.8
Let $\mathcal{L}$ be a residuated lattice with square roots. The following holds for every $a, b \in L$.

(a) $a \leq \sqrt{a}, \qquad a \leq b \quad \textit{implies} \quad \sqrt{a} \leq \sqrt{b}$,

(b) $\sqrt{a} \otimes \sqrt{b} \leq \sqrt{a \otimes b}$,

(c) $\sqrt{a} \rightarrow \sqrt{b} = \sqrt{a \rightarrow b}$,

(d) $\sqrt{a} \wedge \sqrt{b} = \sqrt{a \wedge b}$,

(e) $a \wedge b \leq \sqrt{a} \otimes \sqrt{b} \leq a \vee b$.

Linear ordering and completeness of residuated lattices. Lattice completeness is a property usually required when evaluating quantified formulas. As we will see, the completeness provides also a natural possibility how the residuation operation can be directly defined using the multiplication, or vice-versa (see Lemma 2.9(a),(b)).

DEFINITION 2.9
Let $\mathcal{L}$ be a residuated lattice.

(i) $\mathcal{L}$ *is* linearly ordered *if $a \leq b$ or $b \leq a$ holds for every $a, b \in L$.*

(ii) $\mathcal{L}$ *is* complete *if the underlying lattice is complete.*

(iii) $\mathcal{L}$ *is* divisible *if $a \leq b$ implies existence of $c \in L$ such that $b \otimes c = a$.*

LEMMA 2.9
Let $\mathcal{L}$ be a complete residuated lattice. Then the following identities hold for every $a, b \in L$ and sets $\{a_i \mid i \in I\}$, $\{b_i \mid i \in I\}$ of elements from L over arbitrary set of indices I.

(a) $a \rightarrow b = \bigvee\{x \mid a \otimes x \leq b\}$,

(b) $a \otimes b = \bigwedge\{x \mid a \leq b \rightarrow x\}$.

(c) $(\bigvee_{i \in I} a_i) \otimes b = \bigvee_{i \in I}(a_i \otimes b)$,

(d) $\bigwedge_{i \in I}(a \rightarrow b_i) = a \rightarrow (\bigwedge_{i \in I} b_i)$,

(e) $\bigwedge_{i \in I}(a_i \rightarrow b) = (\bigvee_{i \in I} a_i) \rightarrow b$,

(f) $\bigvee_{i \in I}(a \rightarrow b_i) \leq a \rightarrow (\bigvee_{i \in I} b_i)$,

(g) $\bigvee_{i \in I}(a_i \rightarrow b) \leq (\bigwedge_{i \in I} a_i) \rightarrow b$.

PROOF: Put $K = \{x \mid a \otimes x \leq b\}$.

(a) By adjunction, $x \leq a \rightarrow b$ for every $x \in K$. Hence, $\bigvee_{x \in K} x \leq a \rightarrow b$. Vice-versa, $a \otimes (a \rightarrow b) \leq b$ (by Lemma 2.5 (c)), i.e. $a \rightarrow b \in K$, which gives $a \rightarrow b \leq \bigvee_{x \in K} x$.

(b)–(e) are proved analogously.

(f) and (g) are the consequences of the isotonicity of $\rightarrow$ in the second argument and antitonicity in the first one. □

If the index set I is finite then the properties (c)–(g) of this lemma hold in every residuated lattice.

A special case of residuated lattices are BL-algebras[1], which have been introduced by P. Hájek to develop a kernel logical calculus, which would be included in various kinds of many- valued logical calculi. From the algebraic point of view, BL-algebras enrich residuated lattices by two special properties given in the definition below.

DEFINITION 2.10
Let $\mathcal{L}$ be a residuated lattice. It is called the BL-algebra if the following holds for every $a, b \in L$.

[1]'BL' means "basic logic".

(i) $a \otimes (a \to b) = a \wedge b$,

(ii) $(a \to b) \vee (b \to a) = \mathbf{1}$ *(prelinearity axiom).*

LEMMA 2.10

(a) *A linearly ordered residuated lattice* $\mathcal{L}$ *is a BL-algebra iff it fulfils condition (i) of Definition 2.10.*

(b) *If a linearly ordered residuated lattice* $\mathcal{L}$ *is divisible then it is a BL-algebra.*

(c) *The following identity holds true in any BL-algebra* $a \vee b = ((a \to b) \to b) \wedge ((b \to a) \to a)$.

PROOF: (a) We must check the prelinearity axiom. If L is linearly ordered then either $a \leq b$ or $b \leq a$ holds for every $a, b \in L$. By Lemma 2.5(g) we obtain the assertion.

(b) By (a) it is sufficient to prove the prelinearity. Suppose $a \leq b$, then $a \wedge b = a = a \otimes \mathbf{1} = a \otimes (a \to b)$. Now, let $b < a$. Then $a \wedge b = b$ and $a \otimes (a \to b) \leq b$. Since $\mathcal{L}$ is divisible, there is c such that $a \otimes c = b$. The assertion then follows from the adjunction and the isotonicity of $\otimes$.

(c) The corresponding inequality is obtained by Lemma 2.5(i). We have to prove the opposite one. Indeed, using the prelinearity we have

$$
\begin{aligned}
(((a \to b) \to b) \wedge ((b \to a) \to a)) \otimes ((a \to b) \vee (b \to a)) = \\
= ((a \to b) \otimes (((a \to b) \to b) \wedge ((b \to a) \to a))) \vee \\
\vee ((b \to a) \otimes (((b \to a) \to a) \wedge ((a \to b) \to b))) \leq \\
\leq ((a \to b) \otimes ((a \to b) \to b)) \vee ((b \to a) \otimes ((b \to a) \to a)) \leq b \vee a.
\end{aligned}
$$

□

REMARK 2.2

It is not difficult to deduce from the proof of (b) that the divisibility property is equivalent in residuated lattices with the condition (i) of Definition 2.10.

EXAMPLE 2.11

The Łukasiewicz, Gödel and Goguen algebras from Example 2.5 are BL-algebras. Indeed, by Lemma 2.10(a) it is sufficient to prove the identity $a \otimes (a \to b) = a \wedge b$.

Let $a \leq b$. Then $a \wedge b = a$ and in all three cases $a \to b = \mathbf{1}$ and thus, $a \otimes (a \to b) = a$. Let $b < a$. Then $a \wedge b = b$ and we obtain for the respective algebras the following.

$\mathcal{L}_P$: $a \otimes (a \to b) = a \cdot \frac{b}{a} = b$.

$\mathcal{L}_G$: $a \otimes (a \to b) = a \wedge b = b$.

$\mathcal{L}_Ł$: $a \otimes (a \to b) = a + (1 - a + b) - 1 = b$. □

MV-algebras. The notion of MV-algebra[2] was introduced by C. C. Chang [12] for an algebraic system, which would correspond to the $\aleph_0$-valued propositional calculus. MV-algebras have been developed as generalizations of Boolean

[2] 'MV' means "many-valued".

algebras to be able to transfer the Boolean prime ideal theorem to the general case and thus, to prove completeness of the $\aleph_0$-valued logic. However, it turned out that MV-algebras as well as BL-algebras stand for the algebraic structures of truth values for various non-classical logical calculi including fuzzy logics. As in each example of the algebras presented above, the MV-algebra can also be treated as a special kind of the lattice supplied with two additional algebraic operations called addition and multiplication.

From the algebraic point of view, the MV-algebra differs from the Boolean one by absence of the idempotency law for their algebraic operations (addition and multiplication) and also, by the lack of the law of excluded middle for the lattice operations.

It is also worth noticing that there are some common properties of an MV-algebra and a Boolean ring. For example, each identity from Lemma 2.4 is valid also in an MV-algebra. From this point of view, an MV-algebra restricted only to algebraic operations can also be treated as a generalization of a Boolean ring.

The following definition of an MV-algebra, is not the most economic one. It originates from the definition given by C. C. Chang (see [12]).

DEFINITION 2.11
An MV-algebra is an algebra

$$\mathcal{L} = \langle L, \oplus, \otimes, \neg, \mathbf{1}, \mathbf{0} \rangle \tag{2.29}$$

in which the following identities are valid.

$$a \oplus b = b \oplus a, \qquad a \otimes b = b \otimes a, \tag{2.30}$$

$$a \oplus (b \oplus c) = (a \oplus b) \oplus c, \qquad a \otimes (b \otimes c) = (a \otimes b) \otimes c, \tag{2.31}$$

$$a \oplus \mathbf{0} = a, \qquad a \otimes \mathbf{1} = a, \tag{2.32}$$

$$a \oplus \mathbf{1} = \mathbf{1}, \qquad a \otimes \mathbf{0} = \mathbf{0}, \tag{2.33}$$

$$a \oplus \neg a = \mathbf{1}, \qquad a \otimes \neg a = \mathbf{0}, \tag{2.34}$$

$$\neg(a \oplus b) = \neg a \otimes \neg b, \qquad \neg(a \otimes b) = \neg a \oplus \neg b, \tag{2.35}$$

$$a = \neg\neg a, \qquad \neg\mathbf{0} = \mathbf{1}, \tag{2.36}$$

$$\neg(\neg a \oplus b) \oplus b = \neg(\neg b \oplus a) \oplus a. \tag{2.37}$$

Most of these identities have obvious interpretation. Note that the identity (2.34) is the law of the excluded middle, the identity (2.35) is the de Morgan law for the operations $\otimes$ and $\oplus$ and the first identity of (2.36) is the law of double negation. The last identity (2.37) is specific and it states that the supremum operation is commutative, as can be seen from the definition (2.41) below.

The following are the examples of an MV-algebra.

EXAMPLE 2.12 (ŁUKASIEWICZ ALGEBRA)
The Łukasiewicz algebra $\mathcal{L}_{\text{Ł}}$ from Example 2.8 is an MV- algebra. It can be written as

$$\mathcal{L}_{\text{Ł}} = \langle [0, 1], \oplus, \otimes, \neg, 0, 1 \rangle \tag{2.38}$$

where $\otimes$ is the Łukasiewicz conjunction defined in (2.19), $\oplus$ is called the Łukasiewicz disjunction defined by

$$a \oplus b = 1 \wedge (a + b) \tag{2.39}$$

and $\neg$ is the negation operation defined by $\neg a = 1 - a$. $\square$

EXAMPLE 2.13 (MV-ALGEBRA OF [0,1]-VALUED FUNCTIONS)
The residuated lattice $\mathcal{L}^{func}(X)$ of $[0,1]$-valued functions on a nonempty X from Example 2.9 on page 25 is an MV-algebra. $\square$

EXAMPLE 2.14 (MV-ALGEBRA OF DYADIC NUMBERS)
Let $D \subseteq [0,1]$ be the set of all rational dyadic numbers on $[0,1]$, i.e.

$$D = \bigcup_{n \in \mathbb{N}} \left\{ \frac{i}{2^n} \mid i \in \{0, 1, \ldots, 2^n\} \right\}.$$

Put

$$\begin{aligned} a \oplus b &= 1 \wedge (a + b), \\ a \otimes b &= \max(0, a + b - 1), \\ \neg a &= 1 - a. \end{aligned}$$

for all $a, b \in D$. Then $\mathcal{D} = \langle D, \oplus, \otimes, \neg, 0, 1 \rangle$ is an MV-algebra, which has the square roots defined in Example 2.10. In particular, it is an MV-subalgebra of the Łukasiewicz algebra $\mathcal{L}_Ł$. $\square$

DEFINITION 2.12
Let $\mathcal{L}$ be an MV-algebra given by (2.29). $\mathcal{L}$ is locally finite if to every $a \neq \mathbf{1}$ there is $n \in \mathbb{N}$ such that $a^n = \mathbf{0}$, or equivalently, to every $a \neq \mathbf{0}$ there is $n \in \mathbb{N}$ such that $na = \mathbf{1}$. (In other terminology this property is called a nilpotentness.)

For example, Łukasiewicz MV-algebra $\mathcal{L}_Ł$ is locally finite. In the sequel, suppose that we are given some MV-algebra $\mathcal{L}$. Put

$$a \leq b \quad \text{iff} \quad \neg a \oplus b = \mathbf{1} \tag{2.40}$$

and verify the properties of partial ordering. Indeed, the reflexivity easily follows from (2.34), the antisymmetry is justified by assuming that $\neg a \oplus b = 1$ and $\neg b \oplus a = 1$ and deducing that

$$\begin{aligned} a &= \neg\neg((\neg a \oplus b) \otimes a) = \neg(\neg(\neg a \oplus \neg\neg b) \oplus \neg a) = \neg(\neg(\neg b \oplus \neg\neg a) \oplus \neg b) = \\ &= (\neg b \oplus a) \otimes b = b. \end{aligned}$$

The transitivity is left to the reader.

The lattice operations can be introduced by

$$a \vee b = \neg(\neg a \oplus b) \oplus b = (a \otimes \neg b) \oplus b, \tag{2.41}$$

$$a \wedge b = \neg(\neg a \vee \neg b) = (a \oplus \neg b) \otimes b. \tag{2.42}$$

Moreover, put $a \to b = \neg a \oplus b$ and see that this is a residuation w.r.t. $\otimes$.

THEOREM 2.2

(a) *Every MV-algebra is a residuated lattice.*

(b) *A residuated lattice $\mathcal{L}$ is an MV-algebra iff*

$$(a \to b) \to b = a \vee b \tag{2.43}$$

holds for every $a, b \in L$.

(c) *A residuated lattice $\mathcal{L}$ is an MV-algebra iff it is divisible and keeps the law of double negation (2.36).*

(d) *A BL-algebra is MV-algebra iff it fulfils the law of double negation.*

PROOF: (a) It is necessary to show that the operation $\to$ has the properties of residuation. We will demonstrate only adjunction.

Using (2.31), (2.35) and (2.36), the inequality $a \otimes b \leq c$ is equivalent to $a \otimes b \otimes \neg c = \mathbf{0}$. From the other side, the inequality $a \leq b \to c$ is equivalent to $a \otimes \neg(\neg b \oplus c) = \mathbf{0}$. Using (2.35), (2.36) and (2.31) we conclude that the latter is also equivalent to $a \otimes b \otimes \neg c = \mathbf{0}$.

(b) If a residuated lattice $\mathcal{L}$ fulfils condition (2.43) then by (2.23) we have $a = a \vee \mathbf{0} = (a \to \mathbf{0}) \to \mathbf{0} = \neg\neg a$, which gives (2.36). The other conditions of MV-algebra can be easily verified. The opposite implication is given by (2.41).

(c) see [49].

(d) follows from (c) and Remark 2.2 after Lemma 2.10. □

COROLLARY 2.3

An MV-algebra is a Boolean algebra if the operation $\otimes$ (or, equivalently, $\oplus$) is idempotent.

PROOF: Since MV-algebra is a residuated lattice then by Lemma 2.6 on page 26, it is sufficient to prove that $a \vee \neg a = \mathbf{1}$. Indeed,

$$a \vee \neg a = (a \otimes \neg\neg a) \oplus \neg a = (a \otimes a) \oplus \neg a = a \oplus \neg a = \mathbf{1}.$$

□

Observe that every MV-algebra is a BL-algebra. By Theorem 2.2 (b) both Gödel $\mathcal{L}_G$ as well Goguen $\mathcal{L}_P$ algebras are not MV-algebras. Indeed, let $a < 1$ and $b = 0$. Then in both cases $a \vee b = a$ but $(a \to b) \to b = 1$.

The following lemma demonstrates some characteristic, commonly known properties of MV-algebras. In the formulation of them, we use Theorem 2.2(a), according to which the MV-algebra is a residuated lattice.

LEMMA 2.11

The following are identities in every MV-algebra.

(a) $(a \to b) \vee (b \to a) = \mathbf{1}$,

(b) $(a \wedge b) \to c = (a \to c) \vee (b \to c)$,

(c) $a \to (b \vee c) = (a \to b) \vee (a \to c)$,

(d) $a \to b = \neg b \to \neg a$.

(e) $a \to (a \otimes b) = \neg a \vee b$.

PROOF: We prove (d) and (e).

(d) both inequalities $a \to b \leq \neg b \to \neg a$ the opposite one follow from the definition of $\leq$, see (2.40), and the identity (2.34).

(e) Using (2.41) we obtain

$$\neg a \vee b = (b \otimes \neg\neg a) \oplus \neg a = a \to (a \otimes b).$$

□

As usual, an MV-algebra is complete if the underlying lattice is complete.

LEMMA 2.12

Let $\mathcal{L}$ be a complete MV-algebra. Then the following holds true for every $a, b \in L$ and sets $\{a_i \mid i \in I\}$, $\{b_i \mid i \in I\}$ of elements from L over arbitrary set of indices I.

(a) $\bigvee_{i\in I} a_i = \neg \bigwedge_{i\in I} \neg a_i, \qquad \bigwedge_{i\in I} a_i = \neg \bigvee_{i\in I} \neg a_i,$

(b) $a \wedge (\bigvee_{i\in I} b_i) = \bigvee_{i\in I}(a \wedge b_i), \qquad a \vee (\bigwedge_{i\in I} b_i) = \bigwedge_{i\in I}(a \vee b_i),$

(c) $a \otimes (\bigwedge_{i\in I} b_i) = \bigwedge_{i\in I}(a \otimes b_i),$

(d) $\bigvee_{i\in I}(a_i \to b) = (\bigwedge_{i\in I} a_i) \to b,$

(e) $\bigvee_{i\in I}(a \to b_i) = a \to (\bigvee_{i\in I} b_i).$

PROOF: (a) The inequality $b_i \leq \bigvee_{i\in I} b_i$ for all $i \in I$ implies $\neg \bigvee_{i\in I} b_i \leq \neg b_i$ due to (2.28). From it follows that $\neg \bigvee_{i\in I} b_i \leq \bigwedge_{i\in I} \neg b_i$. The opposite inequality is obtained similarly from $\bigwedge_{i\in I} \neg b_i \leq \neg b_i$. The second equality of (a) is left to the reader.

(b) The inequality $\bigvee_{i\in I}(a \wedge b_i) \leq a \wedge \bigvee_{i\in I} b_i$ is obvious. Vice-versa, using Lemma 2.9(c) and (2.42) we have

$$a \wedge \bigvee_{i\in I} b_i = \left(a \oplus \neg \bigvee_{i\in I} b_i\right) \otimes \bigvee_{i\in I} b_i = \bigvee_{i\in I}\left(\left(a \oplus \neg \bigvee_{i\in I} b_i\right) \otimes b_i\right) \leq$$
$$\leq \bigvee_{i\in I}((a \oplus \neg b_i) \otimes b_i) = \bigvee_{i\in I}(a \wedge b_i).$$

where the inequality follows from $\neg \bigvee_{i\in I} b_i \leq \neg b_i$ by (2.28). The second equality of (b) follows from (a).

(c) By Lemma 2.9(d) and Lemma 2.11(e) together with just obtained (b), we come to the following

$$a \to \bigwedge_{i\in I}(a \otimes b_i) = \bigwedge_{i\in I}(a \to (a \otimes b_i)) = \bigwedge_{i\in I}(\neg a \vee b_i) = \neg a \vee \bigwedge_{i\in I} b_i.$$

Let us now multiply both sides by a. By Theorem 2.2(d), $\mathcal{L}$ is also a BL-algebra, then the multiplied left- hand side is

$$a \otimes \left(a \to \bigwedge_{i \in I} (a \otimes b_i) \right) = a \wedge \left(\bigwedge_{i \in I} (a \otimes b_i) \right) = \bigwedge_{i \in I} (a \otimes b_i).$$

By Lemma 2.5(h), the multiplied right-hand side is

$$a \otimes \left(\neg a \vee \bigwedge_{i \in I} b_i \right) = (a \otimes \neg a) \vee \left(a \otimes \bigwedge_{i \in I} b_i \right) = a \otimes \bigwedge_{i \in I} b_i.$$

(d) First, note that $\neg(a \to b) = a \otimes \neg b$. Then, using (c) we have

$$\bigwedge_{i \in I} (a_i \otimes \neg b) = \left(\bigwedge_{i \in I} a_i \right) \otimes \neg b,$$

i.e. $\bigwedge_{i \in I} \neg(a_i \to b) = \neg\left((\bigwedge_{i \in I} a_i) \to b \right)$ which gives (d) using the law of double negation (2.36) and (a).

(e) is left to the reader.

□

2.2 Filters and Representation Theorems

In the center of interest of this section there are three representation theorems of MV-algebras: representation by subdirect product of linearly ordered MV-algebras, representation by fuzzy sets and representation by the Łukasiewicz algebra.

Whenever possible, lemmas and theorems are formulated for the case of general residuated lattice. Since every MV-algebra is also the residuated lattice (cf. Theorem 2.2 on page 33), every property of the latter is obviously valid also for the former. The results presented in this section heavily rely on the results presented by C. C. Chang in [12], U. Höhle in [48] and R. Cignolli, I. M. L. D'Ottaviano and D. Mundici in [84].

Filters and congruences. Filter theory provides one of the fundamental algebraic techniques, which makes it possible to investigate deep and principal characteristics of the underlying algebras. We introduce few basic definitions and also demonstrate a close connection between filters and congruences.

DEFINITION 2.13

Let $\mathcal{L}$ be a residuated lattice. A subset $F \subseteq L$ is a filter of $\mathcal{L}$ if it fulfils the following conditions:

(i) $\mathbf{1} \in F$,

(ii) if $a \in F$ *and* $a \leq b$ *then* $b \in F$,

(iii) if $a, b \in F$ *then* $a \otimes b \in F$.

A filter F is called proper if it does not contain $\mathbf{0}$.

Note that a dual concept of *ideal* of $\mathcal{L}$ can be introduced as well. In our case, it is a matter of preference which one to choose. Thus, we will base our constructions on filters. In the sequel, we suppose that $\mathcal{L}$ is a residuated lattice and F is a proper filter of it. Since an MV-algebra is a particular case of a residuated lattice, it inherits all the properties of the latter. In some specific cases, which will be specially indicated, the symbol $\mathcal{L}$ will be used to denote an MV- algebra as well.

A proper filter F is called *prime* if

$$a \vee b \in F \quad \text{implies} \quad a \in F \quad \text{or} \quad b \in F.$$

It can be easily demonstrated that F is prime iff either $a \to b \in F$ or $b \to a \in F$. A proper filter F is called *maximal* if for any other filter G, $F \subseteq G$ implies $F = G$. We will call a maximal filter an *ultrafilter*.

The following is the classical lemma concerning the existence of ultrafilters.

LEMMA 2.13
Every proper filter F can be extended to an ultrafilter containing it.

PROOF: Let $\mathcal{F} = \{F_i \mid i \in I\}$ be a chain of filters containing a proper filter F. Obviously, $\bigcup \mathcal{F}$ is a filter containing F. Then use Zorn's lemma. □

The lemma below is a good example of the filter based technique. It has also it's own importance in the proof of the first representation theorem.

LEMMA 2.14
Let $\mathcal{L}$ be an MV-algebra and $a \in L$, $a \neq \mathbf{1}$. Then there is a prime filter F of $\mathcal{L}$ not containing a.

PROOF: Let $a \in L$ and $a \neq \mathbf{1}$. By $\mathcal{F}(a)$ we denote the set of all filters, which do not contain a. Obviously, $\{\mathbf{1}\} \in \mathcal{F}(a)$ and therefore, $\mathcal{F}(a)$ is nonempty. Using Zorn's lemma, there is a maximal element F in $\mathcal{F}(a)$ w.r.t. the ordinary inclusion. We show that F is prime.

Suppose the opposite. Let $b_1 \vee b_2 \in F$ and $b_1, b_2 \notin F$. Since F is a maximal element in $\mathcal{F}(a)$, there are $c_1, c_2 \in F$ and natural numbers n_1, n_2 such that $b_1^{n_1} \otimes c_1 \leq a$ and $b_2^{n_2} \otimes c_2 \leq a$. (Indeed, otherwise we could construct filter F' containing b_1 or b_2 and not containing a such that $F \subset F'$ which contradicts with maximality of F.) Consequently,

$$(b_1 \vee b_2)^{n_1+n_2} \otimes c_1 \otimes c_2 \leq a,$$

which follows that $a \in F$ — a contradiction. □

DEFINITION 2.14
Let $\mathcal{L}$ be a residuated lattice given by (2.14). $\cong$ is a congruence relation on $\mathcal{L}$ if it is an equivalence relation on $\mathcal{L}$ and additionally, if $a, b, c, d \in L$ are such

that $a \cong c$, $b \cong d$ *then*

$$a \vee b \cong c \vee d, \qquad a \wedge b \cong c \wedge d,$$
$$a \otimes b \cong c \otimes d, \qquad a \to b \cong c \to d.$$

Obviously, the congruence relation on $\mathcal{L}$ preserves the additional operations $\oplus$, $\leftrightarrow$, and $\neg$ on $\mathcal{L}$ as well. In the following theorem, we show that filters and congruences in residuated lattices (and, consequently, in MV-algebras) can be treated equally.

THEOREM 2.3
Let $\mathcal{L}$ be a residuated lattice.

(a) *For any congruence $\cong$ on $\mathcal{L}$, the set*

$$F = \{a \in L \mid a \cong \mathbf{1}\}$$

is a filter of $\mathcal{L}$.

(b) *For any filter F of $\mathcal{L}$, the relation $\cong_F$ on $\mathcal{L}$ defined by*

$$a \cong_F b \quad \text{iff} \quad a \leftrightarrow b \in F$$

is a congruence on $\mathcal{L}$.

(c) *The assertions (a) and (b) establish a bijection between the set of all filters of $\mathcal{L}$ and the set of all congruences on $\mathcal{L}$.*

PROOF: (a) Let $a, b \in F$. Then $a \cong \mathbf{1}$ and $b \cong \mathbf{1}$, which gives $a \otimes b \cong \mathbf{1}$, i.e. $a \otimes b \in F$. Let $a \in F$ and $a \leq b$. Then $a \vee b = b$ and we obtain

$$b = a \vee b \cong \mathbf{1} \vee b = \mathbf{1}.$$

(b) Let $a \cong_F c$ and $b \cong_F d$. Then $a \leftrightarrow c \in F$ and $b \leftrightarrow d \in F$. Using Lemma 2.7(a)–(f) we easily obtain that $\cong_F$ is a congruence.

(c) Let $F \subseteq L$ be a filter, $\cong_F$ a corresponding congruence and $F_{\cong_F}$ a filter corresponding to $\cong_F$. We show that $F_{\cong_F} = F$.

Let $a \in F$. Then $a = a \leftrightarrow \mathbf{1}$, i.e. $a \cong_F \mathbf{1}$ and, hence, $a \in F_{\cong_F}$. Vice-versa, let $a \in F_{\cong_F}$. Then $a \cong_F \mathbf{1}$, i.e. $a \leftrightarrow \mathbf{1} \in F$, which means that $a \in F$. □

Let $\mathcal{L}$ be a residuated lattice and $\cong$ be a congruence on it. Given an element $a \in L$, we denote the corresponding equivalence class by $[a]$ and by $L|\cong$ the factor set given by $\cong$. Furthermore, we define on $L|\cong$ the operations of a residuated lattice by putting

$$[a] \vee [b] = [a \vee b] \qquad [a] \wedge [b] = [a \wedge b],$$
$$[a] \otimes [b] = [a \otimes b] \qquad [a] \to [b] = [a \to b],$$
$$\neg[a] = [\neg a].$$

Thus, we have defined the algebra

$$\mathcal{L}|\cong = \langle L|\cong, \vee, \wedge, \otimes, \to, [\mathbf{0}], [\mathbf{1}]\rangle$$

of the same type as is $\mathcal{L}$, which is called the *factor algebra* of $\mathcal{L}$. The following lemma demonstrates that the factor-algebra of an MV-algebra is an MV-algebra as well.

THEOREM 2.4
Let $\mathcal{L}$ be an MV-algebra, F be a proper filter and $\cong_F$ a corresponding congruence on $\mathcal{L}$. Then

(a) *The factor algebra $\mathcal{L}|\cong_F$ is an MV-algebra.*

(b) *$\mathcal{L}|\cong_F$ is linearly ordered iff F is a prime filter.*

PROOF: (a) It is sufficient to verify the identities (2.30)–(2.37) of an MV-algebra over $\mathcal{L}|\cong_F$. For the demonstration let us choose (2.35). We have to prove that $\neg([a] \oplus [b]) = \neg[a] \otimes \neg[b]$. Indeed, $\neg([a] \oplus [b]) = [\neg(a \oplus b)] = [\neg a \otimes \neg b] = \neg[a] \otimes \neg[b]$.

(b) Let $\mathcal{L}|\cong_F$ be linearly ordered. Then either $[a] \le [b]$ or $[b] \le [a]$ holds for every $a, b \in L$. But then either $a \to b \in F$ or $b \to a \in F$, i.e. F is prime. Vice-versa, let F be a prime filter of MV-algebra $\mathcal{L}$. Then $a \to b \in F$, i.e. $[a] \le [b]$ or $b \to a \in F$, i.e. $[b] \le [a]$ and consequently, $\mathcal{L}|\cong_F$ is linearly ordered. □

The factor-algebra $\mathcal{L}|\cong_F$ w.r.t. a filter F of $\mathcal{L}$ will also be denoted by $\mathcal{L}|F$. The following lemma characterizes $\mathcal{L}|F$ when F is an ultrafilter of $\mathcal{L}$.

LEMMA 2.15
Let $\mathcal{L}$ be a residuated lattice and F be a proper filter of $\mathcal{L}$. Then F is an ultrafilter iff $\mathcal{L}|F$ is locally finite.

PROOF: Let F be an ultrafilter $\mathcal{L}$, $a \in L$ and $[0] < [a] < [1]$. By Theorem 2.3, $a \notin F$. At first we show that there are $b \in F$ and $n \in \mathbb{N}$ such that $a^n \otimes b = \mathbf{0}$. Indeed, in the opposite case we can construct a filter F' such that $F \cup \{a\} \subset F'$ — a contradiction with maximality of F. By the adjunction, $b \le a^n \to \mathbf{0}$, hence $a^n \to \mathbf{0} \in F$, which follows $[a]^n = [\mathbf{0}]$.

Vice-versa, let $\mathcal{L}|F$ be locally finite, but F be not maximal. Then there are a proper filter G of $\mathcal{L}$ and an element $a \in L$ such that $F \subset G$ and $a \in G \backslash F$. Due to the local finiteness of $\mathcal{L}|F$, there is n such that $[a]^n = [\mathbf{0}]$ and thus, $a^n \to \mathbf{0} \in F$. But then $a^n, a^n \to \mathbf{0} \in G$ and by Definition 2.13(iii) of filter, we obtain $\mathbf{0} \in G$ — a contradiction. Hence, F is maximal. □

DEFINITION 2.15
Let $\mathcal{L}$ be a residuated lattice. We say that $\mathcal{L}$ is semisimple if the intersection of all ultrafilters of $\mathcal{L}$ coincides with the trivial filter $\{\mathbf{1}\}$. We say that $\mathcal{L}$ is simple if its only filters are $\{\mathbf{1}\}$ and L.

Obviously, nilpotent (locally finite) residuated lattices are simple and, of course, the Łukasiewicz MV-algebra $\mathcal{L}_{\text{Ł}}$ is simple as well.

The following lemma presents a certain criterion of semisimplicity. Its proof can be found in [84].

LEMMA 2.16
Let $\mathcal{L}$ be an MV-algebra and F be a filter of it. Then $\mathcal{L}|F$ is semisimple iff F is an intersection of all ultrafilters of $\mathcal{L}$.

Semisimplicity of MV-algebras is a property required in the representation theorems given below, which generalize the representation theorem for Boolean algebras. At first, we present a theorem, by which semisimple MV-algebras are strongly connected with the divisibility property.

THEOREM 2.5
(a) *Every divisible locally finite residuated lattice is an MV-algebra.*

(b) *Given a semisimple residuated lattice. Then it is divisible iff it is an MV-algebra.*

PROOF: See Lemma 4.11 and Theorem 4.10 from [48]. □

Representation theorems for MV-algebras. As stressed many times, MV-algebras are generalization of Boolean ones. Thus, it is expected to have at disposal analogies of the main properties of the latter. In the center of them, there is a representation theorem, which gives a canonical image of any Boolean algebra. In this subsection, we will show how this theorem can be generalized and suggest three different analogies of it for the cases of MV-algebras with some additional properties.

We begin with definitions of all the necessary terms.

DEFINITION 2.16
Let $\mathcal{L}_1 = \langle L_1, \oplus_1, \otimes_1, \neg_1, \mathbf{1}_1, \mathbf{0}_1\rangle$ and $\mathcal{L}_2 = \langle L_2, \oplus_2, \otimes_2, \neg_2, \mathbf{1}_2, \mathbf{0}_2\rangle$ be MV-algebras. A homomorphism from $\mathcal{L}_1$ to $\mathcal{L}_2$ is a function $h : L_1 \longrightarrow L_2$ preserving the operations, i.e. it fulfils the equalities

$$h(x \otimes_1 y) = h(x) \otimes_2 h(y), \qquad h(x \oplus_1 y) = h(x) \oplus_2 h(y),$$
$$h(\neg_1 x) = \neg_2 h(x), \qquad h(\mathbf{0}_1) = \mathbf{0}_2, \qquad h(\mathbf{1}_1) = \mathbf{1}_2.$$

If h is a surjection, injection or bijection then we call it epimorphism, monomorphism *or* isomorphism, *respectively.*

DEFINITION 2.17
Let $\{\mathcal{A}_j \mid j \in J\}$ be a system of algebras of the same type where J is some index set. Their direct product is an algebra $\mathcal{A}$ whose support is the Cartesian product $A = \prod_{j\in J} A_j$ with projections $p_j : A \longrightarrow A_j$, and the operations on A are defined so that if g is an n-ary operation on A_j, $j \in J$ then for every $a_1, \dots, a_n \in A$ we can define $g(a_1, \dots, a_n)$ using its projections

$$p_j(g(a_1, \dots, a_n)) = g(p_j(a_1), \dots, p_j(a_n)), \qquad j \in J.$$

The direct product of the algebras $\{\mathcal{A}_j \mid j \in J\}$ will be denoted by $\prod_{j\in J}\{\mathcal{A}_j \mid j \in J\}$.

Before we formulate the first representation theorem for MV-algebras let us recall the original one for Boolean algebras using the introduced terminology. It states that every complete and completely distributive Boolean algebra $\mathcal{L}$ is isomorphic with a direct product $\prod_A \mathcal{L}_B$ of Boolean algebras for classical logic $\mathcal{L}_B$ (see Example 2.1) where A is the set of atoms of $\mathcal{L}$.

THEOREM 2.6 (REPRESENTATION BY LINEARLY ORDERED MV-ALGEBRAS) *Every MV-algebra is isomorphic with a subalgebra of a direct product of linearly ordered MV-algebras.*

PROOF: Let $\mathcal{L}$ be an MV-algebra and $\mathcal{F} = \{F \mid F \subseteq L \text{ is a prime filter}\}$ be the set of all its prime filters. By Theorem 2.4(b), $\mathcal{L}|F$ is linearly ordered for every $F \in \mathcal{F}$. Put

$$\mathcal{L}^* = \prod_{F\in\mathcal{F}} \mathcal{L}|F$$

and define a function $f : \mathcal{L} \longrightarrow \mathcal{L}^*$ by

$$f(a) = \langle [a]_F \mid F \in \mathcal{F}\rangle$$

where $[a]_F \in L|F$ is the equivalence class w.r.t. the congruence $\cong_F$. It is easy to verify that $f(L)$ is a subalgebra of $\mathcal{L}^*$. It remains to prove that f is an injection.

Let $a \neq b$. Then taking into account prelinearity, either $a \to b < \mathbf{1}$ or $b \to a < \mathbf{1}$. Assume the first. By Lemma 2.14, there is a prime filter $F \in \mathcal{F}$ not containing $a \to b$. By the proof of Theorem 2.4(b), $[a]_F \not\leq [b]_F$, i.e. $[a]_F \neq [b]_F$ and thus, $f(a) \neq f(b)$. □

The following theorem demonstates the distinguished role of Łukasiewicz algebra among other infinite and complete MV-algebras. In Chapter 4 we will see that the assumptions about an MV-algebra, which lead to the isomorphism with Łukasiewicz algebra are quite reasonable for the structure of truth values of FLn.

THEOREM 2.7 (REPRESENTATION BY ŁUKASIEWICZ ALGEBRA)

(a) *Every infinite, locally finite, complete MV-algebra is isomorphic with the Łukasiewicz algebra $\mathcal{L}_L$.*

(b) *Let M be the set of all ultrafilters of an MV-algebra $\mathcal{L}$. Then $\mathcal{L}$ is semisimple iff it is isomorphic with a subalgebra of a direct product of Łukasiewicz algebras $\prod_M \mathcal{L}_L$.*

The proof can be found in [48].

The third representation theorem is also a generalization of the similar one (cf. Theorem 2.1) for Boolean algebras. It can be easily seen from the reformulation of the latter that the direct product of Boolean algebras $\prod_A \mathcal{L}_B$ has the support consisting of all the functions defined on A and taking values from $\{0, 1\}$.

THEOREM 2.8 (REPRESENTATION BY FUZZY SETS)
Let X be a compact Hausdorff space and $\mathcal{L}_C^{func}(X)$ be a subalgebra of $\mathcal{L}^{func}(X)$ consisting of all continuous functions on X. Then every semisimple MV-algebra $\mathcal{L}$ is isomorphic with an algebra of continuous functions $\mathcal{L}_C^{func}(X)$.

The proof can be found in [84].

Theorem 2.8[3] is called the "representation by fuzzy sets" because, as defined in Chapter 1, fuzzy sets are identified with special functions $A : X \longrightarrow L$. Thence, we may conclude that the representation for MV-algebras has a similar role for fuzzy sets as the general Stone representation theorem has for sets. Thus, we have still another demonstration that fuzzy sets can indeed be taken as generalization of sets.

Injective algebras. At the end of this section, we will, in addition, present the concept of the injectivity property, which will be used in Chapter 6.

DEFINITION 2.18
An MV-algebra $\mathcal{L}$ is injective if given an MV-subalgebra $\mathcal{K}$ of $\mathcal{L}$, MV-algebra $\mathcal{M}$ and a homomorphism $g : K \longrightarrow M$, it is possible to extend g uniquely to a homomorphism $g' : L \longrightarrow M$.

The proof of the following theorem can be found in [30, 48].

THEOREM 2.9

The Łukasiewicz MV-algebra $\mathcal{L}_Ł$ is injective.

2.3 Elements of the theory of t-norms

The previous two sections focused on the basic structures for truth values of fuzzy and many-valued logic. The central role is played by the residuated lattice, which is a lattice endowed by two more operations of multiplication and the corresponding residuation. We have presented three distinguished examples, namely, Łukasiewicz, Gödel and Goguen algebras, which are residuated lattices based on $[0, 1]$ differing from each other by the chosen multiplication (and residuation). However, the class of such examples is significantly wider. In this section we will study the general class of multiplications known as *triangular norms* (briefly t-norms). These are binary operations $\mathbf{t} : [0, 1]^2 \longrightarrow [0, 1]$, which have been introduced by K. Menger[4] in [77] and later elaborated into the up-to-date form by B. Schweizer and A. Sklar in [122]. They became

[3]This theorem holds for semisimple MV-algebras only. Though this class is wide enough to cover many important examples of MV-algebras, it would be interesting to know whether there is a representation theorem for all the MV-algebras. The following more general proposition due to A. di Nola can be found in [20].

Every MV-algebra $\mathcal{L}$ is isomorphic with an algebra of functions $\mathcal{L}_F^*$ with values in the interval $[0, 1]^*$ of nonstandard reals.

[4]His motivation for introduction of the triangular norms was not logical. The main goal was to generalize the concept of the triangular inequality.

interesting for fuzzy logic because they preserve the fundamental properties of the conjunction 'and' (to hold at the same time), namely commutativity, monotonicity, associativity and boundedness and thus, they serve as a natural generalization of the classical conjunction to many valued reasoning systems.

A concept associated with the t-norm is the triangular conorm (t-conorm) $\mathbf{s} : [0,1]^2 \longrightarrow [0,1]$. This corresponds to the behaviour of truth values when joined by the connective 'or'. The other two concepts studied in this section are the residuation operation, which we already know from Section 2.1, and the negation operation $\mathbf{n} : [0,1] \longrightarrow [0,1]$ generalizing behaviour of the logical negation.

Since t-norms can also be seen as functions on $[0,1]$, they can be studied not only from the algebraic point of view, but also as real functions. In this section, we will mostly confine to the properties of the continuous t-norms because these are especially interesting due to the continuity property of the vagueness phenomenon.

Some of the results presented in this section are taken especially from the monograph [56] in preparation by E. P. Klement, R. Mesiar and E. Pap, and also from the monograph [57] by G. J. Klir and B. Yuan. However, they may be found also elsewhere in the extensive literature on t-norms.

Basic properties of triangular norms and conorms.

DEFINITION 2.19
A t-norm is a binary operation $\mathbf{t} : [0,1]^2 \longrightarrow [0,1]$ such that the following axioms are satisfied for all $a, b, c \in [0,1]$:

(i) commutativity

$$a \,\mathbf{t}\, b = b \,\mathbf{t}\, a,$$

(ii) associativity

$$a \,\mathbf{t}\, (b \,\mathbf{t}\, c) = (a \,\mathbf{t}\, b) \,\mathbf{t}\, c,$$

(iii) monotonicity

$$a \leq b \quad \textit{implies} \quad a \,\mathbf{t}\, c \leq b \,\mathbf{t}\, c,$$

(iv) boundary condition

$$1 \,\mathbf{t}\, a = a.$$

Note that due to the commutativity, we also have $a \,\mathbf{t}\, 1 = a$. Moreover, every t-norm fulfils the additional boundary condition $0 \,\mathbf{t}\, a = a \,\mathbf{t}\, 0 = 0$ for all $a \in [0,1]$. Indeed, for all $a \in [0,1]$ we have $a \leq 1$ and thus, $a \,\mathbf{t}\, 0 \leq 1 \,\mathbf{t}\, 0 = 0$. Note also that all t-norms coincide on $\{0,1\}$.

EXAMPLE 2.15
The most important t-norms are *minimum* '$\wedge$', *product* '$\cdot$' and *Łukasiewicz conjunction* '$\otimes$' (2.19), which are known already from Section 2.1. Other examples of t-norms are *drastic product*

$$a\,\mathbf{t}_W\,b = \begin{cases} \min(a,b) & \text{if } \max(a,b) = 1, \\ 0 & \text{otherwise,} \end{cases}$$

or *nilpotent minimum* (also called Fodor t-norm)

$$a\,\mathbf{t}_F\,b = \begin{cases} \min(a,b) & \text{if } a+b > 1, \\ 0 & \text{otherwise.} \end{cases}$$

□

The following lemma follows from the commutativity and monotonicity.

LEMMA 2.17
Each t-norm is jointly monotone, i.e. $a \le a'$ and $b \le b'$ imply $a\,\mathbf{t}\,b \le a'\,\mathbf{t}\,b'$ for every $a, b, a', b' \in [0,1]$.

The t-norms can be (partially) ordered as functions, i.e. for every t-norms $\mathbf{t}_1, \mathbf{t}_2$ we can put

$$\mathbf{t}_1 \le \mathbf{t}_2 \quad \text{iff} \quad a\,\mathbf{t}_1\,b \le a\,\mathbf{t}_2\,b, \qquad a, b \in [0,1]. \tag{2.44}$$

If (2.44) holds then the t-norm $\mathbf{t}_1$ is weaker than $\mathbf{t}_2$ or, equivalently, t-norm $\mathbf{t}_2$ is stronger than $\mathbf{t}_1$. We write $\mathbf{t}_1 < \mathbf{t}_2$ if $\mathbf{t}_1 \le \mathbf{t}_2$ and $\mathbf{t}_1 \ne \mathbf{t}_2$.

It is not difficult to prove the following lemma.

LEMMA 2.18
For every t-norm $\mathbf{t}$ it holds that

$$\mathbf{t}_W \le \mathbf{t} \le \wedge.$$

This means that the drastic product $\mathbf{t}_W$ is the weakest and $\wedge$ the strongest t-norm. Furthermore, the following inequality holds for the t-norms introduced above:

$$\mathbf{t}_W < \otimes < \cdot < \wedge,$$

Given a t-norm $\mathbf{t}$, we put

$$a^n_{\mathbf{t}} = \underbrace{a\,\mathbf{t}\cdots\mathbf{t}\,a}_{n-\text{times}}.$$

A dual concept to t-norm is that of t-conorm.

DEFINITION 2.20
A t-conorm is a binary operation $\mathbf{s} : [0,1]^2 \longrightarrow [0,1]$*, which fulfils the axioms (i)–(iii) from Definition 2.19, and for all* $a \in [0,1]$ *it fulfils the following boundary condition:*

$$0 \, \mathbf{s} \, a = a. \tag{2.45}$$

A t-conorm is dual to the given t-norm $\mathbf{t}$ *if*

$$a \, \mathbf{s} \, b = 1 - (1-a) \, \mathbf{t} \, (1-b)$$

holds for all $a, b \in [0,1]$.

Note that due to the commutativity, $a \, \mathbf{s} \, 0 = a$ holds for all $a \in [0,1]$. Moreover, every t-conorm fulfils the additional boundary condition $1 \, \mathbf{s} \, a = a \, \mathbf{s} \, 1 = 1$.

EXAMPLE 2.16
The most important t-conorms dual to the t-norms from Example 2.15 are the following. Dual to minimum is *maximum* '$\vee$', dual to product is *probabilistic sum*

$$a \, \mathbf{s}_P \, b = a + b - a \cdot b,$$

dual to Łukasiewicz conjunction is *Łukasiewicz disjunction*

$$a \oplus b = \min(1, a+b),$$

dual to drastic product is *drastic sum*

$$a \, \mathbf{s}_W \, b = \begin{cases} \max(a,b) & \text{if } \min(a,b) = 0, \\ 1 & \text{otherwise,} \end{cases}$$

and dual to nilpotent minimum is *nilpotent maximum*

$$a \, \mathbf{t}_F \, b = \begin{cases} \max(a,b) & \text{if } a+b < 1, \\ 1 & \text{otherwise.} \end{cases}$$

□

Analogously as in the case of t-norms, we can introduce ordering of t-conorms. Then, similarly as in Lemma 2.18, we have the following lemma.

LEMMA 2.19
For every t-conorm $\mathbf{s}$ *it holds that*

$$\vee \leq \mathbf{s} \leq \mathbf{s}_W.$$

This means that maximum is the weakest and the drastic sum the strongest t-conorm. For the t-conorms introduced above, we have

$$\vee < \mathbf{s}_P < \oplus < \mathbf{s}_W.$$

Given a t-conorm $\mathbf{s}$, we put

$$na_{\mathbf{s}} = \underbrace{a \, \mathbf{s} \cdots \mathbf{s} \, a}_{n\text{-times}}.$$

Triangular norms with special properties. The set of t-norms can be classified into various, partly overlapping groups according to their specific properties. We will focus especially to three classes, namely continuous, Archimedean and non-Archimedean t-norms.

To be able to speak about properties of t-norms, we need to represent the latter also as real functions on $[0, 1]$. In the sequel, both representations will be equivalently used.

LEMMA 2.20

(a) *A t-norm* $\mathbf{t}$ *is continuous iff it is continuous in its first component.*

(b) *A t-norm* $\mathbf{t}$ *is left(right)-continuous iff it is left(right)-continuous in its first component. This means that for each* $b \in [0, 1]$ *and each sequence* $(a_n)_{n\in\mathbb{N}}$ *of elements from* $[0, 1]$ *either*

$$\sup_{n\in\mathbb{N}}(a_n \,\mathbf{t}\, b) = \left(\sup_{n\in\mathbb{N}} a_n\right) \mathbf{t}\, b \tag{2.46}$$

for the left-continuity, and

$$\inf_{n\in\mathbb{N}}(a_n \,\mathbf{t}\, b) = \left(\inf_{n\in\mathbb{N}} a_n\right) \mathbf{t}\, b \tag{2.47}$$

for the right-continuity.

EXAMPLE 2.17

The minimum, product and Łukasiewicz conjunction are continuous and, hence, left- as well as right-continuous. An example of a non-continuous left-continuous t-norm is the nilpotent minimum $\mathbf{t}_F$. □

DEFINITION 2.21

(i) *A t-norm* $\mathbf{t}$ *is called strictly monotone if*

$$a \,\mathbf{t}\, b < c \,\mathbf{t}\, b$$

holds for every $a, b, c \in (0, 1)$ *and* $a < c$. *It is strict if it is strictly monotone and continuous.*

(ii) *A t-norm* $\mathbf{t}$ *is Archimedean if for every* $a, b \in (0, 1)$ *there is* $n \in \mathbb{N}$ *such that*

$$a^n_{\mathbf{t}} < b.$$

(iii) *A t-norm* $\mathbf{t}$ *is idempotent if* $a \,\mathbf{t}\, a = a$ *holds for all* $a \in [0, 1]$.

(iv) *A t-norm* $\mathbf{t}$ *is nilpotent if for every* $a \in (0, 1)$ *there is* n *such that* $a^n_{\mathbf{t}} = 0$.

An element $a \in [0, 1]$ is called an *idempotent* if $a \,\mathbf{t}\, a = a$. An element $a \in (0, 1)$ is *nilpotent* if there is n such that $a^n_{\mathbf{t}} = 0$. We may see that a t-norm is Archimedean if it has no other idempotents except for 0 and 1.

EXAMPLE 2.18
The product, Łukasiewicz conjunction, and drastic product are Archimedean while minimum is not. The Łukasiewicz conjunction is nilpotent and minimum is idempotent. The product is strict. □

The following theorem demonstrates that continuous Archimedean t-norms can be divided into exactly two disjoint classes, namely nilpotent and strict.

THEOREM 2.10
Let $\mathbf{t}$ *be a continuous Archimedean t-norm. Then the following is equivalent:*

(a) $\mathbf{t}$ *is nilpotent.*

(b) There exists some nilpotent element of $\mathbf{t}$*.*

(c) $\mathbf{t}$ *is not strict.*

Generated t-norms. An interesting calls of t-norms is formed by those having generators. Let $f : [0,1] \longrightarrow [0,\infty]$ (note that ∞ is included) be a continuous strictly decreasing function such that $f(1) = 0$. The function $f^{(-1)} : [0,\infty] \longrightarrow [0,1]$ such that

$$f^{(-1)}(y) = \begin{cases} f^{-1}(y) & \text{if } y \in [0, f(0)] \\ 0 & \text{if } y \in (f(0), \infty] \end{cases}$$

is called the *pseudoinverse* of f.

DEFINITION 2.22
Let $f : [0,1] \longrightarrow [0,\infty]$ *be a continuous strictly decreasing function such that* $f(1) = 0$*. Then* f *is an* additive generator of *of a t-norm* $\mathbf{t}$ *if*

$$a \,\mathbf{t}\, b = f^{(-1)}(\min(f(0), f(a) + f(b))) \tag{2.48}$$

holds for all $a, b \in [0,1]$.

Additive continuous generators are determined uniquely up to a positive multiplicative constant. It is also possible to define a multiplicative generator. Moreover, the additive generator exists also for t-conorms.

EXAMPLE 2.19
The product t-norm is generated by the additive generator $f_P(x) = -\ln x$. The Łukasiewicz t-norm is generated by the additive generator $f_{\text{L}}(x) = 1 - x$. □

THEOREM 2.11
A function $\mathbf{t} : [0,1]^2 \longrightarrow [0,1]$ *is a continuous Archimedean t-norm iff it has a continuous additive generator.*

LEMMA 2.21
Let $\mathbf{t}$ *be a continuous Archimedean t-norm and* $f(x)$ *be its additive generator. Then the following holds true.*

(a) *The t-norm* $\mathbf{t}$ *is strict iff* $f(0) = +\infty$.

(b) *The t-norm* $\mathbf{t}$ *is nilpotent iff* $f(0) < +\infty$.

PROOF: (a) Let $f(0) = +\infty$. Then to all $a > 0$ and $b, c \in (0,1)$ such that $b < c$ we have

$$a \mathbf{t} b = f^{(-1)}(f(a) + f(b)) < f^{(-1)}(f(a) + f(c)) = a \mathbf{t} c,$$

i.e. $\mathbf{t}$ is strictly monotone.

Vice-versa, if $f(0) < +\infty$ then for each $a \in (0,1)$ there is n such that $nf(a) \geq f(0)$. Hence, by (2.48) $a_{\mathbf{t}}^n = 0$, i.e. $\mathbf{t}$ is nilpotent and thus, not strictly monotone. This proves (a) as well as the "only if" part of (b).

The "if" part of (b): Let $\mathbf{t}$ be nilpotent. Then for each $a \in (0,1)$ there is n such that $a_{\mathbf{t}}^n = 0$. Hence, by (2.48) we have $0 = a\mathbf{t}a^{n-1} = f^{(-1)}\min(f(0), f(a) + f(a^{n-1}))$. This gives us $f(0) \leq f(a) + f(a^{n-1})$, which implies that $f(0) < +\infty$ because f is strictly decreasing. □

Ordinal sums of t-norms. The concept enabling us to construct new t-norms is that of the ordinal sum. Besides others, it makes us possible to prove that in the study of the continuous t-norms, three basic t-norms, namely product, Łukasiewicz conjunction and minimum, have a distinguished role. This has further consequences especially for the many-valued logic (cf. [41]).

DEFINITION 2.23
Let $(\mathbf{t}_i)_{i\in I}$ *be an arbitrary family of t-norms and* $((u_i, v_i))_{i\in I}$ *be a family of pairwise disjoint open subintervals of* $[0,1]$. *Then the function*

$$a \mathbf{t} b = \begin{cases} u_i + (v_i - u_i) \cdot \left(\frac{a-u_i}{v_i-u_i} \mathbf{t}_i \frac{b-u_i}{v_i-u_i}\right) & \text{if } a, b \in (u_i, v_i), \\ \min(a,b) & \text{otherwise} \end{cases} \tag{2.49}$$

is called the ordinal sum of the t-norms $(\mathbf{t}_i)_{i\in I}$.

THEOREM 2.12
The ordinal sum of t-norms is a t-norm.

The index set I may be finite as well as infinite. Obviously, if $I = \emptyset$ then we obtain the minimum $\wedge$. At the same time, every t-norm can be seen as a trivial ordinal sum of itself. Analogously, we can define also ordinal sum of t-conorms.

The following two theorems provide an interesting classification of the continuous t-norms.

THEOREM 2.13

(a) *A continuous Archimedean t-norm* $\mathbf{t} : [0,1]^2 \longrightarrow [0,1]$ *is nilpotent iff it is isomorphic with the Łukasiewicz t-norm* $\otimes$, *i.e. there is a strictly increasing continuous bijection* $\varphi : [0,1] \longrightarrow [0,1]$ *such that*

$$a \mathbf{t} b = \varphi^{-1}(\varphi(a) \otimes \varphi(b))$$

holds for all $a, b \in [0, 1]$.

(b) *A continuous Archimedean t-norm* $\mathbf{t} : [0,1]^2 \longrightarrow [0,1]$ *is strict iff it is isomorphic with the product t-norm, i.e. there is a strictly increasing continuous bijection* $\varphi : [0,1] \longrightarrow [0,1]$ *such that*

$$a \,\mathbf{t}\, b = \varphi^{-1}(\varphi(a) \cdot \varphi(b))$$

holds for all $a, b \in [0, 1]$.

THEOREM 2.14
A function $\mathbf{t} : [0,1]^2 \longrightarrow [0,1]$ *is a continuous t-norm iff it is an ordinal sum of continuous Archimedean t-norms.*

It follows from Theorem 2.11 and Lemma 2.21 that every continuous Archimedean t-norm is either strict of nilpotent. By Theorem 2.13, this furthermore follows that it is isomorphic either to the product '$\cdot$' or to Łukasiewicz t-norm '$\otimes$'. But then, by Theorem 2.14 we conclude that every continuous t-norm is isomorphic either to product, or to Łukasiewicz t-norm, or it is equal to minimum '$\wedge$', or in the sense of the ordinal sum it is a 'mixture' of all of them. Therefore, we will call these three t-norms the basic ones. However, this result has also an impact on the deliberation about fuzzy logic since it is possible to reduce drastically the variety of possibilities how the interpretation of logical connectives should be chosen[5].

Residuation and biresiduation. It follows from the definition of the t-norm that it is a monoidal operation on $[0, 1]$. Furthermore, $\langle [0,1], \wedge, \vee \rangle$ is a complete lattice. Therefore, we can introduce the residuation operation in the same way as is done in Lemma 2.9(a) on page 29.

DEFINITION 2.24
Let $\mathbf{t}$ *be a t-norm. The residuation operation* $\rightarrow_{\mathbf{t}} : [0,1]^2 \longrightarrow [0,1]$ *is defined by*

$$a \rightarrow_{\mathbf{t}} b = \bigvee \{z \mid a \,\mathbf{t}\, z \leq b\}. \tag{2.50}$$

If $\mathbf{t}$ is generated by a continuous additive generator f then the corresponding residuation is given by

$$a \rightarrow_{\mathbf{t}} b = f^{(-1)}\left(\max(0, f(b) - f(a))\right). \tag{2.51}$$

LEMMA 2.22
Let $\mathbf{t}$ *be a left-continuous t-norm. Then* $a \rightarrow_{\mathbf{t}} b = 1$ *iff* $a \leq b$.

PROOF: Let $a \leq b$. Then $a \,\mathbf{t}\, 1 = a \leq b$, i.e. we have $a \rightarrow_{\mathbf{t}} b = 1$ by (2.50).

Conversely, let $a \rightarrow_{\mathbf{t}} b = 1$ and let $(z_n)_{n \in \mathrm{N}}$ be a sequence of reals from $[0, 1]$ fulfilling (2.50). Using Lemma 2.20 we obtain $a \,\mathbf{t}\, (\sup_{n \in \mathrm{N}} z_n) = a \,\mathbf{t}\, 1 = a \leq b$. □

[5] For the details see [41]

PROPOSITION 2.1
The following are the residuation operations corresponding to the basic t-norms:

(a) *Gödel implication $\rightarrow_G$ given in (2.16) is a residuation operation corresponding to minimum '$\wedge$'.*

(b) *Goguen implication $\rightarrow_P$ given in (2.17) is a residuation operation corresponding to multiplication '$\cdot$'.*

(c) *Łukasiewicz implication $\rightarrow_L$ given in (2.20) is a residuation operation corresponding to Łukasiewicz conjunction '$\otimes$'.*

PROOF: All these t-norms are left-continuous and thus, $a \leq b$ implies $a \rightarrow_{\mathbf{t}} b = 1$ by Lemma 2.22. Let $b < a$. Then the greatest z such that $a \wedge z \leq b$ is $z = b$. Similarly we get $z = b/a$ for the Goguen implication and $z = 1 - a + b$ for the Łukasiewicz implication. □

THEOREM 2.15
Let $\mathbf{t}$ *be a continuous t-norm and* $\rightarrow_{\mathbf{t}}$ *the corresponding residuation. Then* $a\,\mathbf{t}\,(a \rightarrow_{\mathbf{t}} b) = a \wedge b$.

PROOF: Let $a \leq b$. Then $a\,\mathbf{t}\,(a \rightarrow_{\mathbf{t}} b) = a\,\mathbf{t}\,1 = a$. Let $b < a$. Since $\mathbf{t}$ is left-continuous, we have $a\,\mathbf{t} \bigvee\{z \mid a\,\mathbf{t}\,z \leq b\} = \bigvee\{a\,\mathbf{t}\,z \mid a\,\mathbf{t}\,z \leq b\}$. Furthermore, since $\mathbf{t}$ is continuous, there is $z \in [0,1]$ such that $a\,\mathbf{t}\,z = b$. Consequently, $a\,\mathbf{t}\,(a \rightarrow_{\mathbf{t}} b) = b$. □

This theorem and the fact that $[0,1]$ is linearly ordered immediately gives us the following (cf. Lemma 2.10).

COROLLARY 2.4
Let $\mathbf{t}$ *be a continuous t-norm. Then the algebra*

$$\mathcal{L}_{\mathbf{t}} = \langle [0,1], \vee, \wedge, \mathbf{t}, \rightarrow_{\mathbf{t}}, 0, 1\rangle \tag{2.52}$$

is a BL-algebra.

DEFINITION 2.25
Let $\mathbf{t}$ *be a t-norm. The biresiduation operation* $\leftrightarrow_{\mathbf{t}}$ *(cf. (2.24)) is the operation* $\leftrightarrow_{\mathbf{t}}\colon [0,1]^2 \longrightarrow [0,1]$ *defined by*

$$a \leftrightarrow_{\mathbf{t}} b = (a \rightarrow_{\mathbf{t}} b) \wedge (b \rightarrow_{\mathbf{t}} a). \tag{2.53}$$

Since $[0,1]$ is linearly ordered then for the left-continuous $\mathbf{t}$, at least one of the arguments on the right hand side of (2.53) is equal to 1. Hence, the minimum operation can be replaced by an arbitrary t-norm. Furthermore, for the left-continuous $\mathbf{t}$ we have also

$$a \leftrightarrow_{\mathbf{t}} b = (a \vee b) \rightarrow_{\mathbf{t}} (a \wedge b).$$

PROPOSITION 2.2
The following are biresiduation operations corresponding to the above introduced basic t-norms:

(a) *For minimum* $\wedge$*:*

$$a \leftrightarrow_G b = \begin{cases} 1 & \text{if } a = b, \\ a \wedge b & \text{otherwise,} \end{cases} \qquad a, b \in [0,1].$$

(b) *For product* $\cdot$*:*

$$a \leftrightarrow_P b = \frac{a \wedge b}{a \vee b} = \exp\left(-|\ln a - \ln b|\right), \qquad a, b \in [0,1]$$

where $0/0 := 1$*, resp.* $\infty - \infty := 0$.

(c) *For Łukasiewicz conjunction* $\otimes$*:*

$$a \leftrightarrow_L b = 1 - |a - b|, \qquad a, b \in [0,1].$$

If a t-norm $\mathbf{t}$ *is generated by a continuous additive generator* f*, then*

$$a \leftrightarrow_{\mathbf{t}} b = f^{(-1)}(|f(a) - f(b)|), \qquad a, b \in [0,1]. \tag{2.54}$$

Negation. In the definition of t-conorm, we have used the operation $\neg a = 1 - a$, which, as the reader surely have noticed, works as a negation operation. It has originally been used by L. A. Zadeh for the definition of the complement of a fuzzy set. His choice was good since it turned out that this operation is a convenient interpretation of the negation in fuzzy logic. However, it is an interesting question, what properties must be fulfilled by an operation to be considered as a many-valued negation. Since every generalization should preserve the original, it seems that the weakest requirement is merely to reverse 0 to 1 and vice-versa. However, such a negation may hardly be sufficient. Therefore, a more specific definition of the negation operation is necessary. We will focus on it in this subsection.

DEFINITION 2.26
The negation is a non-increasing operation $\mathbf{n} : [0,1] \longrightarrow [0,1]$ *such that* $\mathbf{n}(0) = 1$ *and* $\mathbf{n}(1) = 0$*. The negation is involutive if* $\mathbf{n}(\mathbf{n}(a)) = a$ *holds for every* $a \in [0,1]$*. The negation* $\mathbf{n}$ *is strict if it is continuous and* $a < b$ *implies* $\mathbf{n}(a) > \mathbf{n}(b)$*. It is strong if it is strict and involutive.*

EXAMPLE 2.20
A typical example of the involutive negation is $\mathbf{n}(a) = 1 - a$. This is obtained from Łukasiewicz implication by putting $\mathbf{n}_{\mathrm{L}}(a) = a \rightarrow_{\mathrm{L}} 0$. This negation is even strong.

The negation $\mathbf{n}_G(a) = a \rightarrow_G 0$ is non involutive since $\mathbf{n}_G(0) = 1$ but $\mathbf{n}_G(a) = 0$ for every $a \in (0,1]$. □

THEOREM 2.16
The function $\mathbf{n} : [0,1] \longrightarrow [0,1]$ *is a strong negation iff there is a continuous strictly increasing function* $f : [0,1] \longrightarrow [0,\infty)$ *such that* $f(0) = 0$ *and*

$$\mathbf{n}(a) = f^{-1}(f(1) - f(a)) \tag{2.55}$$

holds for every $a \in [0,1]$.

PROOF: Let $\mathbf{n}$ be a strong negation. Put

$$f(a) = \frac{1}{2}(1 - \mathbf{n}(a) + a).$$

Then $f(a)$ is a continuous strictly increasing function, for which $f(1) = 1$ and

$$f(\mathbf{n}(a)) = \frac{1}{2}(1 - \mathbf{n}(\mathbf{n}(a)) + \mathbf{n}(a)) = \frac{1}{2}(1 - a + \mathbf{n}(a)) = 1 - f(a) = f(1) - f(a).$$

This immediately gives (2.55).

Conversely, if f fulfils the conditions of the theorem then the operation $\mathbf{n}$ defined by (2.55) is strictly decreasing and $\mathbf{n}(0) = 1$ and $\mathbf{n}(1) = 0$. Thus, $\mathbf{n}$ is strict. Furthermore,

$$\mathbf{n}(\mathbf{n}(a)) = f^{-1}(f(1) - f(f^{-1}(f(1) - f(a)))) = f^{-1}(f(a)) = a,$$

i.e. $\mathbf{n}$ is also involutive and we conclude that it is strong. □

The function f from Theorem 2.16 is called a *generator* of the negation $\mathbf{n}$.

Let $\mathbf{t}$ be a t-norm and $\mathbf{n}$ be a negation. Put

$$a \,\mathbf{s}\, b = \mathbf{n}(\mathbf{n}(a) \,\mathbf{t}\, \mathbf{n}(b)), \qquad a, b \in [0,1]. \tag{2.56}$$

This defines an operation $\mathbf{s}$ dual to the t-norm $\mathbf{t}$. However, note that this needs not necessarily be a t-conorm. For example, let $\mathbf{t} = \wedge$ and $\mathbf{n} = \mathbf{n}_G$ from Example 2.20. Then $a \,\mathbf{s}\, 0 = \mathbf{n}(\min(\mathbf{n}(a), \mathbf{n}(0))) = 1$, which means that the boundary condition is harmed.

LEMMA 2.23
Let $\mathbf{n}$ *be an involutive negation.*

(a) *If* $\mathbf{t}$ *is a t-norm then the operation* $\mathbf{s}$ *defined in (2.56) is a t-conorm.*

(b) *If* $\mathbf{s}$ *is a t-conorm then the operation* $\mathbf{t}$ *defined by*

$$a \,\mathbf{t}\, b = \mathbf{n}(\mathbf{n}(a) \,\mathbf{s}\, \mathbf{n}(b)), \qquad a, b \in [0,1] \tag{2.57}$$

is a t-norm.

REMARK 2.3
The operations $\mathbf{t}$, $\mathbf{s}$ and $\mathbf{n}$, where the latter is an involutive negation may form so called de Morgan triples. More precisely, if either (and consequently both) of the equations (2.56) or (2.57) holds then the triple of operations

$$\langle \mathbf{t}, \mathbf{s}, \mathbf{n} \rangle$$

is called the de Morgan triple[6]. It can be proved that there is no de Morgan triple with the operations $\mathbf{t} \neq \wedge$ and $\mathbf{s} \neq \vee$ fulfilling both the distributive law as well as the law of the excluded middle with respect to the operations $\mathbf{t}, \mathbf{s}$.

2.4 Introduction to Topos Theory

Motivation. S. Eilenberg and S. MacLane created categories in the 1940s as a way of relating systems of algebraic structures and systems of topological spaces in algebraic topology. In 1960s, F. W. Lawvere presents purely categorical definitions of many of classical mathematical structures which result in a way of categorical foundation for mathematics. A. Grothendieck and his students introduced new structures constructed from sets by categorical methods for solving several classical problems in geometry and number theory. Among these structures, the structure of a topos turned out to be the most important. A topos is a special kind of a category defined by axioms which allow that certain constructions with sets can be done in this category too. The introduction of the topos theory as a generalization of the set theory is revolutionary. The new idea is based on the fact that for the investigation and definition of some objects, we can use *relations* between objects instead of *elements* of these objects.

In this section we present very shortly some basic notions of the topos theory which will be useful in later sections. We would like to emphasize that topos theory and its first order theory can serve as a foundation of a mathematics, alternative to the common foundation by familiar axiomatizations of the membership relation, as in the axioms of Zermelo-Fraenkel set theory. However, the "internal logic" of such foundation is intuitionistic. From this point of view, various topoi can provide very different "worlds" of mathematics. We want to show that one of such world can be some part of fuzzy set theory.

The reader who is interested in this theory can find further details in [33] or [52]. We suppose that the reader is familiar with some basic notions from category theory and especially with notions of various limits and colimits of functors (like products, sums, equalizers, pullbacks etc.). The reader who needs more information is advised to read at first some of the relevant parts of [67].

Let us remark that in this section as well as in Section 3.4 and the whole Chapter 7 we will use specific arrows typical for the category theory.

2.4.1 Topos theory

Motivation. Topos is a special kind of a category defined by axioms saying that certain constructions well known with sets can be done in this category too. In this sense, topos is a generalized set theory.

The most frequently used topos is the category **Set** of abstract sets and mappings in which the development has been frozen so that morphisms $X \longrightarrow Y$ are completely determined by "external" elements of X, where by an external

[6] Note that in [57], de Morgan triples are called *dual triples*.

element x of an object X we understand a morphism $x : \mathbf{To} \longrightarrow X$ defined over the terminal object **To**.

We begin this short introduction with the notion of an elementary topos which enables us to apply a number of set-theoretic constructions, well known from the category **Set**.

Subobject classifier. In set theory, the power-set construction plays an important role. Let us denote $2 := \{0, 1\}$. Then the resulting object $P(X) = 2^X$, i.e. the set of all subsets of a set X (for any set X) has a significant property, which may be expressed using a bijection between the set $P(X)$ and the set of all mappings $X \longrightarrow \mathbf{2}$. This bijection assigns to any subset $A \subseteq X$ a characteristic map $\chi_A : X \longrightarrow \mathbf{2}$ which is defined by the following rule:

$$(\forall x \in X) \quad \chi_A(x) = \begin{cases} 1, & \text{if } x \in A \\ 0, & \text{if } x \notin A. \end{cases}$$

In the sequel, we will use the symbol **To** to denote the terminal object in the corresponding category (i.e. **To** is the one-element set $\{0\}$ in **Set**), and the symbol ! to denote the unique map $A \longrightarrow \mathbf{To}$. Then it should be observed that the diagram

$$\begin{array}{ccc} A & \rightarrowtail & X \\ {\scriptstyle !}\downarrow & & \downarrow{\scriptstyle \chi_A} \\ \mathbf{To} & \xrightarrow{\text{embedding}} & \mathbf{2} \end{array}$$

is the pullback. This construction of one of the basic properties of the category **Set** motivates the following definition.

DEFINITION 2.27
Let $\mathcal{E}$ be a category with the terminal object **To**. *By a subobject classifier in $\mathcal{E}$ we mean an object $\Omega \in \mathcal{E}$ together with a morphism $\mathbf{To} \xrightarrow{\top} \Omega$ such that the following axiom (called* **Ω – axiom***) is satisfied.*

For any monomorphism $f : a \rightarrowtail d$ there exists the unique morphism (called the characteristic morphism of f) $\chi_f : d \longrightarrow \Omega$ such that the diagram

$$\begin{array}{ccc} a & \rightarrowtail & d \\ {\scriptstyle !}\downarrow & & \downarrow{\scriptstyle \chi_f} \\ \mathbf{To} & \xrightarrow{\top} & \Omega \end{array}$$

is a pullback.

Note, that a subobject classifier, provided that it exists in a category $\mathcal{E}$, is unique up to isomorphism. For any object d of a category $\mathcal{E}$ by $\mathrm{Sub}_{\mathcal{E}}(d)$ (or simply $\mathrm{Sub}(d)$)) we denote the set of all *subobjects* of d. Sometimes, for simplicity, the sentence "f is a subobject of d" means that f is a monomorphism $x \rightarrowtail d$ for some object x. It easily follows that $\mathrm{Sub}(-)$ may be considered as a functor $\mathrm{Sub}(-) : \mathcal{E}^{op} \longrightarrow \mathbf{Set}$. In a category $\mathcal{E}$ with the subobject classifier $\top : \mathbf{To} \to \Omega$ we obtain for any object $d \in \mathcal{E}$ a bijection

$$\sigma_d : \mathrm{Sub}(d) \longrightarrow \mathrm{Hom}_{\mathcal{E}}(d, \Omega),$$

which represents a natural isomorphism of functors $\mathrm{Sub}(-)$ and $\mathrm{Hom}_{\mathcal{E}}(-,\Omega)$, i.e. we can write

$$\sigma : \mathrm{Sub}(-) \longrightarrow \mathrm{Hom}_{\mathcal{E}}(-,\Omega).$$

Exponential objects. Another important property of the category **Set** represents the notion of exponential objects B^A for any sets B, A, i.e.,

$$B^A = \{f \mid f \text{ is a map } A \longrightarrow B\}.$$

To characterize the object B^A using only the properties of morphisms, i.e. using *internal* properties of this object instead of *external* definition by its external elements, we observe that a morphism $ev : B^A \times A \longrightarrow B$ such that $ev((f,x)) = f(x)$ is associated with the set B^A. This morphism is called the *evaluation*.

Moreover, the categorical description of B^A is motivated by the fact that the morphism ev has some universal property among functions g of the type

$$C \times A \xrightarrow{g} B.$$

In fact, for every such function g there exists precisely one map $\widehat{g} : C \longrightarrow B^A$ such that the diagram

$$\begin{array}{ccc} B^A \times A & \xrightarrow{ev} & B \\ {\scriptstyle \widehat{g}\times 1_A}\uparrow & & \| \\ C \times A & \xrightarrow{g} & B \end{array}$$

commutes. Using this universal property we obtain that the functor $\mathrm{Hom}_{\mathbf{Set}}(-\times A, B) : \mathbf{Set} \longrightarrow \mathbf{Set}$ is representable by this exponential object B^A. More generally, a functor $F : \mathcal{E} \longrightarrow \mathbf{Set}$ is called *representable* if there exists an object $a \in \mathcal{E}$ such that there exists a natural isomorphism $\tau : F \longrightarrow \mathrm{Hom}_{\mathcal{E}}(-,a)$. The object a is called the *representing object.* This property of the category **Set** leads us to the following definition.

DEFINITION 2.28
A category $\mathcal{E}$ is Cartesian closed if the functor $\mathrm{Hom}_{\mathcal{E}}(-\times a, b)$ *is representable. The representing object is denoted by b^a and it is called the exponential of b by a.*

Hence, a category $\mathcal{E}$ is Cartesian closed if for any two objects $a, b \in \mathcal{E}$ there exists an object $b^a \in \mathcal{E}$ and a morphism $ev : b^a \times a \longrightarrow b$ such that for any object $c \in \mathcal{E}$ and any morphism $g : c \times a \longrightarrow b$ there exists precisely one morphism $\widehat{g} : c \longrightarrow b^a$ such that the diagram

$$\begin{array}{ccc} b^a \times a & \xrightarrow{ev} & b \\ {\scriptstyle \widehat{g}\times 1_a}\uparrow & & \| \\ c \times a & \xrightarrow{g} & b \end{array}$$

commutes. The map $g \mapsto \widehat{g}$ defines a bijection

$$\mathrm{Hom}_{\mathcal{E}}(c \times a, b) \simeq \mathrm{Hom}_{\mathcal{E}}(c, b^a), \tag{2.58}$$

which is a part of the natural isomorphism

$$\mathrm{Hom}_{\mathcal{E}}(- \times a, b) \simeq \mathrm{Hom}_{\mathcal{E}}(-, b^a).$$

Elementary topos. Using these categorical tools we can now define the notion of an elementary topos.

DEFINITION 2.29
A category $\mathcal{E}$ is an elementary topos if $\mathcal{E}$ has finite limits, a subobject classifier and it is Cartesian closed.

The definition of a topos may be given in an even simpler way. For this modified definition, for any object $a \in \mathcal{E}$ we introduce the notion of the *power object* of a as an object $\mathbf{P}a$ in the category $\mathcal{E}$ which represents the functor

$$\mathrm{Sub}(- \times a) : \mathcal{E} \xrightarrow{-\times a} \mathcal{E}^{op} \xrightarrow{\mathrm{Sub}(-)} \mathbf{Set}.$$

It can be proved that there exists a natural isomorphism

$$\phi : \mathrm{Hom}_{\mathcal{E}}(-, \mathbf{P}a) \longrightarrow \mathrm{Sub}(- \times a).$$

Then the following proposition holds, a proof of which may be found, for example in [3].

THEOREM 2.17
A category $\mathcal{E}$ is a topos iff $\mathcal{E}$ has finite limits and every object of $\mathcal{E}$ has a power object.

Instead of the proof of this theorem we observe only that in a topos $\mathcal{E}$, we have

$$\mathrm{Sub}(b \times a) \simeq \mathrm{Hom}_{\mathcal{E}}(b \times a, \Omega) \simeq \mathrm{Hom}_{\mathcal{E}}(b, \Omega^a),$$

for objects $a, b \in \mathcal{E}$ where the symbol $\simeq$ is an abbreviation for the existence of a bijection. Moreover, these bijections are parts of natural isomorphisms, namely let us consider the following compositions of functors.

$$\mathrm{Sub}(- \times a) : \mathcal{E} \xrightarrow{-\times a} \mathcal{E} \xrightarrow{\mathrm{Sub}(-)} \mathbf{Set},$$

$$\mathrm{Sub}(b \times -) : \mathcal{E} \xrightarrow{b\times -} \mathcal{E} \xrightarrow{\mathrm{Sub}(-)} \mathbf{Set},$$

$$\mathrm{Hom}_{\mathcal{E}}(- \times a, \Omega) : \mathcal{E} \xrightarrow{-\times a} \mathcal{E} \xrightarrow{\mathrm{Hom}_{\mathcal{E}}(-,\Omega)} \mathbf{Set},$$

$$\mathrm{Hom}_{\mathcal{E}}(b \times -, \Omega) : \mathcal{E} \xrightarrow{b\times -} \mathcal{E} \xrightarrow{\mathrm{Hom}_{\mathcal{E}}(-,\Omega)} \mathbf{Set},$$

$$\mathrm{Hom}_{\mathcal{E}}(b, \Omega^-) : \mathcal{E} \xrightarrow{\Omega^-} \mathcal{E} \xrightarrow{\mathrm{Hom}_{\mathcal{E}}(b,-)} \mathbf{Set}.$$

Then the above mentioned bijections are elements of the following natural isomorphisms.

$$\begin{aligned} \mathrm{Sub}(-\times a) &\longrightarrow \mathrm{Hom}_{\mathcal{E}}(-\times a,\Omega) \longrightarrow \mathrm{Hom}_{\mathcal{E}}(-,\Omega^a),\\ \mathrm{Sub}(b\times -) &\longrightarrow \mathrm{Hom}_{\mathcal{E}}(b\times -,\Omega) \longrightarrow \mathrm{Hom}_{\mathcal{E}}(b,\Omega^-). \end{aligned}$$

In any topos $\mathcal{E}$ the object Ω^a is the power object of a, i.e. $\Omega^a = \mathbf{P}a$.

2.4.2 Lattice of subobjects in topos

Motivation. A classical two-valued Boolean logic is closely connected with the two elements set $\{\mathbf{0},\mathbf{1}\}$, which is considered as the truth value set for logical formulas. In general, any truth value in this classical logic can be considered as a subset in this two element set. It means that the important role from this logical point of view has the power object $\mathbf{P}\{\mathbf{0},\mathbf{1}\} = \mathbf{P}(\{\mathbf{0},\mathbf{1}\}) = \mathrm{Sub}(\{\mathbf{0},\mathbf{1}\})$ in the topos **Set**. This power object becomes a *Boolean lattice*, where infimum is the intersection and supremum the union of subsets in $\mathbf{P}\{\mathbf{0},\mathbf{1}\}$. Since a topos is a generalization of a category **Set**, very natural idea is to consider analogical situation in arbitrary topos. To realize this project the first step is to investigate the structure of $\mathrm{Sub}(\Omega)$ which is the analogy of $\mathrm{Sub}(\{\mathbf{0},\mathbf{1}\})$. Surprisingly, for some topoi we will receive a very similar structure as we do for the topos **Set**. On the other hand, for another topoi the structure of $\mathrm{Sub}(\Omega)$ is more complicated and the resulting interpretation is different.

We continue with some results about the structure of subobjects in an elementary topos, i.e about the set $\mathrm{Sub}(a)$. For monomorphisms $f : a \rightarrowtail d$, $g : b \rightarrowtail d$ in a topos $\mathcal{E}$, we set $f \le g$ if there exists a morphism $h : a \longrightarrow b$ such that the diagram

$$\begin{array}{ccc} a & \overset{f}{\rightarrowtail} & d \\ {\scriptstyle h}\downarrow & & \| \\ b & \overset{g}{\rightarrowtail} & d \end{array}$$

commutes.

Epi-mono decomposition of morphisms. Before giving the definition of truth-valued morphisms, we need to introduce another notion in a topos $\mathcal{E}$ which represents the well-known property in the topos **Set**. Namely, we need a generalization of the epi-mono decomposition property of functions in **Set**. An analogous fact also holds in the general elementary topos, as is shown in the following lemma. In this lemma we also introduce a new notation, which will be used in a sequel.

LEMMA 2.24
Let $\mathcal{E}$ be an elementary topos, and $f : a \longrightarrow b$ any morphism. Then there exists an object $\exists_f a$ in $\mathcal{E}$ (called the image of f), epimorphism $f^* : a \twoheadrightarrow \exists_f a$ and a monomorphism $\mathrm{Im}\, f : \exists_f a \rightarrowtail b$ such that $f = \mathrm{Im}\, f \circ f^*$. Such a decomposition is unique up to isomorphism.

The notation $\exists_f a = \exists_f(a)$ does not look natural at the present time, but it will appear as very natural in the sequel when we interpret logical formulas of the form $(\exists y)A$. As we mentioned above, the classical internal logic associated with the topos **Set** is closely connected with truth-valued functions

$$\wedge, \vee, \rightarrow : \mathrm{Sub}(\{\mathbf{0},\mathbf{1}\}) \times \mathrm{Sub}(\{\mathbf{0},\mathbf{1}\}) \longrightarrow \mathrm{Sub}(\{\mathbf{0},\mathbf{1}\}),$$
$$\neg : \mathrm{Sub}(\{\mathbf{0},\mathbf{1}\}) \longrightarrow \mathrm{Sub}(\{\mathbf{0},\mathbf{1}\}).$$

These logical functions are sometimes constructed using truth tables (cf. Example 2.1 on page 17). We now show that even in general topos, we have very similar situation when instead of the structure $\mathrm{Sub}(\{\mathbf{0},\mathbf{1}\})$ we will consider the structure $\mathrm{Sub}(\Omega)$.

Truth-valued morphisms. We proceed now to define some truth-valued morphisms. So, let $\mathcal{E}$ be any elementary topos with the subobject classifier $\mathbf{To} \xrightarrow{\top} \Omega$ where **To** is the terminal object in $\mathcal{E}$.

DEFINITION 2.30
We introduce the following notation:

(a) *Let* **Io** *be the initial object in* $\mathcal{E}$. *By* $\bot$ *we denote the characteristic morphism of the unique morphism* $! : \mathbf{Io} \rightarrow \mathbf{To}$.

(b) *By* $\neg : \Omega \rightarrow \Omega$ *we denote the characteristic morphism of the subobject* $\bot : \mathbf{To} \longrightarrow \Omega$.

(c) *By* $\cap : \Omega \times \Omega \longrightarrow \Omega$ *we denote the characteristic morphism of the product* $\langle \top, \top \rangle : \mathbf{To} \longrightarrow \Omega \times \Omega$.

(d) *Let* $\top_\Omega$ *be the following composition*

$$\top_\Omega : \Omega \xrightarrow{!} \mathbf{To} \xrightarrow{\top} \Omega,$$

and let α *be the image of the morphism*

$$\langle \top_\Omega, 1_\Omega \rangle + \langle 1_\Omega, \top_\Omega \rangle : \Omega + \Omega \longrightarrow \Omega \times \Omega,$$

where $+$ *is the symbol for coproduct (= sum) of objects and the corresponding morphisms. Then, by* $\cup : \Omega \times \Omega \longrightarrow \Omega$ *we denote the characteristic morphism of* α *(it is necessary to verify that* α *is a monomorphism).*

(e) *Let* $e : \leq \longrightarrow \Omega \times \Omega$ *be the equalizer of morphisms* $\cap$ *and* pr_1 *(the first projection). Then the characteristic morphism of* e *is denoted by*

$$\rightarrow_\Omega : \Omega \times \Omega \longrightarrow \Omega.$$

Example of truth-valued morphisms. Note that for any topos $\mathcal{E}$, the set $\mathrm{Sub}(\Omega)$ can be considered as an overset of $\{\mathbf{Io}, \mathbf{To}\}$. In fact, we can define a map $\{\mathbf{Io}, \mathbf{To}\} \longrightarrow \mathrm{Sub}(\Omega)$ such that $\mathbf{Io} \mapsto \bot$ and $\mathbf{To} \mapsto \top$. In the following theorem we show that this embedding preserves all the classical truth-valued functions.

THEOREM 2.18
In any elementary topos $\mathcal{E}$ the truth-valued morphisms $\top$ and $\bot$ satisfy the following composition properties.

$\neg$	
$\top$	$\bot$
$\bot$	$\top$

$\cap$	$\top$	$\bot$
$\top$	$\top$	$\bot$
$\bot$	$\bot$	$\bot$

$\cup$	$\top$	$\bot$
$\top$	$\top$	$\top$
$\bot$	$\top$	$\bot$

$\to$	$\top$	$\bot$
$\top$	$\top$	$\bot$
$\bot$	$\top$	$\top$

Structure of subobjects in topos. We can now investigate the order structure of the set $\mathrm{Sub}(a)$. As mentioned above, in the category **Set** the partially ordered set $(\mathrm{Sub}(A), \leq)$ is a Boolean lattice for any set A.

In a topos, the situation is more complicated. Nevertheless, the following principal theorem describes this structure completely.

THEOREM 2.19
For any elementary topos $\mathcal{E}$ and any object d of $\mathcal{E}$, the set $\mathrm{Sub}(d)$ *with classical ordering of subobjects is Heyting algebra.*

It is worthwhile to mention that the *pseudo-complement* $-f$ of an element $f : a \longrightarrow d$ in $\mathrm{Sub}(d)$ is the pullback of $\top$ along $\neg \circ \chi_f$, i.e.,

$$\begin{array}{ccc} -a & \xrightarrow{-f} & d \\ \downarrow & & \downarrow{\scriptstyle \neg \circ \chi_f} \\ \mathbf{To} & \xrightarrow{\top} & \Omega \end{array}$$

where the truth-valued morphism $\neg$ is defined in Definition 2.30.

Internal lattice objects. We have investigated structure $\mathrm{Sub}(d)$ in a topos $\mathcal{E}$. Although there is an analogy between $\mathrm{Sub}(d)$ for object $d \in \mathcal{E}$ and $\mathrm{Sub}(D)$ for a set $D \in$ **Set**, there is a great difference between these two objects. Namely, although $\mathrm{Sub}(D)$ is also an object in the category **Set**, the structure $\mathrm{Sub}(d)$ is not an object in the topos $\mathcal{E}$. Hence, $\mathrm{Sub}(d)$ can be considered an "external" version of this lattice of subobjects. Nevertheless, in any topos $\mathcal{E}$ an internal version of this "lattice of subobjects" elements can be defined as well. We now present the some facts about this object.

Let $\mathcal{E}$ be an elementary topos. In general, by a *lattice object* in a topos $\mathcal{E}$ we understand an object $L \in \mathcal{E}$ together with two morphisms

$$\wedge : L \times L \longrightarrow L, \qquad \vee : L \times L \longrightarrow L,$$

which have to satisfy some commutative diagrams, derived from classical definitions of a lattice. For example, let us consider the absorption law

$$x \wedge (y \vee x) = x.$$

The variables x, y from this law can be expressed as the corresponding projections of some products of the object L. Namely, the left-hand side of this law can be considered as the sequence of maps and projections

$$(x,y) \xrightarrow{\langle\Delta,1_y\rangle} ((x,x),y) \xrightarrow{\cong} (x,(x,y)) \xrightarrow{\langle 1_x,p2\rangle}$$
$$\xrightarrow{\langle 1_x,p2\rangle} (x,(y,x)) \xrightarrow{\langle 1_x,\vee\rangle} (x, y\vee x) \xrightarrow{\wedge} x\wedge(y\vee x),$$

which corresponds to the following sequence of morphism in the category $\mathcal{E}$.

$$L\times L \xrightarrow{\langle\Delta,1_L\rangle} (L\times L)\times L \xrightarrow{\cong} L\times(L\times L) \xrightarrow{\langle 1_L,p2\rangle}$$
$$\xrightarrow{\langle 1_L,p2\rangle} L\times(L\times L) \xrightarrow{\langle 1_L,\vee\rangle} L\times L \xrightarrow{\wedge} L$$

where Δ is the diagonal morphism, $\cong$ is the isomorphism of objects $L\times(L\times L)$ and $(L\times L)\times L$ and $p2$ is the twist morphism interchanging the factors of the product. Analogously, the right-hand side of this law can be expressed as the sequence

$$(x,y) \xrightarrow{\mathrm{pr}_x} x,$$

which corresponds to the following morphism

$$L\times L \xrightarrow{\mathrm{pr}_1} L.$$

Putting together these sequences we obtain the following commutative diagram, which expresses the above mentioned law:

$$\begin{array}{ccccccc} L\times L & \xrightarrow{\langle\Delta,1_L\rangle} & (L\times L)\times L & \xrightarrow{\langle 1_L,p2\rangle\circ\cong} & L\times(L\times L) & \xrightarrow{\langle 1_L,\wedge\rangle} & L\times L \\ \downarrow{\scriptstyle pr_1} & & & & & & {\scriptstyle\vee}\downarrow \\ L & & & =\!=\!=\!=\!= & & & L \end{array}$$

Furthermore, we say that the object $L\in\mathcal{E}$ is a *Heyting algebra object* in $\mathcal{E}$ or an *internal Heyting algebra*, if there exists an additional binary operation $\to\colon L\times L\to L$ satisfying the identities expressing the fact that the pair $\langle\to,\wedge\rangle$ is the adjoint pair. For example, the following lemma presents such a system of identities.

LEMMA 2.25
Let L be a lattice with 0 and 1 and a binary operation $\to$. Then L is a Heyting algebra such that $\langle\to,\wedge\rangle$ is adjoint pair iff the operation $\to$ satisfies the identities

$$x\wedge(x\to y)=x\wedge y, \qquad y\wedge(x\to y)=y,$$
$$x\to(y\wedge z)=(x\to y)\wedge(x\to z), \qquad (x\to x)=1.$$

for all the elements $x,y,z\in L$.

Using this "internal" definition of the Heyting algebra in a topos we can obtain the following "internal" version of Theorem 2.19.

THEOREM 2.20
For any object a in a topos $\mathcal{E}$, the power object $\mathbf{P}a$ is the internal Heyting algebra. Moreover, for each $b \in \mathcal{E}$, the internal structure on $\mathbf{P}a$ makes $\mathrm{Hom}_{\mathcal{E}}(b, \mathbf{P}a)$ *an external Heyting algebra so that the canonical morphism*

$$\mathrm{Sub}_{\mathcal{E}}(a \times b) \cong \mathrm{Hom}_{\mathcal{E}}(b, \mathbf{P}a)$$

is an isomorphism of the external Heyting algebras.

3 LOGICAL CALCULI AND MODEL THEORY

In this chapter, we present a brief overview of classical logical calculi and model theory. Our aim is twofold: first, we remind some of its main concepts and properties, and introduce notation. Second, our focus is directed especially to those properties which will be furthermore generalized due to the principle of the graded approach assumed in this book. In our opinion, this enables the reader to see the origins and way of the development of fuzzy logic. In endeavour for global point of view, we present general principles of formal logical systems (cf. A. Tarski [124]) in Section 3.3. They can be applied also to non-classical approaches and will be important for further explanation in Chapter 4. The interpretation, however, is there mostly a direct generalization of the classical approaches. A very abstract interpretation based on the category theory may have a special interest for fuzzy logic and it will be the topic of Chapter 7. A brief reminder of some fundamental notions of the model theory in categories is presented in Section 3.4. The reader who wishes to learn more about the topic of this chapter is referred to the special literature, for example [17, 121, 68, 76].

3.1 Classical Logic

The fundamental concept of classical two-valued logic is that of a *proposition*. This is any kind of statement, which can be *unambiguously* decided to be *true* or *false*. For example, "37 is an odd number" is a proposition which is apparently true. In classical logic, no other possibility than true or false propositions are accepted. This works quite well in mathematics where all the statements are precisely formulated. However, note that classical logic is often used in situations where the statements contain expressions of natural language, for example "the cost is big". To be able to decide unambiguously, we must give a precise definition, e.g. "big means more than 10,000\$". A more realistic solution taking into account vagueness of such propositions is offered by fuzzy logic presented in the next chapter. However, in classical logic, vagueness is completely avoided.

To express whether a proposition is true or false, we use the algebra for classical logic $L = \{\mathbf{0}, \mathbf{1}\}$ (cf. Example 2.1). Then we identify *false* with the element $\mathbf{0}$ and *true* with $\mathbf{1}$. These elements are called the *truth values*.

3.1.1 Propositional logic

This logic is the simplest kind of classical logic. The objects of its study are propositions, their interrelations and semantics. Propositional logic has two constituents: syntax and semantics. The first one is a formal deduction theory also called a propositional calculus while the second is its interpretation.

Syntax. A propositional calculus is a special case of a formal logical system, which has been elaborated for study mathematical theories as objects of more general metatheory. Each formal theory consists of a language, a set of formulas, axioms and inference rules. In what follows we introduce all the components of the formal system for the propositional calculus.

The formal *language* J of a propositional calculus is a set, which consists of:

(i) A countable set of propositional variables $\{p, q, \dots\}$.

(ii) Symbol for the logical binary connective of implication $\Rightarrow$.

(iii) The symbol $\perp$ for falsity.

(iv) Various kinds of brackets as auxiliary symbols.

Propositional variables are used to denote variable elementary propositions. The symbol $\perp$ is also called the *logical constant* and it represents a proposition which is always false, for example "1 is not equal to 1".

Formulas are used to denote any kind of propositions: elementary and compound ones. The latter are formed from the former with help of logical connectives. In propositional calculus, formulas are the special finite sequences of symbols from J defined using the following inductive definition.

DEFINITION 3.1

(i) A propositional variable or the logical constant $\perp$ is a (atomic) formula.

(ii) If A, B are formulas then $A \Rightarrow B$ is a formula.

Formulas constructed using this definition are sometimes called *well formed formulas*. The set of all the well formed formulas of the given language J is denoted by F_J. In the sequel, we will usually speak about formulas omitting the attribute "well formed". In the notation of formulas we will use brackets for the indication the order in using connectives.

EXAMPLE 3.1
The expressions $p, q, p \Rightarrow q, (p \Rightarrow q) \Rightarrow \bot$ are well formed formulas in the language J while $p \Rightarrow, Ap \Leftrightarrow q, \bot \wedge \vee s$ are not. □

Except for the basic connective of implication $\Rightarrow$, we may also consider other connectives, namely *negation* ($\neg$), *conjunction* ($\wedge$), *disjunction* ($\vee$) and *equivalence* ($\Leftrightarrow$). They may be introduced either as basic connectives being elements of the language J, or (in our case) defined as the derived ones, i.e. as shorts for special formulas, namely

$$\neg A := A \Rightarrow \bot, \qquad \text{negation} \tag{3.1}$$

$$A \wedge B := \neg(A \Rightarrow \neg B), \qquad \text{conjunction} \tag{3.2}$$

$$A \vee B := \neg A \Rightarrow B, \qquad \text{disjunction} \tag{3.3}$$

$$A \Leftrightarrow B := (A \Rightarrow B) \wedge (B \Rightarrow A). \qquad \text{equivalence} \tag{3.4}$$

We may also define the second logical constant $\top := \neg\bot$ for 'true', as a formula which denotes *truth* and can be regarded as an always true proposition, such as "1 is equal to 1".

The next step in the description of the propositional calculus is choosing some formulas called (logical) *axioms.*

DEFINITION 3.2
Let $A, B, C \in F_J$ be formulas. The set LAx *of axioms consists of the following formulas:*

(CL1) $\vdash A \Rightarrow (B \Rightarrow A)$.

(CL2) $\vdash (\neg B \Rightarrow \neg A) \Rightarrow (A \Rightarrow B)$.

(CL3) $\vdash (A \Rightarrow (B \Rightarrow C)) \Rightarrow ((A \Rightarrow B) \Rightarrow (A \Rightarrow C))$.

Let us stress that these are *schemes of axioms.* This means that we obtain an axiom $D \in$ LAx whenever we substitute some concrete formula for A, B, and C in (CL1)–(CL3). Every such concrete axiom is called an *instance* of the axiom.

The final step in the description of a propositional calculus is definition of inference (deduction) rules, which can be regarded as special partial operations on the set F_J. In formal logical systems, inference rules are treated purely formally, which means "blind" derivation of formulas without special attention to their meaning. Classical propositional calculus has the single inference rule known as *modus ponens* (or the detachment rule). This rule has the form

$$r_{MP} : \frac{A, A \Rightarrow B}{B}. \tag{3.5}$$

This rule is the crucial logical rule, which can be explained semantically as follows: knowing that the propositions represented by the formulas A and $A \Rightarrow B$ are true, we infer that the proposition represented by the formula B is true as well.

Deduction.

DEFINITION 3.3
A proof in the propositional calculus is a finite sequence of formulas

$$w := B_1, \dots, B_n$$

where each formula B_i is either an instance of the logical axiom (CL1)–(CL3), or it is derived from some formulas B_j, B_k, $j, k < i$, using the rule of modus ponens. We say that the formula $A := B_n$ is proved using the proof w.

If a formula A is proved using some proof then we write $\vdash A$ and say that A is *provable* in the propositional calculus, or that it is a formal *theorem* of it.

The propositional calculus can be considered as the basic object among other formal logical systems. The specificity of the latter is reflected in the extension of the set of logical axioms LAx by the set of special axioms, $\text{SAx} \subset F_J$. Special axioms, together with the logical ones and the inference rule of modus ponens determine a *propositional theory* T.

We say that a formula A is proved in T if there exists a proof w such that

$$w := B_1, \dots, B_n$$

where $A := B_n$ and each formula B_i is either an instance of logical axiom (CL1)–(CL3), or a special axiom, or it is derived from some formulas B_j, B_k, $j, k < i$, using the rule of modus ponens. In this case we write $T \vdash A$ and say that A is provable in the propositional theory T (or, it is a theorem of T).

Let $\Gamma \subseteq F_J$ be a set of formulas and T be a theory with a set of special axioms SAx. Then by $T' = T \cup \Gamma$ we denote a theory whose set of special axioms is $\text{SAx} \cup \Gamma$. The theory T' is called the *extension* of the theory T.

Extension of a theory is characterized in the following theorem, being one of the basic theorems of classical propositional logic.

THEOREM 3.1 (DEDUCTION)
Let T be a theory and $A, B \in F_J$. Then

$$T \vdash A \Rightarrow B \quad \textit{iff} \quad T \cup \{A\} \vdash B.$$

Semantics. Semantics in logic means to say, which propositions are true and which are false. More formally, we assign a truth value from $\{\mathbf{0}, \mathbf{1}\}$ to the formulas from F_J. This procedure will be called the *truth valuation* of formulas from F_J.

The truth valuation of formulas is heavily based on the principle of the *truth functionality*. This means that the truth value of a formula is computed from the truth values of its parts using special operations assigned to the logical connectives. With respect to this, we come to the following definition.

DEFINITION 3.4
A truth valuation of formulas in the propositional calculus is a function $\mathcal{D}$: $F_J \longrightarrow \{\mathbf{0}, \mathbf{1}\}$ defined as follows. If A is a propositional variable p say, then

without any restrictions $\mathcal{D}(p) \in \{\mathbf{0}, \mathbf{1}\}$. *Otherwise,*

$$\mathcal{D}(\perp) = \mathbf{0},$$
$$\mathcal{D}(A \Rightarrow B) = \mathcal{D}(A) \to \mathcal{D}(B),$$

and for the derived connectives

$$\mathcal{D}(\neg A) = \neg\mathcal{D}(A), \qquad \mathcal{D}(A \vee B) = \mathcal{D}(A) \vee \mathcal{D}(B),$$
$$\mathcal{D}(A \wedge B) = \mathcal{D}(A) \wedge \mathcal{D}(B), \qquad \mathcal{D}(A \Leftrightarrow B) = \mathcal{D}(A) \leftrightarrow \mathcal{D}(B)$$

where the operations on $[0,1]$ "$\to, \neg, \vee, \wedge, \leftrightarrow$" *have been defined in Example 2.1 using the truth tables.*

DEFINITION 3.5

(i) *A formula A is a tautological consequence of formulas $B_1, \ldots, B_n$ if in every truth valuation $\mathcal{D} : F_J \longrightarrow \{\mathbf{0}, \mathbf{1}\}$ such that $\mathcal{D}(B_i) = \mathbf{1}$, $i = 1, \ldots, n$ it holds that $\mathcal{D}(A) = \mathbf{1}$.*

(ii) *A formula A is a tautology if it is true for all truth valuations, i.e. $\mathcal{D}(A) = \mathbf{1}$ holds for all $\mathcal{D} : F_J \longrightarrow \{\mathbf{0}, \mathbf{1}\}$. If A is a tautology then we write $\models A$.*

A formula is a tautology if it is a tautological consequence of the empty set of formulas. Note that all the formulas (CL1)–(CL3) are tautologies.

The above definition is a special case of the following.

DEFINITION 3.6

Let T be a theory with a set SAx *of special axioms. A truth valuation $\mathcal{D}$ is a model of the theory T if $\mathcal{D}(A) = \mathbf{1}$ holds for all formulas $A \in$* SAx*. This relation is denoted by $\mathcal{D} \models T$.*

A formula A is true in T, $T \models A$, if it is true in all models of T, i.e. if $\mathcal{D}(A) = \mathbf{1}$ for all $\mathcal{D} \models T$.

In Subsection 2.1.1, the concept of normal form in Boolean algebras has been introduced. Its consequence in the propositional logic are the concepts of conjunctive and disjunctive normal forms. These may serve as a useful technical simplification for dealing with more complicated formulas since, as seen from Theorem 3.2 below, they make possible to transform each formula into one of two standard shapes. In Chapter 5, we will extensively develop the theory of normal forms in fuzzy logic.

DEFINITION 3.7

Let B_i, $i = 1, \ldots, n$, be atomic formulas or their negations. A conjunction $B_{i_1} \wedge \cdots \wedge B_{i_r}$, $1 \le r \le n$ of pairwise disjoint components is called an elementary conjunction. Analogously, a disjunction $B_{i_1} \vee \cdots \vee B_{i_r}$, $1 \le r \le n$ of pairwise disjoint components is called an elementary disjunction.

(i) *A formula $A^{\vee} := C_1 \vee \cdots \vee C_s$, $s \ge 1$, is in the conjunctive normal form.*

(ii) *A formula $A^{\wedge} := C_1 \wedge \cdots \wedge C_s$, $s \ge 1$, is in the disjunctive normal form.*

THEOREM 3.2 (NORMAL FORMS)

To every formula A there is a conjunctive normal form $A^{\wedge}$ and a disjunctive normal form $A^{\vee}$ such that $\models A \Leftrightarrow A^{\wedge}$ and $\models A \Leftrightarrow A^{\vee}$.

Consistency and completeness. One of the fundamental problems when dealing with some formal theory is to decide, whether it is consistent or contradictory. Both concepts are complementary as can be seen from the following definition.

DEFINITION 3.8
A theory T is contradictory if $T \vdash A$ holds for every formula $A \in F_J$. Otherwise it is consistent. A theory T is maximally consistent if every theory T' such that $T \subset T'$ is contradictory.

THEOREM 3.3 (CONSISTENCY)
A theory T is contradictory iff there is a formula A such that $T \vdash A \wedge \neg A$.

The formula $A \wedge \neg A$ is called the *contradiction.*

LEMMA 3.1
A theory $T \cup \{\neg A\}$ is contradictory iff $T \vdash A$.

The following lemma will be proved in a more general setting in Section 3.3 (Lemma 3.6).

LEMMA 3.2
Every consistent theory T is contained in some maximally consistent theory.

Note that both provability as well as truth in the model are two different ways how to characterize true formulas. The principal balance between them is reflected by the soundness property.

THEOREM 3.4 (SOUNDNESS)

$$T \vdash A \quad \textit{implies} \quad T \models A$$

holds for every formula $A \in F_J$.

A stronger property is completeness which states that provable formulas are just those being true. In other words, both ways of characterization of true formulas lead to the same result.

THEOREM 3.5 (WEAK COMPLETENESS)

$$\vdash A \quad \textit{iff} \quad \models A$$

holds for every tautology $A \in F_J$.

Since there are formulas which are not tautologies (for example, $(A \Rightarrow B) \Rightarrow B$), it follows from this and the consistency theorem that classical propositional logic is consistent.

The above theorem characterizes completeness only for tautologies. A more general result is obtained in the following theorem.

THEOREM 3.6 (COMPLETENESS)

$$T \vdash A \quad \textit{iff} \quad T \models A$$

holds for every formula $A \in F_J$.

The proof of the weak completeness[1] follows immediately from the completeness when putting SAx $= \emptyset$. There are also direct proofs of the weak completeness (cf. [17, 121]).

3.1.2 Predicate logic

The expressive power of propositional calculus is not sufficient to describe more complex mathematical concepts since it deals only with whole propositions and has no possibility to work with constituent them objects. The way out is the extension of propositional calculus to predicate one. This section gives a brief overview of classical *first-order* predicate calculus, which makes possible to quantify objects. Predicate calculus is the other example of a formal logical system and thus, we will introduce it analogously to propositional calculus.

Syntax. The formal *language* J of classical predicate calculus consists of:

(i) Countable set of object variables $x, y, \ldots$.

(ii) Finite or countable set of object constants $\mathbf{u}_1, \mathbf{u}_2, \ldots$.

(iii) Finite or countable set of n-ary[2] functional symbols $f, g, \ldots$.

(iv) Nonempty finite or countable set of n-ary predicate symbols $P, Q, \ldots$.

(v) Symbol (logical constant) for the falsity $\bot$.

(vi) Symbol for the logical binary connective of implication $\Rightarrow$.

(vii) Symbol for the general quantifier $\forall$.

(viii) Various kinds of brackets as auxiliary symbols.

Terms in predicate calculus are used to denote objects.

DEFINITION 3.9

(i) A variable x *or constant* $\mathbf{u}$ *is a (atomic) term.*

(ii) Let f *be an* n*-ary functional symbol and* $t_1, \ldots, t_n$ *terms. Then the expression* $f(t_1, \ldots, t_n)$ *is a term.*

Formulas of classical predicate calculus are defined as follows.

---

[1]Let us remark that the terminology is not well-established. Some authors speak about "strong completeness" instead of "completeness". J. R. Shoenfield in [121] calls weak completeness the "tautology theorem".

[2]To simplify the notation, we will generally write arity of various symbols as n without specific stress that it may differ from symbol to symbol, but always keeping this fact on mind.

DEFINITION 3.10

(i) *The logical constant $\bot$ is an (atomic) formula.*

(ii) *Let P be an n-ary predicate symbol and $t_1, \dots, t_n$ be terms. Then the expression $P(t_1, \dots, t_n)$ is a (atomic) formula.*

(iii) *If A, B are formulas then $A \Rightarrow B$ is a formula.*

(iv) *If x is a variable and A a formula then $(\forall x)A$ is a formula.*

The derived connectives are defined in the same way as in propositional calculus, i.e. as shorts for formulas (3.1)–(3.4). The existential quantifier is defined as a short for the formula $(\exists x)A := \neg(\forall x)\neg A$.

Given a formula $(\forall x)A$. Then the *scope of the quantifier* $(\forall x)$ is the formula A. If the variable x is in the scope of $(\forall x)$ then it is *bound* in A. If x is not in the scope of any quantifier then it is *free*. The set of free variables $FV(t)$ of the term t consists of all distinct variables occurring in t. Analogously we can define the set $BV(A)$ of bound variables of the formula A. Note that, in general, $FV(A) \cap BV(A) \neq \emptyset$. A term t is closed if $FV(t) = \emptyset$ and open otherwise.

If $FV(A) \subseteq \{x_1, \dots, x_n\}$ then we will often write $A(x_1, \dots, x_n)$. A formula A is *closed* (a *sentence*) if $FV(A) = \emptyset$. It is *open* if it contains no quantifiers.

Let $A(x)$ be a formula and t a term. By $A_x[t]$ we denote a formula in which all the free occurrences of the variable x are replaced by the term t. The term t is *substitutible* for the variable $x \in FV(A)$ if there is no variable $y \in FV(t)$ such that x is in the scope of the quantifier $\forall y$ in A, i.e. that y would become bound in $A_x[t]$.

EXAMPLE 3.2

The formula $P(x, f(x, y))$ is atomic. The scope of $(\forall x)$ in $A := Q(x) \vee (\forall x)P(x, f(x, y))$ is the formula $P(x, f(x, y))$. The variable x is bound in $(\forall x)P(x, f(x, y))$ but both free as well as bound in A. The variable y is free in A. If t is the term then $A_y[t]$ is $(\forall x)P(x, f(x, t)) \vee Q(x)$. If, for example, $t := h(z, u)$ then it is substitutible for y in A, but $t := g(x, z)$ is not. □

An *instance* of a formula $A(x_1, \dots, x_n)$ is a formula $A_{x_1, \dots, x_n}[t_1, \dots, t_n]$. A *closure* of a formula $A(x_1, \dots, x_n)$ is a formula $(\forall x_1) \cdots (\forall x_n)A$.

Let $A, B, C \in F_J$ be formulas and t be a term. The set LAx of *logical axioms* of classical predicate calculus are all formulas of the form (CL1)–(CL3) (i.e., all propositional logical axioms are also predicate logical axioms) as well as formulas of the form

(CL4) $\vdash (\forall x)A(x) \Rightarrow A_x[t]$, provided that t is substitutible for x in A.

(CL5) $\vdash (\forall x)(A \Rightarrow B) \Rightarrow (A \Rightarrow (\forall x)B)$ where x is not free in A.

The formula of the form (CL4) is the *substitution axiom*.

Basic inference (deduction) rules of first-order predicate calculus is the *rule of modus ponens* (3.5) and the *rule of generalization*

$$r_G : \frac{A}{(\forall x)A}. \tag{3.6}$$

The first-order formal theory is a set of formulas $T \subseteq F_J$, which can be characterized as a triple

$$T = \langle \mathrm{LAx}, \mathrm{SAx}, \{r_{MP}, r_G\}\rangle.$$

If SAx $= \emptyset$ then T is (classical) *predicate calculus*. The notions of a proof and a proof in a theory are defined in the same way as in the propositional calculus.

THEOREM 3.7 (DEDUCTION)
Let T be a theory and $A, B \in F_J$ formulas such that A is closed. Then

$$T \vdash A \Rightarrow B \quad \textit{iff} \quad T \cup \{A\} \vdash B.$$

The first-order theory with equality. Let us extend the language J of predicate calculus by the binary predicate symbol $=$. Furthermore, the atomic formula based on this symbol will be shortly written as $t_1 = t_2$ where t_1, t_2 are some terms.

The following are (logical) *equality axioms*.

(EQ1) $\vdash x = x$ (*identity axiom*)

(EQ2) Let f be an n-ary functional symbol. Then

$$\vdash (x_1 = y_1) \Rightarrow (\cdots \Rightarrow ((x_n = y_n) \Rightarrow (f(x_1, \ldots, x_n) = f(y_1, \ldots, y_n)) \cdots).$$

(EQ3) Let P be an n-ary predicate symbol. Then

$$\vdash (x_1 = y_1) \Rightarrow (\cdots \Rightarrow ((x_n = y_n) \Rightarrow (P(x_1, \ldots, x_n) \Rightarrow P(y_1, \ldots, y_n)) \cdots).$$

THEOREM 3.8 (EQUALITY)

(i) *Let the term s' be obtained from the term s by means of replacement of some occurrences of terms $t_1, \ldots, t_n$ in s by terms $t'_1, \ldots, t'_n$, respectively. If $T \vdash t_1 = t'_1$, ..., $T \vdash t_n = t'_n$ then $T \vdash s = s'$.*

(ii) *Let the formula A' be obtained from the formula A by means of replacement of some occurrences of terms $t_1, \ldots, t_n$ (containing no bound variables) in A by terms $t'_1, \ldots, t'_n$, respectively. If $T \vdash t_1 = t'_1$, ..., $T \vdash t_n = t'_n$ then $T \vdash A \Leftrightarrow A'$.*

Semantics. Semantics of predicate calculus is determined within a certain structure, which gives a meaning to symbols from the language.

DEFINITION 3.11
Let J be the language of predicate calculus. The structure for J is

$$\mathcal{D} = \langle D, P_D, \ldots, f_D, \ldots, u, \ldots\rangle$$

*where D is a nonempty set, $P_D \subseteq D^n$, ... are n-ary relations assigned to each n-ary predicate symbol $P, \ldots$, and $f_D : D^n \longrightarrow D$ are n-ary functions on D assigned to each n-ary functional symbol f. Finally, the $u, \ldots \in D$ are designated elements of D assigned to each constant **u**, ... of J.*

There are several equivalent ways how we can define interpretation of terms and formulas. In this book, we will employ the following.

Let $\mathcal{D}$ be a structure for the language J. We extend J to the language $J(\mathcal{D})$ by new constants being names for all the elements from D. These constants will be denoted by the corresponding bold-face letter, i.e. if $d \in D$ then $\mathbf{d}$ is a constant in $J(\mathcal{D})$ denoting it. Hence, $J(\mathcal{D}) = J \cup \{\mathbf{d} \mid d \in D\}$.

DEFINITION 3.12

(i) Interpretation of closed terms:

$$\begin{aligned} \mathcal{D}(\mathbf{u}_i) &= u_i, \quad \mathbf{u}_i \in J, u_i \in D, \\ \mathcal{D}(\mathbf{d}) &= d, \quad d \in D, \\ \mathcal{D}(f(t_1, \dots, t_n)) &= f_D(\mathcal{D}(t_1), \dots, \mathcal{D}(t_n)). \end{aligned}$$

(ii) Interpretation of closed formulas: let $t_1, \dots, t_n$ be closed terms. Then

$$\begin{aligned} \mathcal{D}(\bot) &= \mathbf{0}, \\ \mathcal{D}(P(t_1, \dots, t_n)) &= P_D(\mathcal{D}(t_1), \dots, \mathcal{D}(t_n)), \\ \mathcal{D}(A \Rightarrow B) &= \mathcal{D}(A) \to \mathcal{D}(B), \\ \mathcal{D}((\forall x)A) &= \inf\{\mathcal{D}(A_x[\mathbf{d}]) \mid d \in D\}, \end{aligned}$$

and in the case of the derived connectives

$$\begin{aligned} \mathcal{D}(A \vee B) &= \mathcal{D}(A) \vee \mathcal{D}(B), & \mathcal{D}(A \wedge B) &= \mathcal{D}(A) \wedge \mathcal{D}(B), \\ \mathcal{D}(A \Leftrightarrow B) &= \mathcal{D}(A) \leftrightarrow \mathcal{D}(B) & \mathcal{D}((\exists x)A) &= \sup\{\mathcal{D}(A_x[\mathbf{d}]) \mid d \in D\}. \end{aligned}$$

Note that the same element $d \in D$ may be assigned to two constants, namely to the additional constant $\mathbf{d}$ being its name, and, provided it is one of the designated elements $d = u_i$ of the structure $\mathcal{D}$, to the constant $\mathbf{u}_i$ of the language J. If $\mathcal{D}(A) = \mathbf{1}$ then we will write $\mathcal{D} \models A$.

In the previous definition, only closed formulas (sentences) have been considered. If $A(x_1, \dots, x_n)$ is not a closed formula, we must first define an *evaluation* $e : FV(A) \longrightarrow D$ of its free variables $x_1, \dots, x_n$ such that $e(x_1) = d_1, \dots, e(x_n) = d_n$. Then the interpretation of A can be given by the interpretation of $A_{x_1,\dots,x_n}[\mathbf{d}_1, \dots, \mathbf{d}_n]$, which is a closed formula.

DEFINITION 3.13

A formula $A(x_1, \dots, x_n)$ is satisfied *in $\mathcal{D}$ by the evaluation e, $e(x_1) = d_1, \dots, e(x_n) = d_n$, if $\mathcal{D}(A_{x_1,\dots,x_n}[\mathbf{d}_1, \dots, \mathbf{d}_n]) = \mathbf{1}$.*

A formula $A(x_1, \dots, x_n)$ is satisfiable *in $\mathcal{D}$ if it is satisfied for some evaluation e.*

A formula A is true *in $\mathcal{D}$, $\mathcal{D} \models A$, if it is satisfied by all evaluations e.*

A formula A is false *in $\mathcal{D}$ if it is not satisfied by all evaluations e.*

Some properties of first-order theories. Let T be a theory given by a set of special axioms SAx. The structure $\mathcal{D}$ is a *model* of T, $\mathcal{D} \models T$, if all its special axioms $A \in \mathrm{SAx}$ are true in $\mathcal{D}$, i.e. if $\mathcal{D} \models A$ holds for all $A \in \mathrm{SAx}$. Then $T \models A$ if A is true in all models of T.

Theorems 3.2, 3.3, 3.4 can be proved also in predicate logic. We will furthermore list some other theorems whose proofs can be found in the special literature. Let us remark that these are generalized also in fuzzy logic (see Chapter 4).

DEFINITION 3.14

(i) *A formula A is in prenex form if*

$$A := (Qx_1) \cdots (Qx_n)B \tag{3.7}$$

where Q is either of the quantifiers $\forall$ or $\exists$, and B is an open formula.

(ii) *A formula A' is a variant of A if A' is obtained from A by successive replacements of its subformulas of the form $(\forall x)B(x)$ by the formulas $(\forall y)B_x[y]$, provided that y is not free in B.*

The formula B from (3.7) is called *matrix* and $(Qx_1) \cdots (Qx_n)$ its *prefix*.

THEOREM 3.9

(a) *For every $A \in F_J$ there is $A' \in F_J$ in prenex form such that $\vdash A \Leftrightarrow A'$.*

(b) *If A' is a variant of A then $\vdash A' \Leftrightarrow A$.*

DEFINITION 3.15

Let T be a theory in the language J, $J' \supset J$ an extension of J and $\Gamma \subset F_{J'}$ some set of formulas of the language J'. Then the theory $T' = T \cup \Gamma$ is called an extension of T. The extension T' is simple if $J' = J$. It is conservative if $T' \vdash A$ implies $T \vdash A$ holds for every formula $A \in F_J$.

Obviously, if T' is an extension of T then $T \vdash A$ implies $T' \vdash A$.

THEOREM 3.10 (CONSTANTS)

Let T' be an extension of T by adding new constants, i.e. $J(T') = J(T) \cup \{\mathbf{c}_1, \ldots, \mathbf{c}_n\}$. Then $T \vdash A$ iff $T' \vdash A_{x_1,\ldots,x_n}[\mathbf{c}_1, \ldots, \mathbf{c}_n]$ holds true, i.e. extension of a theory by new constants is conservative.

DEFINITION 3.16

Theories T and T' are equivalent if one is extension of the other one, i.e. they have the same language and $T \vdash A$ iff $T' \vdash A$ holds for every formula.

LEMMA 3.3

The theories T and T' are equivalent iff they have the same models.

A theory T is called *Henkin* if for every sentence $(\forall x)A(x)$ there is a constant $\mathbf{u_A}$ such that

$$T \vdash A_x[\mathbf{u_A}] \Rightarrow (\forall x)A(x). \tag{3.8}$$

Formula (3.8) is also called *Henkin axiom*. The constant $\mathbf{u_A}$ is called *special* for $(\forall x)A(x)$. It can be proved that we can conservatively extend every theory to the Henkin one.

A formula A is *decidable* in T if either $T \vdash A$ or $T \vdash \neg A$. A theory T is *complete* if every formula $A \in F_J$ is decidable in it. Note that, in general, there exist undecidable formulas. However, it can be proved that every consistent theory has a simple complete extension.

The following two important theorems of classical logic are usually referred to as Gödel's completeness theorems.

THEOREM 3.11 (COMPLETENESS II)
A theory T is consistent iff it has a model.

THEOREM 3.12 (COMPLETENESS I)

$$T \vdash A \quad \textit{iff} \quad T \models A$$

holds for every formula $A \in F_J$.

DEFINITION 3.17
A formula A is a quasitautology if there are some formulas $B_1, \dots, B_n$ being instances of the equality axioms (EQ1)– (EQ3) such that

$$\models B_1 \Rightarrow (\cdots \Rightarrow (B_n \Rightarrow A) \cdots).$$

EXAMPLE 3.3
Let t, s be terms. The formula $(t = s) \Rightarrow (P_x[t] \Rightarrow (Q \Rightarrow P_x[s]))$ is a quasitautology since it is a tautological consequence of the instances of the equality axioms $(t = s) \Rightarrow ((t = t) \Rightarrow ((t = t) \Rightarrow (s = t)))$ and $(t = s) \Rightarrow (P_x[t] \Rightarrow P_x[s])$. □

The following is the well-known Hilbert-Ackermann theorem.

THEOREM 3.13 (CONSISTENCY)
Open theory T (i.e. a theory with special axioms being open formulas) is contradictory iff there is a quasitautology

$$\neg B_1 \vee \cdots \vee \neg B_n$$

where $B_1, \dots, B_n$ are instances of special axioms of the theory T.

A formula A in prenex form is existential if it contains no general quantifiers.

DEFINITION 3.18
Let A be a formula. We assign it a formula A_H as follows. If A is existential then $A_H := A$. Otherwise $A := (\exists x_1) \cdots (\exists x_n)(\forall y)B$. We extend J with a new n-ary functional symbol f and put

$$A^* := (\exists x_1) \cdots (\exists x_n) B_y[f(x_1, \dots, x_n)].$$

This procedure is repeated until the result is an existential formula A'. Then $A_H := A'$. We will call A_H a Herbrand variant of the formula A.

THEOREM 3.14 (HERBRAND)
Let A be a closed formula and A_H its Herbrand variant. Then $\vdash A$ iff there is a quasitautology

$$B_1 \vee \cdots \vee B_n$$

where $B_1, \ldots, B_n$ are instances of the matrix of the formula A_H.

3.1.3 Many-sorted predicate logic

A modification of predicate logic is the many-sorted one. It enables us to describe the behaviour of more complicated relations among objects of different kinds. In the syntax, these various kinds of objects are represented by variables of different *sorts*. It turns out that the modification of the language is not significant and such logic preserves all the properties of classical one. Moreover, it can be demonstrated that many-sorted logic can be represented within a single-sorted one.

We introduce an abstract set Υ of sorts. Any sequence $\langle \iota_1, \ldots, \iota_n \rangle$ of sorts from Υ is called a *type*. The formal *language* J of the many-sorted predicate logic consists of variables $x_\iota, \ldots$ of various sorts $\iota \in \Upsilon$, constants $\mathbf{u}_\iota, \ldots$ of various sorts $\iota \in \Upsilon$, predicate symbols $P, Q, \ldots$ of various types, and functional symbols $f, g, \ldots$ of various types. The other symbols are the same as in classical predicate logic, which can be understood also as "one-sorted predicate logic". Therefore, terms and formulas are defined in the same way as for predicate logic with the exception that we have terms of different types. At the same time, we must always say, which sorts of variables can occur on the given place in the formula. Obviously, the type $\langle \iota \rangle$ is identified with the sort ι.

Semantics of formulas of the many-sorted predicate logic is also slightly modified.

DEFINITION 3.19
The structure for the language J of the many-sorted predicate logic is

$$\mathcal{D} = \langle \{D_\iota \mid \iota \in \Upsilon\}, P_D, \ldots, f_D, \ldots, u, \ldots \rangle$$

where D_ι are nonempty sets corresponding to the respective sorts $\iota \in \Upsilon$. If P is a predicate symbol of the type $\langle \iota_1, \ldots, \iota_n \rangle$ then it is assigned a relation $P_D \subseteq D_{\iota_1} \times \cdots \times D_{\iota_n}$. Similarly, If f is a functional symbol of the type $\langle \iota_1, \ldots, \iota_{n+1} \rangle$ then it is assigned a function $f_D \subseteq D_{\iota_1} \times \cdots \times D_{\iota_n} \longrightarrow D_{\iota_{n+1}}$. Finally, if $\mathbf{u}$ is a constant of the sort ι then it is assigned an element $u \in D_\iota$.

All the other concepts in predicate logic are defined in the same way or slightly modified accordingly. Also all theorems of predicate logic are valid for many-sorted logic.

3.2 Classical Model Theory

Model theory can be taken as a part of the formal logical theory. It investigates special properties of models, especially of predicate logic (the concept of model of the propositional logic is very simple). Let us stress that by *model* in the model theory, we understand the *structure* introduced in Subsection 3.1.2.

Model theory has important applications. For example, big parts of algebra can be considered to belong to the model theory. Hence, all structures discussed in Section 2.1 and elsewhere are models. The difference is that in model theory, the stress is put also on the role of the language, which in algebra is often neglected. This section contains a brief overview of some of the main concepts and results. More details and proofs can be found in [8, 15, 17, 121] and elsewhere.

Some basic concepts of model theory. Models, i.e. structures for the given language J of classical predicate logic, will be denoted by the script letters $\mathcal{D}, \ldots$. Recall that by means of them, we interpret terms as well as formulas, i.e. each term t is assigned an element $\mathcal{D}(t) \in D$ and each formula A is assigned a truth value $\mathcal{D}(A) \in \{\mathbf{0}, \mathbf{1}\}$.

DEFINITION 3.20
Two models $\mathcal{D}$ and $\mathcal{D}'$ are isomorphic, $\mathcal{D} \cong \mathcal{D}'$, if there is a bijection $g : D \longrightarrow D'$ such that the following holds for all $d_1, \ldots, d_n \in D$:

(i) For each couple of functions f_D in $\mathcal{D}$ and $f_{D'}$ in $\mathcal{D}'$ assigned to a functional symbol $f \in J$

$$g(f_D(d_1, \ldots, d_n)) = f_{D'}(g(d_1), \ldots, g(d_n)).$$

(ii) For each couple of relations P_D in $\mathcal{D}$ and $P_{D'}$ in $\mathcal{D}'$ assigned to a predicate symbol $P \in J$

$$\langle d_1, \ldots, d_n\rangle \in P_D \quad \text{iff} \quad \langle g(d_1), \ldots, g(d_n)\rangle \in P_{D'}.$$

(iii) For each couple of elements u in $\mathcal{D}$ and u' in $\mathcal{D}'$ assigned to a constant symbol $\mathbf{u} \in J$, $g(u) = u'$.

DEFINITION 3.21
Two models $\mathcal{D}$ and $\mathcal{D}'$ are elementary equivalent, $\mathcal{D} \equiv \mathcal{D}'$, if

$$\mathcal{D} \models A \quad \text{iff} \quad \mathcal{D}' \models A \tag{3.9}$$

holds for every formula $A \in F_J$.

A model $\mathcal{D}$ is a *submodel* of $\mathcal{D}'$, $\mathcal{D} \subset \mathcal{D}'$, if $D \subseteq D'$, $f_D = f_{D'}|D^n$, and $P_D = P_{D'} \cap D^n$ (recall that n denotes the corresponding arity of f and P, respectively) and for each pair of constants u in $\mathcal{D}$ and u' in $\mathcal{D}'$ assigned to a constant symbol $\mathbf{u} \in J$ is $u = u'$. The submodel $\mathcal{D}$ is an *elementary submodel* of $\mathcal{D}'$, $\mathcal{D} \prec \mathcal{D}'$, if (3.9) holds for every formula $A_{x_1,\ldots,x_n}[\mathbf{d}_1, \ldots, \mathbf{d}_n]$ and $d_1, \ldots, d_n \in D$. The model $\mathcal{D}'$ is an *extension* or *elementary extension* of $\mathcal{D}$, respectively.

LEMMA 3.4
Let $\mathcal{D}$ and $\mathcal{D}'$ be models.

(a) *If $\mathcal{D} \cong \mathcal{D}'$ then $\mathcal{D} \equiv \mathcal{D}'$.*

(b) *If $\mathcal{D} \prec \mathcal{D}'$ then $\mathcal{D} \equiv \mathcal{D}'$.*

Compactness, elementary chains and cardinality. Compactness property is in general a characterization of the infinite object by means of the finite ones. If it is true then it may significantly simplify various provability techniques. We have compactness also in model theory.

THEOREM 3.15 (COMPACTNESS)
A theory T has model iff each its finite subtheory $T' \subseteq T$ has a model.

Let us consider the language $J(\mathcal{D})$ (defined on page 70). The expanded model (or *expansion* of the model $\mathcal{D}$) is the model

$$\mathcal{D}_D = \langle \mathcal{D}, \{\mathbf{d} \mid d \in D\} \rangle, \tag{3.10}$$

i.e. we add to $\mathcal{D}$ the constants $\mathbf{d} \in J(\mathcal{D})$ being names for all the elements of D.

DEFINITION 3.22

(i) *A diagram of the model $\mathcal{D}$ is a set $\Delta_{\mathcal{D}}$ of atomic formulas B or their negations $\neg B$, $B \in F_{J(\mathcal{D})}$ true in $\mathcal{D}_D$.*

(ii) *A positive diagram is a set $\Delta^+_{\mathcal{D}}$ of atomic formulas $B \in F_{J(\mathcal{D})}$ true in $\mathcal{D}_D$.*

(iii) *A model $\mathcal{E}$ is isomorphically embedded into $\mathcal{D}$ if there is a submodel $\mathcal{C} \subset \mathcal{D}$ such that $\mathcal{E} \cong \mathcal{C}$.*

LEMMA 3.5
Let $\mathcal{D}, \mathcal{E}$ be models of the language J.

(a) *The model $\mathcal{E}$ is isomorphically embedded in $\mathcal{D}$ iff there is an expansion of $\mathcal{D}$ being a model of the diagram of $\mathcal{E}$.*

(b) *The model $\mathcal{D}$ is elementary extension of $\mathcal{E}$ iff $\mathcal{E} \subset \mathcal{D}$ and $\mathcal{E}_D \equiv \mathcal{D}_D$.*

A set of models $\{\mathcal{D}_\alpha \mid \alpha < \xi\}$ where ξ is some ordinal number is an *elementary chain* if $\mathcal{D}_\alpha \prec \mathcal{D}_\beta$ for each $\alpha < \beta < \xi$.

THEOREM 3.16 (ELEMENTARY CHAINS)
Let $\mathcal{D}_0 \prec \mathcal{D}_2 \prec \ldots \prec \mathcal{D}_\alpha \prec \ldots$ be an elementary chain of models, $\alpha < \xi$ for some ordinal number ξ. Then

$$\mathcal{D} = \bigcup_{\alpha<\xi} \mathcal{D}_\alpha$$

is an elementary extension of each model $\mathcal{D}_\alpha$ from this chain.

Ultraproducts. An important method in the model theory is construction of models using filters. Let us be given a set of models $\{\mathcal{D}_i \mid i \in I\}$ of the language J where I is some index set and let $F \subset P(I)$ be a filter. We construct a new model $\mathcal{D}_F$ as follows. Put

$$D = \prod_{i \in I} D_i,$$

which is a Cartesian product of the supports D_i of the corresponding models $\mathcal{D}_i$, $i \in I$. The element $d \in D$ is a tuple $d = \langle (d)_i \mid i \in I \rangle$ where $(d)_i$ is i-th component of d. Put

$$d \sim d' \quad \text{iff} \quad \{i \in I \mid (d)_i = (d')_i\} \in F.$$

It can be demonstrated that $\sim$ is an equivalence on D. Then the *filter product* with respect to the filter F of the sets D_i, $i \in I$, is the factor set $D| \sim$ denoted by

$$D_F = \prod_F D_i = D| \sim . \tag{3.11}$$

Elements of this factor set will be denoted by $[d] \in D_F$ where $d \in D$.

Let $f \in J$ be an n-ary functional symbol. Then it is assigned a function $f_{D_F} : D_F^n \longrightarrow D_F$ defined by

$$f_{D_F}([d_1], \ldots , [d_n]) = [\langle f_{D_i}((d_1)_i, \ldots , (d_n)_i) \mid i \in I \rangle]. \tag{3.12}$$

where f_{D_i} are functions assigned to f in $\mathcal{D}_i$.

Let $P \in J$ be an n-ary predicate symbol. Then it is assigned a relation $P_{D_F} \subseteq D_F^n$ defined by

$$\langle [d_1], \ldots , [d_n] \rangle \in P_{D_F} \quad \text{iff} \quad \{i \in I \mid \langle (d_1)_i, \ldots , (d_n)_i \rangle \in P_{D_i}\} \in F \tag{3.13}$$

where P_{D_i} are relations assigned to P in $\mathcal{D}_i$.

Let $\mathbf{u} \in J$ be a constant. Then it is assigned an element

$$u_F = [\langle (u)_i \mid i \in I \rangle] \in D_F \tag{3.14}$$

where $(u)_i \in D_i$ are elements assigned to $\mathbf{u}$ in $\mathcal{D}_i$.

DEFINITION 3.23
Let F be an ultrafilter on I. An ultraproduct of the set of models $\{\mathcal{D}_i \mid i \in I\}$ of the language J is the structure

$$\mathcal{D}_F = \prod_F \mathcal{D}_i = \langle D_F, P_{D_F}, \ldots , f_{D_F}, \ldots , u_F, \ldots \rangle \tag{3.15}$$

where D_F, P_{D_F}, f_{D_F} and u_F are defined in (3.11)–(3.14), respectively.

THEOREM 3.17 (ŁOŚ)
Let $\mathcal{D}_F$ be an ultrapoduct $\prod_F \mathcal{D}_i$ and I an index set.

(a) *For every term t it holds that*

$$\mathcal{D}_F(t) = [d] = [\langle \mathcal{D}_i(t) \mid i \in I \rangle],$$

where the i-th component $(d)_i = \mathcal{D}_i(t)$ is an interpretation of the term t in $\mathcal{D}_i$.

(b) *For every formula $A(x_1, \ldots, x_n)$ and elements $[d_1], \ldots, [d_n]$ such that $\mathcal{D}_F(\mathbf{d}_1) = [d_1], \ldots, \mathcal{D}_F(\mathbf{d}_n) = [d_n]$ and $\mathcal{D}_i(\mathbf{d}_1) = (d_1)_i, \ldots, \mathcal{D}_i(\mathbf{d}_n) = (d_n)_i$ it holds*

$$\mathcal{D}_F \models A[\mathbf{d}_1, \ldots, \mathbf{d}_n] \quad \textit{iff} \quad \{i \in I \mid \mathcal{D}_i \models A[\mathbf{d}_1, \ldots, \mathbf{d}_n]\} \in F.$$

3.3 Formal Logical Systems

Classical logic presented in the previous sections is the most elaborated example of the so called *formal logical system.* Formal logical systems represent a general formalization of common principles, which are used as the frame for formulation of other formal logical systems. Besides classical logic we can name, for example intuitionistic, modal, temporal and other logics. The aim of this section is to describe these principles so that they could be used in the explanation of the formal system of fuzzy logic in the subsequent chapters.

Language and syntax. To characterize a formal logical system, we must first define its formal language J. This is a set consisting of:

(i) Symbols specific for the given formal system.

(ii) Finite set of symbols $\square_i$, $i = 1, \ldots, r$ for n_i-ary logical connectives.

(iii) Finite (possibly empty) set Q_i, $i = 1, \ldots, s$ of symbols for quantifiers.

(iv) Symbols $\bot, \top$ for *logical constants.*

(v) Various kinds of brackets as auxiliary symbols.

If the language J contains quantifiers then we suppose it to contain also variables $(x, y, \ldots)$ denoting objects.

Syntax is given by the set of formulas and inference rules. *Formulas* are precisely defined *finite* sequences of symbols from J derived using a *formation sequence* defined below.

First, we specify certain finite sequences of symbols from J, which we call *atomic formulas.* These depend on the concrete formal system. For example, in the propositional logic, atomic formulas are propositional variables, in the predicate logic these are expressions of the form $p(t_1, \ldots, t_n)$, etc. Besides this, the logical constants $\bot, \top$ are also atomic formulas. Then we may define the formation sequence in the following way.

DEFINITION 3.24
A sequence $A_1, \ldots A_n$ is a formation sequence of $A := A_n$ if for all $i \leq n$ either of the following holds:

(i) *A_i is an atomic formula,*

(ii) $A_i := \Box(A_{k_1}, \dots, A_{k_n})$ *for certain* $k_1, \dots, k_n < i$ *and an* n*-ary connective* $\Box \in J$,

(iii) $A_i := (Qx)A_k$ *for certain* $k < i$ *and a quantifier* $Q \in J$ *where* x *is a variable denoting objects.*

A well formed formula is the expression having a formation sequence. The set of all the well formed formulas is denoted by F_J. Note that if the connective $\Box$ is unary or binary then we usually write $\Box A_k$ or $A_{k_1} \Box A_{k_2}$, respectively.

Some chosen formulas from F_J are called axioms. They are of two kinds: *logical axioms* forming a set LAx $\subset F_J$ and *special axioms* forming a set SAx $\subseteq F_J$. Logical axioms are always fixed in the given formal system while special axioms depend on the specific problem we want to describe.

A concept enabling us to derive new formulas from the other ones in the syntax is that of inference rule[3].

DEFINITION 3.25
An inference rule r *is a partial operation on* F_J *represented by a scheme*

$$r : \frac{A_1, \dots, A_n}{r(A_1, \dots, A_n)} \tag{3.16}$$

where $A_1, \dots, A_n \in F_J$ *are premises of* r *and* $r(A_1, \dots, A_n) \in F_J$ *is a formula being its conclusion.*

In a given formal logical system, we work with some fixed set R of inference rules. All the syntactical derivations are then realized with respect to R using formal proofs.

DEFINITION 3.26
Let $X \subseteq F_J$ *be a set of formulas. A formal* proof *of a formula* $A \in F_J$ *from the set* X *is a* finite *sequence of formulas*

$$w := B_1, \dots, B_n \tag{3.17}$$

where $A := B_n$ *and for each formula* B_i, $i = 1, \dots, n$ *the following holds: either* $B_i \in X$ *or it is derived from some formulas* B_j, $j < i$, *using an inference rule* $r \in R$*; in symbols,* $B_i = r(B_{j_1}, \dots, B_{j_n})$ *where* $j_1, \dots, j_n < i$.

We say that a formula $A \in F_J$ is *provable* from the set $X \subseteq F_J$ if there is a formal proof (3.17) such that $A := B_n$.

Syntactic consequence. A crucial role in every formal logical system is played by the *consequence operation*. This is an operation on the power set of formulas being a *closure operation* $\mathcal{C} : P(F_J) \longrightarrow P(F_J)$, i.e. it fulfils the conditions

(a) $X \subseteq \mathcal{C}(X)$,

[3]The form of the inference rules used in this book has been proposed by D. Hilbert.

(b) $X \subseteq Y$ implies $\mathcal{C}(X) \subseteq \mathcal{C}(Y)$,

(c) $\mathcal{C}(X) = \mathcal{C}(\mathcal{C}(X))$

holds for all $X, Y \subseteq F_J$.

There are two kinds of the consequence operation in the formal logical systems, namely the syntactic and semantic ones. Intuitively, the syntactic consequence operation assigns to a set of formulas all the formulas derivable from the former, i.e. those to which it is possible to find a proof. The semantic consequence operation assigns to a set of formulas all the formulas being true provided that the former formulas are true as well.

First, we will characterize basic behaviour of sets of formulas with respect to the inference rules.

DEFINITION 3.27
Let $V \subseteq F_J$ be a set of formulas and r be an inference rule. We say that V is closed with respect to the inference rule r if $A_1, \ldots, A_n \in V$ implies $r(A_1, \ldots, A_n) \in V$ for all formulas $A_1, \ldots, A_n \in \mathrm{Dom}(r)$.

Using the concept of sets closed with respect to inference rules, we define the syntactic consequence operation as follows.

DEFINITION 3.28
The syntactic consequence operation is an operation $\mathcal{C}^{syn}$ on $P(F_J)$ which assigns to every $X \subseteq F_J$ the set of formulas

$$\mathcal{C}^{syn}(X) = \bigcap \{V \subseteq F_J \mid X \subseteq V, V \textit{ is closed w.r.t. all } r \in R\}. \qquad (3.18)$$

It is easy to prove that the operation $\mathcal{C}^{syn}$ is a closure operation on $P(F_J)$.

Recall that the intuitive idea behind the concept of the syntactic consequence operation is the requirement to find a proof to every formula being a consequence of formulas from X. The following theorem demonstrates that Definition 3.28 indeed fits this requirement.

THEOREM 3.18
Given a set $X \subseteq F_J$. Then $A \in \mathcal{C}^{syn}(X)$ iff it is provable from X.

PROOF: Put

$$D = \{A \mid \text{there is a formal proof of } A \text{ from } X\}.$$

Obviously, $X \subseteq D$ since if $A \in X$ then $w := A$ is a proof of A. We show that D is closed with respect to all $r \in R$. Indeed, let $B_1, \ldots, B_n \in D$ and consider $r \in R$. Then there are proofs $w_{B_1}, \ldots, w_{B_n}$ of these formulas from X. Consequently,

$$w := w_{B_1}, \ldots, w_{B_n}, r(B_1, \ldots, B_n)$$

is a proof of the formula $A := r(B_1, \ldots, B_n)$ from X and we conclude that $A \in D$.

Now, let $A \in \mathcal{C}^{syn}(X)$. Since $X \subseteq D$ and D is closed with respect to all $r \in R$, we have $A \in D$, which means that there exists its formal proof of A from X.

Vice-versa, we have to show that every provable formula, $A \in D$, belongs to $\mathcal{C}^{syn}(X)$, i.e. that $D \subseteq V$ for all V such that $X \subseteq V$ and closed w.r.t. all $r \in R$. We will proceed by induction on the length of the proof.

Let $w := A$ where $A \in X$. Obviously $A \in V$ by the definition of V. Let us consider now the proof $w := w_{B_1}, \dots, w_{B_n}, A$ where $A := r(B_1, \dots, B_n)$ and $w_{B_1}, \dots, w_{B_n}$ are proofs of some formulas $B_1, \dots, B_n$, respectively. By the induction step, $B_1, \dots, B_n \in V$, but since V is closed, we have $A \in V$. □

By this theorem, the set of syntactic consequences of the set X consists of all formulas provable from X. Moreover, it states that the nature of the syntactic consequence operation is finitistic. This is important since we thus have a possibility to characterize true formulas algorithmically.

Semantics. Now we will define *semantics* of the formal logical systems. This is realized by assignment of truth values to formulas. Hence, we must first characterize a set L of truth values. This is some special algebra, which we will suppose to be some of those discussed in Chapter 2. Furthermore, we distinguish a set $G \subset L$ of *designated truth values* . It is assumed that $\mathbf{1} \in G$ and we often simply put $G = \{\mathbf{1}\}$.

Recall that the *truth functionality* means that logical connectives and quantifiers are assigned operations on the set of truth values L. More precisely, each n-ary logical connective $\square \in J$ is assigned an n-ary function $h : L^n \longrightarrow L$ and each quantifier $Q \in J$ is assigned a generalized operation $H : P(L) \longrightarrow L$. If not stated otherwise, we will work with truth functional formal logical systems only in this book.

DEFINITION 3.29

The truth valuation of formulas is a function $\mathcal{D} : F_J \longrightarrow L$ such that:

(i) $\mathcal{D}(\bot) = \mathbf{0}$ *and* $\mathcal{D}(\top) = \mathbf{1}$.

(ii) Let $\square \in J$ be an n-ary connective having assigned an n-ary function $h : L^n \longrightarrow L$. Then

$$\mathcal{D}(\square(A_1, \dots, A_n)) = h(\mathcal{D}(A_1), \dots, \mathcal{D}(A_n)).$$

(iii) Let $Q \in J$ be a quantifier having assigned a generalized function $H : P(L) \longrightarrow L$. Furthermore, we assume that $\mathcal{D}(A(x))$ is a truth value assigned to the formula A for an object corresponding to the variable x. Then

$$\mathcal{D}((Qx)A) = H\{\mathcal{D}(A(x)) \mid \textit{for all objects corresponding to } x\}.$$

A formula A for which $\mathcal{D}(A) \in G$ holds for every truth valuation $\mathcal{D}$ is called a *tautology*. We require every logical axiom $A \in \text{LAx}$ to be a tautology.

Moreover, special axioms $A \in \mathrm{SAx}$ are always assigned some of the designated truth values.

Recall that the semantic consequence operation assigns to a set of formulas all the formulas being true provided that the former are true, as well. The precise formulation is given in the following definition.

DEFINITION 3.30
Let $X \subseteq F_J$ and $\mathcal{I}_X$ be the set of all truth valuations $\mathcal{D}_X : F_J \longrightarrow L$ such that $\mathcal{D}_X(A) \in G$ for all $A \in X$. The semantic consequence operation $\mathcal{C}^{sem} : P(F_J) \longrightarrow P(F_J)$ is a function which assigns to every set $X \subseteq F_J$ the set of formulas

$$\mathcal{C}^{sem}(X) = \{B \in F_J \mid \mathcal{D}_X \in \mathcal{I}_X \text{ implies } \mathcal{D}_X(B) \in G\}. \tag{3.19}$$

Analogously as in the case of the syntactic consequence operation, the operation $\mathcal{C}^{sem}$ is a closure operation on $P(F_J)$.

Formal theories. A formal *theory* is a set $T \subseteq F_J$ of all formulas provable from the set of axioms $\mathrm{LAx} \cup \mathrm{SAx} \subset F_J$, i.e.

$$T = \mathcal{C}^{syn}(\mathrm{SAx} \cup \mathrm{LAx}). \tag{3.20}$$

The theory is equivalently determined as a triple

$$T = \langle \mathrm{LAx}, \mathrm{SAx}, R \rangle.$$

where R is a set of inference rules. Since the sets of logical axioms and inference rules are always fixed, the theory T is uniquely defined by its set of special axioms SAx.

If a formula A is provable in T, i.e. $A \in \mathcal{C}^{syn}(\mathrm{LAx} \cup \mathrm{SAx})$, we write

$$T \vdash A \tag{3.21}$$

and say that A is a *theorem* of the formal theory T. If $\mathrm{SAx} = \emptyset$, we write $\vdash A$. Given a theory T, we denote its language by $J(T)$.

Occasionally, SAx may be extended by some set of formulas $\Gamma \subseteq F_J$. Then by $T \cup \Gamma$ we understand an extended theory $T \cup \Gamma = \mathcal{C}^{syn}(\mathrm{LAx} \cup \mathrm{SAx} \cup \Gamma)$.

DEFINITION 3.31

(i) *A theory T is contradictory if $T \vdash \bot$. Otherwise it is consistent.*

(ii) *A theory T is maximally consistent if every theory T' such that $T \subset T'$ is contradictory.*

Recall Definition 3.9 where we defined a contradictory theory to be such that $\mathcal{C}^{syn}(\mathrm{LAx} \cup \mathrm{SAx}) = F_J$. In classical logic, both definitions are equivalent. However, our definition is general enough to prove the following lemma (the proof of analogous Lemma 3.2 requires precise definition of the syntax).

LEMMA 3.6
Every consistent theory T is contained in some maximally consistent theory $\overline{T}$.

PROOF: Let $\{T_\alpha \mid \alpha < \beta\}$ (β is an ordinal number), $T_\alpha \subseteq T_{\alpha'}$ for $\alpha < \alpha'$ be a chain of consistent theories such that $T_0 := T$. Let $\bigcup_{\alpha<\beta} T_\alpha$ be contradictory. Then there is T_α such that $T_\alpha \vdash \bot$, i.e. T_α is contradictory — a contradiction with the assumption. Then we apply Zorn's lemma and obtain a maximally consistent theory $\overline{T} \supseteq T$. □

Given a theory T. Its *model* is a truth valuation $\mathcal{D}$ such that $\mathcal{D}(A) \in G$ holds for every $A \in \mathrm{SAx}$. If $\mathcal{D}$ is a model of T then we write $\mathcal{D} \models T$.

We say that a formula $A \in F_J$ is *true* in T if $A \in \mathcal{C}^{sem}(\mathrm{LAx} \cup \mathrm{SAx})$ and write $T \models A$. If $\mathrm{SAx} = \emptyset$, we write $\models A$.

Since $\mathcal{D}(A) \in G$ holds for every logical axiom $A \in \mathrm{LAx}$ and a model $\mathcal{D} \models T$, we immediately obtain that

$$\mathcal{C}^{sem}(\mathrm{LAx} \cup \mathrm{SAx}) = \{A \in F_J \mid \mathcal{D} \models T \quad \text{and} \quad \mathcal{D}(A) \in G\}. \tag{3.22}$$

Semantic consequence and completeness. Both syntactic as well as semantic consequence operations can be considered as two *different characterizations of the truth of formulas.* A natural question arises what is the balance between them. To answer it, we must assume that all the inference rules are sound.

DEFINITION 3.32
The inference rule $r \in R$ is sound if $\mathcal{D}(r(A_1, \ldots, A_n)) \in G$ whenever $\mathcal{D}(A_1), \ldots, \mathcal{D}(A_n) \in G$ holds for every truth valuation $\mathcal{D}$ and all formulas $A_1, \ldots, A_n$.

Thus, using sound inference rules, we always derive true formulas from the true ones. Apparently, it is reasonable to work with sound inference rules only.

We say that the formal logical system is *sound* if the set of syntactic consequences of any set of formulas is a subset of its semantic consequences, i.e. if

$$\mathcal{C}^{syn}(X) \subseteq \mathcal{C}^{sem}(X) \tag{3.23}$$

holds for every $X \subseteq F_J$. Using the concept of the formal theory, we may write (3.23) equivalently as

$$T \vdash A \quad \text{implies} \quad T \models A$$

holds for every theory T and a formula $A \in F_J$.

The logical system is *complete* if

$$\mathcal{C}^{syn}(X) = \mathcal{C}^{sem}(X) \tag{3.24}$$

holds for every set $X \subseteq F_J$. Equivalently, the system is complete if

$$T \vdash A \quad \text{iff} \quad T \models A \tag{3.25}$$

holds for every theory T and a formula $A \in F_J$. If (3.25) holds for tautologies only then we say that the logical system is *weakly complete*.

3.4 Model Theory in Categories

Motivation. The development of the theory of topoi shows the existence of constructions and interpretations of logical operations, which enable any elementary topos to be considered as a generalization of the category of sets. In order to do this, it is necessary to provide a language in which these definitions and constructions of the category of sets may be described, together with an interpretation of this language within a topos. The aim of this section is to introduce this language and its interpretation in a topos. The reader could find a number of examples and further details in [68].

In Section 3.1, we introduced the concept of many-sorted predicate logic, which is based on the many-sorted language J with a set Υ of sorts, a set $\mathcal{P}$ of predicate symbols of various types (or sorts) (= finite sequences of types $\iota \in \Upsilon$) and a set $\mathcal{F}$ of functional symbols of various types. We show shortly how this structure can be interpreted in a topos.

Interpretation of language in topos. Using an analogy of interpretations of formulas in the category **Set** we define two "step-by-step" methods that enable us to interpret any formula in a topos $\mathcal{E}$. These two methods will now be summarized.

Definition 3.33 (Interpretation by subobjects)
Let $\mathcal{E}$ be a topos. Then the interpretation by subobjects is a method which consists of the following steps:

(i) For any sort $\iota \in \Upsilon$ we define an object $M(\iota) \in \mathcal{E}$.

(ii) For any predicate symbol $P \in \mathcal{P}$ of a type $\iota_1 \times \cdots \times \iota_n$ we have to define a subobject $M(P)$ of $\prod_{i=1}^n M(\iota_i)$.

(iii) For any operational symbol $f \in \mathcal{F}$ of a type $\iota_1 \times \cdots \times \iota_n \longrightarrow \iota$ we have to define a morphism $M(f) : \prod_{i=1}^n M(\iota_i) \longrightarrow M(\iota)$.

(iv) For any term $t(t_1, \ldots, t_n)$ of a type $\iota_1 \times \cdots \times \iota_n \longrightarrow \iota$ we have to define a morphism $[t]^{\mathcal{D}} : \prod_{i=1}^n M(\iota_i) \longrightarrow M(\iota)$.

(v) For any formula $A(x_1, \ldots, x_n)$ (x_i is of the sort ι_i) we have to define $[A]^{\mathcal{D}}$ as a subobject of $\prod_{i=1}^n M(\iota_i)$.

Definition 3.34 (Interpretation by morphisms)
Let $\mathcal{E}$ be a topos. Then the interpretation by morphisms is a method which consists of the following steps:

(i) The same as (i) above.

(ii) For any predicate symbol $P \in \mathcal{P}$ of a type $\iota_1 \times \cdots \times \iota_n$ we have to define a morphism $M(P) : \prod_{i=1}^n M(\iota_i) \longrightarrow \Omega$.

(iii) The same as (iii) above.

(iv) The same as (iv) above.

(v) For any formula $A(x_1, \ldots, x_n)$ we have to define $||A||^{\mathcal{D}}$ as a morphism $\prod_{i=1}^n M(\iota_i) \longrightarrow \Omega$.

To realize these two methods we have to specify exactly the steps (iv) and (v) of both methods. To do this, we need some notations and notions.

Image of subobjects. For simplicity, suppose that $\mathcal{E}$ is an elementary topos although some constructions could also be done in a more general category. We say that a morphism $a \xrightarrow{f} y$ is *surjective* if whenever $b \longrightarrow y$ is a monomorphism such that f factors through $b \longrightarrow y$, then $b \longrightarrow y$ is the maximal subobject, i.e., $b \longrightarrow y$ is an isomorphism. The *image of* $x \longrightarrow a$ *under* f is the subobject $\exists_f x \rightarrowtail b$ such that there exists a surjective morphism g which makes the following diagram commutative.

$$\begin{array}{ccc} x & \xrightarrow{i} & a \\ {\scriptstyle g}\downarrow & & \downarrow{\scriptstyle f} \\ \exists_f x & \longrightarrow & b \end{array}$$

LEMMA 3.7
Let $\mathcal{E}$ be an elementary topos. Then for any subobject $x \rightarrowtail a$ and any morphism $a \xrightarrow{f} b$ in $\mathcal{E}$ there exists $\exists_f x$.

Now we are in the position to introduce necessary details of the categorical interpretation for an elementary topos $\mathcal{E}$.

DEFINITION 3.35
*An $\mathcal{E}$-*valued interpretation *(or $\mathcal{E}$-*structure*) of a language J is a structure*

$$\mathcal{D} = \langle \{M(\iota) \mid \iota \in \Upsilon\}, \{M(P) \mid P \in \mathcal{P}\}, \{M(f) \mid f \in \mathcal{F}\} \rangle,$$

such that

(i) For every sort $\iota \in \Upsilon$, $M(\iota)$ is an object of $\mathcal{E}$.

(ii) For every predicate symbol $P \in \mathcal{P}$, $P \subset \iota_1 \times \cdots \times \iota_n$, $M(P)$ is a subobject of $\prod_{i=1}^{n} M(\iota_i)$ in $\mathcal{E}$.

(iii) For every functional symbol $f \in \mathcal{F}$, $\iota_1 \times \cdots \times \iota_n \xrightarrow{f} \iota$, $M(f)$ is a morphism $M(\iota_1) \times \cdots \times M(\iota_n) \longrightarrow M(\iota)$ in $\mathcal{E}$.

Interpretation by subobjects. We will show how to interpret terms in any $\mathcal{E}$-structure $\mathcal{D}$ of J. This definition is based inductively on the structure of terms. We notice here that this definition depends on the set X of distinct free variables which contains all the free variables of the term under consideration. Hence, by the symbol $M(X)$ we denote the product (in $\mathcal{E}$)

$$M(X) = \prod_{\iota \in S} M(\iota),$$

where $S = \{\iota_x \in \Upsilon \mid x \in X, x \text{ is of a sort } \iota_x\}$. Then for a term t of sort ι, such that $FV(t) \subseteq X$, the interpretation $[t]^{\mathcal{D}}(X)$ will be defined as a morphism $M(X) \longrightarrow M(\iota)$ as follows:

(i) Let $t := x \in X$. Then $[x]^{\mathcal{D}}(X)$ is the canonical projection

$$M(X) \longrightarrow M(\iota_x).$$

(ii) Let $t := f(t_1, \dots, t_n)$, where $f \in \mathcal{F}$, $\iota_1 \times \cdots \times \iota_n \xrightarrow{f} \iota$. By the induction principle, $[t_i]^{\mathcal{D}}(X) : M(X) \longrightarrow M(\iota_i)$ are already defined and $[t]^{\mathcal{D}}(X)$ is then the following composition

$$M(X) \xrightarrow{\alpha} \prod_{i=1}^{n} M(\iota_i) \xrightarrow{M(f)} M(\iota)$$

where $\alpha = \langle [t_i]^{\mathcal{D}}(X) \rangle_{i=1}^{n}$.

Now, we will turn our attention to the interpretation of formulas. Let A be a formula such that $FV(A) \subseteq X$. Then $[A]^{\mathcal{D}}(X)$ will be a subobject of $M(X)$ defined as follows.

(i) Let $A := t_1 = t_2$, and let ι be the sort of t_1 and t_2. The morphisms

$$[t_i]^{\mathcal{D}}(X) : M(X) \longrightarrow M(\iota)$$

for $i = 1, 2$, have already been defined by the induction principle, and so by $[A]^{\mathcal{D}}(X)$ we denote the equalizer of $[t_1]^{\mathcal{D}}(X)$ and $[t_2]^{\mathcal{D}}(X)$,

$$[A]^{\mathcal{D}}(X) \rightarrowtail M(X) \underset{[t_2]^{\mathcal{D}}(X)}{\overset{[t_1]^{\mathcal{D}}(X)}{\rightrightarrows}} M(\iota).$$

(ii) Let $A := P(t_1, \dots, t_n)$ where $P \in \mathcal{P}$, $P \subseteq \iota_1 \times \cdots \times \iota_n$. Then $[A]^{\mathcal{D}}(X)$ is the pullback

$$\begin{array}{ccc} [A]^{\mathcal{D}}(X) & \rightarrowtail & M(X) \\ \downarrow & & \downarrow \langle [t_1], \dots, [t_n] \rangle \\ M(P) & \rightarrowtail & \prod_{i=1}^{n} M(\iota_i) \end{array}$$

(Recall that a pullback of monomorphisms is also a monomorphism.)

The connectives $\wedge$, $\vee$, $\neg$, and $\Rightarrow$ are interpreted as the corresponding operations $\wedge$, $\vee$, $\neg$, and $\rightarrow$ in the corresponding lattice $(\mathrm{Sub}(d), \leq)$. Thus we have the following special cases.

(iii) Let $A := \bigwedge \Phi$ (i.e. a conjunction of a set Φ of formulas). Then for every $B \in \Phi$, a subobject $[B]^{\mathcal{D}}(X)$ is already defined and $[A]^{\mathcal{D}}(X)$ is then an infimum of these subobjects in the lattice $(\mathrm{Sub}(M(X)), \leq)$,

$$[A]^{\mathcal{D}}(X) = \left[\bigwedge \Phi\right]^{\mathcal{D}}(X) := \bigwedge \left\{ [B]^{\mathcal{D}}(X) \middle| \, B \in \Phi \right\}.$$

(iv) Let $A := \bigvee \Phi$ (i.e. a disjunction of a set Φ of formulas). Then for every $B \in \Phi$ a subobject $[B]^{\mathfrak{D}}(X)$ is already defined and we put

$$[A]^{\mathfrak{D}}(X) = \left[\bigvee \Phi\right]^{\mathfrak{D}}(X) := \bigvee \{ [B]^{\mathfrak{D}}(X) |\ B \in \Phi\}$$

where the supremum is taken from the lattice $(\mathrm{Sub}(M(X)), \leq)$.

(v) Let $A := B \Rightarrow C$. Then we set

$$[A]^{\mathfrak{D}}(X) = [B \Rightarrow C]^{\mathfrak{D}}(X) = [B]^{\mathfrak{D}}(X) \to [C]^{\mathfrak{D}}(X).$$

The quantifiers $\forall$ and $\exists$ are interpreted by the maps

$$\forall_f, \exists_f : \mathrm{Sub}(a) \longrightarrow \mathrm{Sub}(b)$$

where $f : a \longrightarrow b$ is a morphism. The map $\exists_f$ is introduced according to Lemma 3.7. Let us recall that for any morphism $f : a \longrightarrow b$ in a topos $\mathcal{E}$ there exists a map $f^{-1} : \mathrm{Sub}(b) \longrightarrow \mathrm{Sub}(a)$ such that for any $x \rightarrowtail b$ the following diagram is pullback:

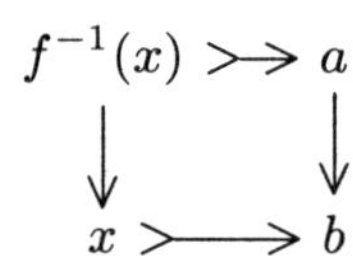

Now, the map $\forall_f : \mathrm{Sub}(a) \longrightarrow \mathrm{Sub}(b)$ is defined such that for any $x \rightarrowtail a$, the subobject $\forall_f(x)$ is the largest subobject $y \rightarrowtail b$ of b such that the pullback $f^{-1}(y)$ factors through $x \rightarrowtail a$, as can be seen from the diagram

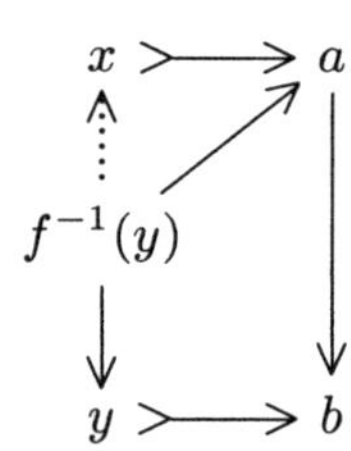

Since the ordered set $(\mathrm{Sub}(a), \leq)$ is a Heyting algebra, it follows that such object $\forall_f(x)$ exists. Then we have

(vi) Let $A := (\exists y)B$. Then, by the induction principle, there exists the interpretation $[B]^{\mathfrak{D}}(X \cup \{y\}) \rightarrowtail M(X \cup \{y\})$. (We may suppose without loss of generality that $y \notin X$.) Let p_X be the canonical projection $M(X \cup \{y\}) \longrightarrow M(X)$. Then $[A]^{\mathfrak{D}}(X)$ is the image of $[B]^{\mathfrak{D}}(X \cup \{y\})$ under p_X,

$$[(\exists y)B]^{\mathfrak{D}}(X) := \exists_{p_X} \left([B]^{\mathfrak{D}}(X \cup \{y\})\right).$$

(This explains the symbol of image.)

(vii) Let $A \equiv (\forall y)B$. Then, as in (v), we put

$$[(\forall y)B]^{\mathfrak{D}}(X) := \forall_{p_X} \left([B]^{\mathfrak{D}}(X \cup \{y\})\right).$$

If A is any formula with $FV(A) \subseteq X$, we can say that A is *true* in $\mathcal{D}$, or that $\mathcal{D}$ is an $\mathcal{E}$-*model* of A, denoted by $\mathcal{D} \models A$, if

$$[A]^{\mathcal{D}}(X) = M(X) \xrightarrow{1_{M(X)}} M(X),$$

that is, if $[A]^{\mathcal{D}}(X)$ is the maximal subobject of $M(X)$. Note also that $\mathcal{D} \models A$ if and only if the characteristic morphism of $[A]^{\mathcal{D}}(X)$ is $\top \circ !$.

Interpretation by morphisms. We will continue to define the interpretation according to Definition 3.34. Such an $\mathcal{E}$-valued interpretation in a topos $\mathcal{E}$ is an analogous structure $\mathcal{D}$, which has to satisfy conditions (i) and (iii) of Definition 3.35, whilst the condition (ii) is replaced by the condition

(ii') For every predicate symbol $P \in \mathcal{P}$, $P \subseteq \iota_1 \times \cdots \times \iota_n$, $M(P)$ is a morphism $M(\iota_1) \times \cdots \times M(\iota_n) \longrightarrow \Omega$ (recall that Ω is the subobject classifier).

Next we concentrate on the step (v) of Definition 3.34, which differs from the corresponding condition (v) of Definition 3.33. We use the notation of the above-mentioned interpretation, i.e. we use the symbols $[t]^{\mathcal{D}}(X)$, $M(X)$, etc.

Let A be a formula such that $FV(A) \subseteq X$. Then the interpretation of A in $\mathcal{E}$-structure $\mathcal{D}$ (according to Definition 3.34) will be the morphism

$$\|A\|^{\mathcal{D}}(X) : M(X) \longrightarrow \Omega$$

defined inductively. To define this morphism, we first need some notation.

Let a be an object of $\mathcal{E}$. By $\delta_a : a \times a \longrightarrow \Omega$ we denote the characteristic morphism of the subobject $a \xrightarrow{\langle 1_a, 1_a \rangle} a \times a$. In Section 2.4 we mentioned the morphism $ev : b^a \times a \longrightarrow b$ which is called an evaluation and which establishes the bijection

$$\mathrm{Hom}_{\mathcal{E}}(c \times a, \Omega) \cong \mathrm{Hom}_{\mathcal{E}}(c, \Omega^a), \quad g \mapsto \widehat{g}.$$

Let $ev : \Omega^a \times a \longrightarrow \Omega$. According to the bijection

$$\mathrm{Hom}_{\mathcal{E}}(\Omega^a \times a, \Omega) \simeq \mathrm{Sub}(\Omega^a \times a),$$

there exists a subobject $\in : \in_a \hookrightarrow \Omega^a \times a$ such that $ev = \chi_{\in}$. Now, let us consider the image of $\in_a$ under the projection $\mathrm{pr} : \Omega^a \times a \longrightarrow \Omega^a$.

$$\begin{array}{ccc} \in_a & \xrightarrow{\in} & \Omega^a \times a \\ \downarrow & & \downarrow \mathrm{pr}\ V \\ \exists_{\mathrm{pr}}(\in_a) & \longrightarrow & \Omega^a \end{array}$$

Finally, by $\exists_a$ we denote the characteristic morphism of subobject $\exists_{\mathrm{pr}}(\in_a)$.

$$\begin{array}{ccc} \exists_{\mathrm{pr}}(\in_a) & \longrightarrow & \Omega^a \\ !\downarrow & & \downarrow \exists_a V \\ \mathbf{T}\mathbf{o} & \xrightarrow{\top} & \Omega \end{array}$$

Analogously, we will define the morphism $\forall_a : \Omega^a \longrightarrow \Omega$. Let a be an object of a topos $\mathcal{E}$. For the largest subobject $1_a : a \longrightarrow a$ there exists its characteristic morphism $\chi_{1_a} : a \longrightarrow \Omega$ which makes the following diagram to be the pullback

$$\begin{array}{ccc} a & \xrightarrow{1_a} & a \\ \big\downarrow & & \big\downarrow \chi_{1_a} = \top_a \\ \mathbf{To} & \xrightarrow{\top} & \Omega \end{array}$$

We can use the bijection $g \mapsto \widehat{g}$ presented above (where $c = \mathbf{To}$) and for a morphism $\top_a$ we can take a morphism $\widehat{\top}_a : \mathbf{To} \longrightarrow \Omega^a$. It is clear that this morphism is monomorphism and it follows that it has also a characteristic morphism $\chi_{\widehat{\top}_a} : \Omega^a \longrightarrow \Omega$. Then this characteristic morphism will be denoted by $\forall_a$. Hence, the following diagram is the pullback.

$$\begin{array}{ccc} \mathbf{To} & \xrightarrow{\widehat{\top}_a} & \Omega^a \\ !\big\downarrow & & \big\downarrow \forall_a \\ \mathbf{To} & \xrightarrow{\top} & \Omega \end{array}$$

Now we may define the interpretation $||A||$.

(i) Let $A := t_1 = t_2$. The morphisms $[t_i]^{\mathcal{D}}(X) : M(X) \longrightarrow M(\iota)$ are already defined by the induction principle and, hence, by $||A||^{\mathcal{D}}(X)$ we may denote the composition of the morphisms

$$||A||^{\mathcal{D}}(X) : M(X) \xrightarrow{\langle [t_1],[t_2] \rangle} M(\iota) \times M(\iota) \xrightarrow{\delta_{M(\iota)}} \Omega.$$

(ii) Let $A := P(t_1, \dots, t_n)$ where $P \in \mathcal{P}$, $P \subseteq \iota_1 \times \cdots \times \iota_n$. Then $||A||^{\mathcal{D}}(X)$ will be the composition of the following morphisms

$$||A||^{\mathcal{D}}(X) : M(X) \xrightarrow{\langle [t_i] \rangle_{i=1}^n} \prod_{i=1}^n M(\iota_i) \xrightarrow{M(P)} \Omega.$$

(iii) Let $A := B_1 \wedge B_2$. Then for every i we have a morphism

$$||B_i||^{\mathcal{D}}(X) : M(X) \longrightarrow \Omega$$

and we put

$$||A||^{\mathcal{D}}(X) = \cap \circ \langle ||B_1||^{\mathcal{D}}(X), ||B_2||^{\mathcal{D}}(X) \rangle$$

where $\cap$ was defined in Definition 2.30 on page 57.

(iv) Let $A := B_1 \vee B_2$. Then, analogously, we put

$$||A||^{\mathcal{D}}(X) = \cup \circ \langle ||B_1||^{\mathcal{D}}(X), ||B_2||^{\mathcal{D}}(X) \rangle .$$

(v) Let $A := (\exists y)B$. Then by the induction principle we have already defined the morphism $||B||^{\mathcal{D}}(X \cup \{y\}) : M(X) \times M(\iota_y) \longrightarrow \Omega$. It follows from the previous considerations that $\mathrm{Hom}_{\mathcal{E}}(M(X) \times M(\iota_y), \Omega) \simeq \mathrm{Hom}_{\mathcal{E}}(M(X), \Omega^{M(\iota_y)})$, and so there exists a morphism

$$\widehat{||B||} : M(X) \longrightarrow \Omega^{M(\iota_y)}$$

associated with $||B||^{\mathcal{D}}(X \cup \{y\})$ under this bijection. Then by $||A||^{\mathcal{D}}(X)$ we denote the following composition

$$||(\exists y)B||^{\mathcal{D}}(X) : M(X) \xrightarrow{\widehat{||B||}} \Omega^{M(\iota_y)} \xrightarrow{\exists_{M(\iota_y)}} \Omega.$$

(vi) Let $A := (\forall y)B$. Then analogously we have already defined a morphism $||B||^{\mathcal{D}}(X \cup \{y\}) : M(X) \times M(\iota_y) \longrightarrow \Omega$ and there exists a morphism $\widehat{||B||} : M(X) \longrightarrow \Omega^{M(\iota_y)}$. Then by $||A||^{\mathcal{D}}(X)$ we denote the following composition

$$||(\forall y)B||^{\mathcal{D}}(X) : M(X) \xrightarrow{\widehat{||B||}} \Omega^{M(\iota_y)} \xrightarrow{\forall_{M(\iota_y)}} \Omega.$$

We say that $\mathcal{D}$ is an $\mathcal{E}$-*model of* A, and we write $\mathcal{D} \models A$ if $||A||^{\mathcal{D}}(X) = \top_{M(X)}$.

Relations between two types of interpretation. In the previous subsections, we have introduced two types of interpretation. A natural question arises, whether these two procedures may lead to identical subobjects. According to the first procedure, for any formula A, $[A]^{\mathcal{D}}(X)$ is a subobject of $M(X)$, whilst with respect to the other procedure, $||A||^{\mathcal{D}}(X)$ is a morphism. It is therefore reasonable to ask whether the characteristic morphism of $[A]^{\mathcal{D}}(X)$ is identical with $||A||^{\mathcal{D}}(X)$. In the following we will try to answer this question.

First, we introduce the following definition which establishes the connection of the two types of interpretation.

DEFINITION 3.36
Let

$$\mathcal{D} = \langle M(\iota) \mid \iota \in \Upsilon\}, \{M(P) \mid P \in \mathcal{P}\}, \{M(f) \mid f \in \mathcal{F}\}\rangle$$

be an $\mathcal{E}$-structure of J defined by using subobjects. Then the $\mathcal{E}$-structure

$$\mathcal{C} = \langle\{N(\iota) \mid \iota \in \Upsilon\}, \{N(P) \mid P \in \mathcal{P}\}, \{N(f) \mid f \in \mathcal{F}\}\rangle$$

of J defined by morphisms is said to be associated with *$\mathcal{D}$, provided the following conditions are satisfied.*

(i) For every sort $\iota \in \Upsilon$, $M(\iota) = N(\iota)$ holds.

(ii) For every operational symbol $f \in \mathcal{F}$, $M(f) = N(f)$ holds.

(iii) For every predicate symbol $P \in \mathcal{P}$, $N(P)$ is the characteristic morphism of a subobject $M(P)$.

The following theorem solves completely the above mentioned problem.

THEOREM 3.19
Let $\mathcal{E}$ be an elementary topos. Let

$$\mathcal{D} = \langle\{M(\iota) \mid \iota \in \Upsilon\}, \{M(P) \mid P \in \mathcal{P}\}, \{M(f) \mid f \in \mathcal{F}\}\rangle$$

be an $\mathcal{E}$-structure of J defined by subobjects and let

$$\mathcal{C} = \langle \{N(\iota) \mid \iota \in \Upsilon\}, \{N(P) \mid P \in \mathcal{P}\}, \{N(f) \mid f \in \mathcal{F}\} \rangle$$

be an $\mathcal{E}$-structure of J defined by morphisms which is associated with $\mathcal{D}$. Let A be a formula in J with $FV(A) \subseteq X$. Then $\|A\|^{\mathcal{C}}(X)$ is the characteristic morphism of the subobject $[A]^{\mathcal{D}}(X)$.

Canonical language and its interpretation. For any topos we can define a many sorted language which can describe properties of this category and which can be interpreted naturally in this topos. This language will be called *the canonical language* and it is introduced in the following definition.

DEFINITION 3.37
Let $\mathcal{E}$ be an elementary topos. The canonical language *$J_{\mathcal{E}}$ of $\mathcal{E}$ is a many-sorted language such that*

(i) The sorts in $J_{\mathcal{E}}$ are exactly the objects in $\mathcal{E}$.

(ii) The finitary sorted predicate symbols correspond to subobjects (of respective sorts).

(iii) Only the unary operational symbols correspond to morphisms in $\mathcal{E}$.

For simplicity we identify the symbols in $J_{\mathcal{E}}$ with the corresponding objects and morphisms in $\mathcal{E}$, respectively. Hence, the language $J_{\mathcal{E}}$ is simply obtained by forgetting the composition law of the category $\mathcal{E}$. Now, by a *canonical $\mathcal{E}$-structure* we understand the structure $\mathcal{I}$ which assigns sorts, predicate symbols and unary operation symbols to corresponding (the same, in fact) objects of the category $\mathcal{E}$. For a formula A in $J_{\mathcal{E}}$ and a set X of free variables such that $FV(A) \subseteq X$, by $[A](X)$ we denote the interpretation of A in $\mathcal{I}$, i.e. $[A](X) \longrightarrow \prod_{x \in X} x$.

Gentzen sequents. Recall that a *(Gentzen) sequent* of J is an object of the form $\Phi \Rightarrow \Psi$, where $\Rightarrow$ is a new symbol not contained in J and Φ, Ψ are finite (possibly empty) sets of formulas. A *theory* in J is then a set of sequents of J. Let $\Phi \Rightarrow \Psi$ be a Gentzen sequent of J and let $\mathcal{D}$ be an interpretation of a language J defined by subobjects. Then we say that $\Phi \Rightarrow \Psi$ *holds in* $\mathcal{D}$, or equivalently, that $\mathcal{D}$ *is a model of* $\Phi \Rightarrow \Psi$ (in symbols, $\mathcal{D} \models \Phi \Rightarrow \Psi$), if the relation

$$\bigwedge_{A \in \Phi} [A]^{\mathcal{D}}(X) \leq \bigvee_{B \in \Psi} [B]^{\mathcal{D}}(X)$$

holds in the lattice $(\mathrm{Sub}(M(X)), \leq)$, where $FV(\Phi) \cup FV(\Psi) \subseteq X$. It should be observed that this definition is correct, i.e. it does not depend on the choice of the set X. Further, we say that $\mathcal{D}$ is a model of a sequent $\Rightarrow \Psi$ (with the empty left side), if $\bigvee_{B \in \Psi} [B]^{\mathcal{D}}(X)$ is the greatest subobject in $(\mathrm{Sub}(M(X)), \leq)$, i.e., $\bigvee_{B \in \Psi} [B]^{\mathcal{D}}(X)$ is isomorphic to $M(X)$. Analogously, we say that $\mathcal{D}$ is a model of the sequent $\Phi \Rightarrow$, if $\bigwedge_{A \in \Phi} [A]^{\mathcal{D}}(X)$ is the smallest subobject in $M(X)$.

Topologies on a topos. We show a simple application of the interpretation of formulas in a topos which enables us to introduce some well known notions in any topos. As an example we will take the *topology*, since it represents a very useful and frequently used tool in any algebraic systems. Hence it seems to be worthwhile to construct in a usual way a topology on a topos, as well. A topology on a set can be defined by well known systems of closure axioms. We will describe a topology on a topos by a similar system of axioms, since the process of interpretation of formulas in topos enables us to handle simply with such defined topology.

Let $\mathcal{E}$ be a topos and let us consider the following definition where formulas are considered in the canonical language $J_{\mathcal{E}}$ of $\mathcal{E}$. The canonical interpretation of $J_{\mathcal{E}}$ in $\mathcal{E}$ will be denoted simply by $\mathcal{E}$ (if there is no danger of misunderstanding).

DEFINITION 3.38
A topology *on a topos* $\mathcal{E}$ *is a morphism* $j : \Omega \longrightarrow \Omega$ *such that*

(i) $\mathcal{E} \models \ \Rightarrow (x \Rightarrow j(x))$,

(ii) $\mathcal{E} \models \ \Rightarrow (j(j(x)) = j(x))$,

(iii) $\mathcal{E} \models \ \Rightarrow (j(x \cap y) = j(x) \cap j(y))$.

The symbol $\cap$ in this definition represents morphism $\Omega \times \Omega \longrightarrow \Omega$, which is the characteristic morphism of $\top \times \top$ (see Definition 2.30 on page 57). Using the properties of interpretations of formulas from canonical language $J_{\mathcal{E}}$ in $\mathcal{E}$ it can be proved simply that a morphism $j : \Omega \longrightarrow \Omega$ is a topology on Ω iff it satisfies the following statements:

(i') $j \circ \top = \top$,

(ii') $j \circ j = j$,

(iii') $j \circ \cap = \cap \circ (j \times j)$.

For example, $\mathcal{E} \models \ \Rightarrow (j(j(x)) = j(x))$ represents the fact that $\Omega \xrightarrow{1_\Omega} \Omega$ is the equalizer of morphisms

$$\Omega \underset{j \circ j \circ \mathrm{pr}_2 \circ (1_\Omega \times 1_\Omega)}{\overset{j \circ \mathrm{pr}_1 \circ (1_\Omega \times 1_\Omega)}{\rightrightarrows}} \Omega.$$

And this is equivalent with the condition $j \circ j = j$. Any topology j defines a closure operation as follows. Let a be an object in $\mathcal{E}$ and let ι be a subobject of a, i.e. there exists a monomorphism $i : \iota \rightarrowtail a$. Let us consider the following diagram:

$$\begin{array}{ccccccc}
\bar\iota & \leftarrowtail & \iota & \rightarrowtail & a & \xrightarrow{=} & a\\
\Big\downarrow{\scriptstyle !} & & \Big\downarrow{\scriptstyle !} & & \Big\downarrow{\scriptstyle \chi_\iota} & & \Big\downarrow{\scriptstyle \chi_{\bar\iota}}\\
 & & \mathbf{1} & \xrightarrow{\top} & \Omega & & \\
 & & \Big\downarrow{\scriptstyle =} & & \Big\downarrow{\scriptstyle j} & & \\
\mathbf{1} & \xrightarrow{=} & \mathbf{1} & \xrightarrow{\top} & \Omega & \xrightarrow{=} & \Omega
\end{array}$$

Then the closure $\bar{\iota}$ of ι (sometimes called the j-closure of ι) in this topology j is the subobject of a with the characteristic morphism $j \circ \chi_\iota$. Moreover, the following lemma holds.

LEMMA 3.8
Let $j : \Omega \longrightarrow \Omega$ be a topology on a topos $\mathcal{E}$ and let $J \rightarrowtail \Omega$ be the object in $\mathcal{E}$ with the characteristic morphism $\chi_J = j$. Then for any subobject $\iota \rightarrowtail a$ the diagram

$$\begin{array}{ccc} \bar{\iota} & \rightarrowtail & a \\ {\scriptstyle u}\downarrow & & \downarrow{\scriptstyle \chi_\iota} \\ J & \rightarrowtail & \Omega \end{array}$$

is the pullback.

PROOF: We will show only how to construct the morphism u. Let $i_1 : \iota \rightarrowtail a$ and $i_2 : J \rightarrowtail \Omega$ be monomorphisms corresponding these subobjects. Then $\chi_\iota \circ i_1 : \bar{\iota} \longrightarrow \Omega$ and since J is a pullback, there exists the unique morphism $u : \bar{\iota} \longrightarrow J$ such that $i_2 \circ u = \chi_\iota \circ i_1$. The rest follows simply from corresponding diagrams. □

The closure operation $\iota \mapsto \bar{\iota}$ has various nice properties.

PROPOSITION 3.1
Let j be a topology on a topos $\mathcal{E}$, let s, t be subobjects in a and let $f : b \longrightarrow a$ be a morphism in $\mathcal{E}$.

(a) *$s \leq \bar{s}$ where $\leq$ is ordering in the Heyting algebra Sub(a).*

(b) *$\bar{\bar{s}} = \bar{s}$.*

(c) *$\overline{s \wedge t} = \bar{s} \wedge \bar{t}$ where $\wedge$ is the infimum in the Heyting algebra Sub(a).*

(d) *If $s \leq t$ then $\bar{s} \leq \bar{t}$.*

(e) *The j-closure of $f^{-1}(s)$ is $f^{-1}(\bar{s})$, where for monomorphism $x \rightarrowtail a$ and a morphism $f : b \longrightarrow a$, $f^{-1}(x)$ is the pullback*

$$\begin{array}{ccc} x & \rightarrowtail & a \\ \uparrow & & \uparrow \\ f^{-1}(x) & \rightarrowtail & b. \end{array}$$

PROOF: For example, to prove the first property let us consider the following pullback diagrams:

$$\begin{array}{ccc} s & \overset{i}{\rightarrowtail} & a \\ {\scriptstyle !}\downarrow & & \downarrow{\scriptstyle \chi_s} \\ \mathbf{To} & \overset{\top}{\longrightarrow} & \Omega \end{array} \qquad \begin{array}{ccc} \bar{s} & \overset{j}{\rightarrowtail} & a \\ {\scriptstyle v}\downarrow & & \downarrow{\scriptstyle \chi_s} \\ J & \overset{\alpha}{\rightarrowtail} & \Omega \end{array} \qquad \begin{array}{ccc} J & \overset{\alpha}{\rightarrowtail} & \Omega \\ {\scriptstyle !}\downarrow & & \downarrow{\scriptstyle j} \\ \mathbf{To} & \overset{\top}{\longrightarrow} & \Omega \end{array}$$

Since $j \circ \top = \top$, the following diagram commutes

$$\begin{array}{ccc} s & \xrightarrow{\chi_s \circ i} & \Omega \\ {\scriptstyle !}\downarrow & & \downarrow{\scriptstyle j} \\ \mathbf{To} & \xrightarrow{\top} & \Omega \end{array}$$

and there exists the unique morphism $\beta : s \rightarrowtail J$ such that $\alpha \circ \beta = \chi_s \circ i$. Since $\bar{s}$ is the pullback, there exists the unique morphism $w : s \rightarrowtail \bar{s}$ such that $j \circ w = i$. Hence, $s \leq \bar{s}$. The other properties can be proved analogously. □

4 FUZZY LOGIC IN NARROW SENSE

In this chapter, we will describe in detail fuzzy logic in narrow sense (FLn) with graded syntax. We will focus on its essential part but, as discussed in Chapter 1, understand it as an open system, ready to absorb various useful concepts coming from outside.

Our explanation of FLn will follow the common lines of presentation of any kind of formal logical systems. Namely, we will strictly distinguish between syntax and semantics, which, unlike the classical systems, are both evaluated by degrees. The global point of view on such a system is explicated in Section 4.1 where, in parallel with Section 3.3, the concept of the graded consequence operation is developed. We distinguish the syntactic as well as semantic graded consequence operations, which are both based on the assumption that *axioms* need not be fully true (convincing) and thus, form a *fuzzy set* of formulas. The syntactic consequence operation is determined by inference rules used in the evaluated proof, thus leading to the concepts of a fuzzy theory and a provability degree of a formula, which unlike the classical case requires information about all its possible proofs. We will show that all these concepts are natural generalization of the corresponding classical ones.

Recall that the system of FLn is basically truth functional though, for example, the inference rules can in general be non-truth functional. However, the truth functionality is used in the definition of the model of a fuzzy theory.

Soundness in the graded logical system means that the truth degree of a formula in a fuzzy theory is greater than or equal to the provability. We will show that the opposite inequality, i.e. completeness, requires infinite distributivity of the implication operation with respect both to supremum as well as to infimum.

A question arises, what is an appropriate structure of truth values of fuzzy logic. This question is analyzed in Section 4.2. We give reasons to assume truth values to form a complete, infinitely distributive residuated lattice, or an MV-algebra. A consequence of the mentioned necessity condition for the completeness of FLn is continuity of the implication with respect to the natural topology. Due to the representation theorem for infinite, locally finite MV-algebras presented in Section 2.1, we conclude that the structure of truth values should form the Łukasiewicz algebra $\mathcal{L}_L$.

Having all these preparatory results, we can focus on the detailed explanation of FLn, which can be found in Sections 4.3–4.6. We will confine to the first-order predicate FLn and prove that it is complete. Our goal is, at least, twofold. First, whenever possible, we develop FLn in parallel with classical predicate logic to show that it indeed either keeps or naturally generalizes all the properties of the latter. A consequence is that in principle, we can switch from classical logic directly to FLn without fear of loosing significant results. This enables us to see classical logic from another side, perhaps helping us to understand it better. One of crucial questions to be solved is consistency of fuzzy theories. We show that inconsistent fuzzy theories are identical with the classical ones. An attempt to weaken the inconsistency concept is presented in Subsection 4.3.10. Surprisingly, we mostly obtain again the classical inconsistent theory. Consequently, the consistency concept cannot be weakened too much.

The second goal accords with the objective that FLn should be able to *extend the capability of classical logic* to the directions where the latter cannot provide satisfactory solution, such as the mentioned sorites paradox (see page 10). Hence, we neither want FLn to simply mimic classical logic with slight generalizations, nor to deny the latter. Therefore, we have to seek new interesting questions. It is likely that they will be trivial in classical logic (but nontrivial in FLn), or even have no sense in it. Such questions are connected, e.g. with the concept of fuzzy equality where on the one side, its axioms are straightforward translation of the classical equality ones, but in the semantics we meet important problems concerning characterization of membership, approximation of functions and others. Other group of non-classical problems can be found in fuzzy model theory. For example, the concept of the degree of elementary subvalence and equivalence of models opens a possibility to introduce a kind of Kripke-like semantics into FLn. The third group of completely new questions and problems is connected with natural language, which can be more deeply embedded into models of reasoning than classical logic could offer. Chapter 6 is devoted to this third group.

Besides the mentioned, the reader may find in Sections 4.3–4.6 characterization theorems for various aspects of fuzzy theories, namely their extension, completion, consistency and provability in open ones, elements of the model theory and outline of the recursive properties of fuzzy theories.

It should be noted that the results and concepts presented in this chapter have been initiated and in many respects elaborated by J. Pavelka in [103].

Namely, the concept of the graded consequence operation, inference rules and evaluated proof, reasoning about soundness and completeness with respect to the necessity of the implication operation to be continuous. He also proved completeness of the graded propositional fuzzy logic.

4.1 Graded Formal Logical Systems

In Section 3.3, we have presented a general concept of the formal logical systems. Our exposition concerned both classical as well as many-valued logics. Stemming from the graded approach principle, we will develop a theory of graded formal logical systems in this section. Unlike the usual many-valued logical systems, whose semantics is many-valued but the syntax is classical, in the graded logical systems both semantics as well as syntax are evaluated. This enables us to formulate the concept of completeness of the graded formal logical systems, which generalizes the common concept of completeness both of classical as well as of many-valued logics, but is slightly weaker. As mentioned, there are some limitations for the completeness to hold. We provide general conditions under which it is possible to obtain a complete graded system.

Truth values and the maximality principle. Since we speak about graded logical systems, we have to agree first on the kind of structure which will serve us as a scale from which we take the truth values. Various reasons, which will be discussed in the next section, make us convinced that the set of truth values should form a residuated lattice

$$\mathcal{L} = \langle L, \vee, \wedge, \otimes, \rightarrow, \mathbf{0}, \mathbf{1} \rangle, \tag{4.1}$$

which is, moreover, complete (see Definitions 2.7 and 2.9 on page 29).

As stressed in Section 1.3, in comparison with the other kinds of many-valued logic, there are no designated truth values, which means that all the truth values are equal in their importance.

However, it may happen during our reasoning that we obtain more truth values from L and it is necessary to decide, which of them we should be taken as the result. Therefore, we postulate the *maximality principle* according to which, if the same object is assigned more truth values then its final truth assignment is equal to the *maximum (supremum) of all of them*. The maximality principle is one of the leading principles assumed in this book.

Language and syntax. Similarly as in the classical case, we will deal with the formal language J and a set F_J of well formed formulas. We will define them in the same way as in Section 3.3. However, the language J of the graded formal logical system is supposed to contain, besides the logical constants $\bot, \top$, also logical constants $\mathbf{a}$ for all the other $a \in L \backslash \{\mathbf{0}, \mathbf{1}\}$. Note that we write $\bot$ instead of $\mathbf{0}$ and $\top$ instead of $\mathbf{1}$.

In the case that L is uncountable, J becomes uncountable, too. However, as will be seen later, it is not indispensable in FLn to consider all $a \in L$ and we may confine only to a countable number of $\mathbf{a}$. In the special case, even the only constant $\bot$ for the falsity is sufficient. However, considering logical constants

$\mathbf{a}$ for all $a \in L$ is a useful technical trick and therefore, we will usually suppose this.

All the logical constants in FLn are atomic formulas. Having this on mind, we define F_J as the set of all the expressions having a formation sequence due to Definition 3.24 on page 77.

Important role in the presentation of the graded formal logical systems is played by evaluated formulas. This concept is introduced in accordance with the consistent application of the graded approach discussed in Chapter 1. Evaluated formulas are realization of the need to introduce grades, besides semantics, also into syntax.

DEFINITION 4.1
An evaluated formula is a couple

$$a/A \tag{4.2}$$

where $A \in F_J$ is a formula and $a \in L$ is its syntactic evaluation.

It should be stressed that the syntactic evaluations accompany all the formulas but they *do not belong* to the language J.

Axioms in the graded formal systems are sets of evaluated formulas. Since the evaluations can be interpreted as membership degrees in the fuzzy set, the axioms are at the same time *fuzzy sets of formulas*. Similarly as in the classical case, we distinguish logical and special axioms, namely $\mathrm{LAx}, \mathrm{SAx} \underset{\sim}{\subset} F_J$. Axioms with the membership degree smaller than $\mathbf{1} \in L$, i.e. those having the initial syntactic evaluation different from $\mathbf{1}$, may be considered as not fully convincing.

The possibility to work with not fully convincing axioms is natural. Recall the sorites paradox discussed in Section 1.3 on page 10 where the implication $\mathbb{FN}(x) \Rightarrow \mathbb{FN}(x+1)$ is a typical example of not fully convincing axiom accepted with the initial evaluation close, but not equal to $\mathbf{1}$.

In the sequel, we will often use the fact that fuzzy sets are functions. Hence, if $V : F_J \longrightarrow L$ is a function and $W \underset{\sim}{\subset} F_J$ a fuzzy set then $V \leq W$ means ordering of V, W as functions.

Inference rules in FLn deal with the evaluated formulas.

DEFINITION 4.2
An n-ary inference rule r in the graded logical system is a scheme

$$r : \frac{a_1/A_1, \ldots, a_n/A_n}{r^{evl}(a_1, \ldots, a_n)/r^{syn}(A_1, \ldots, A_n)}, \tag{4.3}$$

using which the evaluated formulas $a_i/A_i, \ldots, a_n/A_n$ are assigned the evaluated formula $r^{evl}(a_1, \ldots, a_n)/r^{syn}(A_1, \ldots, A_n)$. The syntactic operation r^{syn} is a partial n-ary operation on F_J and the evaluation operation r^{evl} is an n-ary lower semicontinuous operation on L (i.e. it preserves arbitrary suprema in all variables).

Given a formal system, we always fix a set R of inference rules in it.

The following definition generalizes Definition 3.27 on page 79.

DEFINITION 4.3
A fuzzy set $V \underset{\sim}{\subset} F_J$ of formulas is closed with respect to r if

$$V(r^{syn}(A_1, \dots, A_n)) \geq r^{evl}(V(A_1), \dots, V(A_n)) \tag{4.4}$$

holds for all formulas $A_1, \dots, A_n \in \mathrm{Dom}(r^{syn})$.

Recall that according to Definition 3.27, if V contains premises of the rule r, then it contains also its conclusion. First, realize that given the membership degrees $V(A_1), \dots, V(A_n)$ of the premises in the fuzzy set V, the conclusion of r is the evaluated formula $r^{evl}(V(A_1), \dots, V(A_n))/r^{syn}(A_1, \dots, A_n)$. Second, remind property (g) of Lemma 2.5 on page 25 stating that residuation, as the interpretation of implication, is equal to unit iff the left argument is smaller than or equal to the right one. The inequality (4.4) then states that the many-valued implication "if $A_1, \dots, A_n$ belong to V with the respective degrees $V(A_1), \dots, V(A_n)$ then the conclusion belongs to V with the degree $r^{evl}(V(A_1), \dots, V(A_n))$" is true.

To generalize consistently syntax into the graded form we must introduce the concept of an evaluated formal proof.

DEFINITION 4.4
An evaluated formal proof of a formula A from the fuzzy set $X \underset{\sim}{\subset} F_J$ is a finite sequence of evaluated formulas

$$w := a_0/A_0,\ a_1/A_1, \dots,\ a_n/A_n \tag{4.5}$$

such that $A_n := A$ and for each $i \leq n$, either there exists an n-ary inference rule r such that

$$a_i/A_i := r^{evl}(a_{i_1}, \dots, a_{i_n})/r^{syn}(A_{i_1}, \dots, A_{i_n}), \quad i_1, \dots, i_n < i,$$

or

$$a_i/A_i := X(A_i)/A_i.$$

The evaluation a_n of the last member in (4.5) is the *value* of the evaluated proof w. We will usually write the value of the proof w as $\mathrm{Val}(w)$. If it is clear from the context that we deal with the evaluated proof, we will usually omit the adjective "evaluated". If w is a proof of a formula A then we will often write w_A.

Graded syntactic consequence operation. After exposition of the main concepts of the graded approach to syntax, we may define the graded consequence operation. Recall from Section 3.3, page 78, that the consequence operation is defined as a closure operation on the power of F_J. In analogy with this, the *graded consequence operation*, is a *closure operation* $\mathcal{C} : \mathcal{F}(F_J) \longrightarrow \mathcal{F}(F_J)$

assigning a fuzzy set $\mathcal{C}(X) \underset{\sim}{\subset} F_J$ to a fuzzy set $X \underset{\sim}{\subset} F_J$, which fulfils the conditions formulated on page 78 $\mathcal{C}(X)(A) = \mathcal{C}(\mathcal{C}(X))(A)$ for all fuzzy sets of formulas $X, Y \underset{\sim}{\subset} F_J$.

DEFINITION 4.5
Let R be a set of inference rules. Then the fuzzy set of syntactic consequences of the fuzzy set $X \underset{\sim}{\subset} F_J$ is given by the membership function

$$\mathcal{C}^{syn}(X)(A) = \bigwedge \{V(A) \mid V \underset{\sim}{\subset} F_J, X \leq V \text{ and } V \text{ is closed w.r.t. to all } r \in R\}. \tag{4.6}$$

Obviously, the syntactic consequence operation $\mathcal{C}^{syn}$ defined in (4.6) is a closure operation on $\mathcal{F}(F_J)$.

The relation between the concepts of the evaluated proof and the syntactic consequences is clarified in the following important theorem, which can also be seen as a generalization of Theorem 3.18 on page 79.

THEOREM 4.1
Let $X \underset{\sim}{\subset} F_J$. Then

$$\mathcal{C}^{syn}(X)(A) = \bigvee \{\mathrm{Val}(w) \mid w \text{ is a proof of } A \text{ from } X\}. \tag{4.7}$$

PROOF: Let $A \in F_J$. We denote by $W \underset{\sim}{\subset} F_J$ the fuzzy set defined by the right hand side of (4.7).

First, we will show by induction on the length of a proof w_A that $\mathrm{Val}\, w_A \leq V(A)$ for every proof w_A and every fuzzy set $V \underset{\sim}{\subset} F_J$ such that $X \subseteq V$, i.e. that $W(A) \leq \mathcal{C}^{syn}(X)(A)$.

Let the length of w_A be 1. Then $A \in \mathrm{Supp}(X)$ and thus, by definition (4.6) $\mathrm{Val}(w_A) = X(A) \leq V(A)$ holds for every above considered fuzzy set $V \underset{\sim}{\subset} F_J$.

Let the length of w_A be greater than 1 and $A = r^{syn}(B_1, \ldots, B_n)$ for some already proved formulas $B_1, \ldots, B_n$ occurring in w_A. Then for every V from (4.6) we obtain

$$\mathrm{Val}(w_A) = r^{evl}(\mathrm{Val}(w_{B_1}), \ldots, \mathrm{Val}(w_{B_n})) \leq r^{evl}(V(B_1), \ldots, V(B_n)) \leq V(A)$$

by the inductive assumption and the fact that V is closed with respect to r.

Vice-versa, since $w_A := X(A)/A$ is a proof with the value $\mathrm{Val}(w_A) = X(A)$, it holds that $X \leq W$. Therefore, by (4.6) it is sufficient to show that W is closed with respect to all the inference rules from R.

Let $r \in R$ and $A = r^{syn}(B_1, \ldots, B_n)$ for some formulas $B_1, \ldots, B_n$. First realize that if $w_{B_1}, \ldots, w_{B_n}$ are proofs of the respective formulas then

$$w_A := \mathrm{Val}(w_{B_1})/B_1, \ldots, \mathrm{Val}(w_{B_n})/B_1, r^{evl}(\mathrm{Val}(w_{B_1}), \ldots, \mathrm{Val}(w_{B_n})/A$$

is a proof of A. Then

$$\begin{aligned}W(A) &= \bigvee_{w_A} \mathrm{Val}(w_A) \geq \\ &\geq \bigvee_{w_{B_1},\dots,w_{B_n}} r^{evl}(\mathrm{Val}(w_{B_1}),\dots,\mathrm{Val}(w_{B_n})) = \\ &= r^{evl}\left(\bigvee_{w_{B_1}} \mathrm{Val}(w_{B_1}),\dots,\bigvee_{w_{B_n}} \mathrm{Val}(w_{B_n})\right) = \\ &= r^{evl}(W(B_1),\dots,W(B_n))\end{aligned}$$

by the semicontinuity of r^{evl}. □

Analogously as in Theorem 3.18, this theorem also provides a certain kind of finitistic characterization of the syntactic consequence operation. However, unlike the classical syntax where finding one finite proof is sufficient, the situation here is not so convenient. In general, a sequence (possibly infinite) of finite proofs must be considered, because maximum of their values does not need to exist, and so, we cannot confine ourselves to one proof only. Due to this deliberation, we rewrite (4.7) as follows.

Let $G \subset F_J$ be a finite set of formulas and $X \underset{\sim}{\subset} F_J$. We put $(X|G)(A) = X(A)$ if $A \in G$ and $(X|G)(A) = \mathbf{0}$ otherwise. Then the following lemma holds.

LEMMA 4.1

$$\mathcal{C}^{syn}(X)(A) = \bigvee\{\mathcal{C}^{syn}(X|G)(A) \mid G \subset F_J, G \textit{ is finite}\}. \tag{4.8}$$

PROOF: Let G_w be a set of all formulas occurring in the proof w. Obviously, G_w is finite. Using Theorem 4.1, we obtain

$$\begin{aligned}\mathcal{C}^{syn}(X)(A) &= \bigvee\{\mathcal{C}^{syn}(X|G_w)(A) \mid w \text{ is a proof of } A\} \leq \\ &\leq \bigvee\{\mathcal{C}^{syn}(X|G)(A) \mid G \subset F_J, G \text{ is finite}\} \leq \\ &\leq \mathcal{C}^{syn}(X)(A).\end{aligned}$$

□

The consequence operation $\mathcal{C}^{syn}$ fulfilling (4.8) will be called the *compact* consequence operation.

Semantics. A *truth valuation* of formulas from F_J is a function $\mathcal{D} : F_J \longrightarrow L$, which fulfils conditions of Definition 3.29 on page 80. If $\mathbf{a}$ is a logical constant for some $a \in L$ then we obviously put $\mathcal{D}(\mathbf{a}) = a$.

A direct generalization of Definition 3.32 on page 82 is the following.

DEFINITION 4.6
An inference rule r is sound if

$$\mathcal{D}(r^{syn}(A_1,\dots,A_n)) \geq r^{evl}(\mathcal{D}(A_1),\dots,\mathcal{D}(A_n)) \tag{4.9}$$

holds for all truth valuations $\mathcal{D}$ of formulas from F_J.

By this definition, a sound inference rule cannot lead to an evaluation of its consequence being greater than a truth value of the latter in any interpretation. This behaviour of the inference rules assures soundness of the whole formal system (see below). Therefore, if not stated otherwise, we will always suppose that all the inference rules contained in the above considered fixed set R are sound.

DEFINITION 4.7
Let $X \underset{\sim}{\subset} F$ be a fuzzy set of formulas. Then the fuzzy set of its semantic consequences is given by the membership function

$$\mathcal{C}^{sem}(X)(A) = \bigwedge\{\mathcal{D}(A) \mid \textit{for all truth valuations } \mathcal{D} : F_J \longrightarrow L, X \leq \mathcal{D}\}. \tag{4.10}$$

Similarly as above, the semantic consequence operation $\mathcal{C}^{sem}$ defined in (4.10) is a closure operation on $\mathcal{F}(F_J)$.

Because all the truth values are equal in their importance, we may generalize the concept of tautology as follows.

DEFINITION 4.8
We say that a formula A is an a-tautology (tautology in the degree a) if

$$a = \mathcal{C}^{sem}(\emptyset)(A) \tag{4.11}$$

and write $\models_a A$. If $a = \mathbf{1}$ then we write simply $\models A$ and call A a tautology.

The fuzzy set of logical axioms in the given graded formal logical system is a certain fixed fuzzy set

$$\text{LAx} \subset \{\, {}^{a}/A \mid A \text{ is an } a\text{-tautology}\}.$$

EXAMPLE 4.1
Let the set of truth values be the Łukasiewicz algebra $\mathcal{L}_{\text{L}}$. Then the formula of the form $A \vee \neg A$ is a 0.5-tautology. Indeed, in any truth valuation $\mathcal{D}$ we have $\mathcal{D}(A \vee \neg A) = \mathcal{D}(A) \vee \neg\mathcal{D}(A) \geq 0.5$. □

Of course, due to the principle of equal importance of truth values, we have also $\mathbf{0}$-tautologies. However, it may hardly have any sense to speak about them.

Formal fuzzy theories. Having defined the graded syntactic and semantic consequence operations, we can introduce the concept of formal fuzzy theory, which is a graded counterpart to the classical concept of the formal theory.

DEFINITION 4.9
A formal fuzzy theory T in the language J of FLn is a triple

$$T = \langle \text{LAx}, \text{SAx}, R\rangle$$

where LAx $\underset{\sim}{\subset} F_J$ *is a fuzzy set of logical axioms,* SAx $\underset{\sim}{\subset} F_J$ *is a fuzzy set of special axioms, and* R *is a set of sound inference rules.*

Note that a fuzzy theory can also be viewed as a fuzzy set $T = \mathcal{C}^{syn}(\mathrm{LAx} \cup \mathrm{SAx}) \underset{\sim}{\subset} F_J$. We will usually omit the adjectives "formal" and "fuzzy" when speaking about formal fuzzy theories and no danger of misunderstanding impends.

Given a fuzzy theory T. We denote its language by $J(T)$. If w is a proof in T then its value is denoted by $\mathrm{Val}_T(w)$.

DEFINITION 4.10
Let T *be a fuzzy theory and* $A \in F_J$ *a formula.*

(i) *If* $\mathcal{C}^{syn}(\mathrm{LAx}\cup\mathrm{SAx})(A) = a$ *then we write* $T \vdash_a A$ *and say that the formula* A *is a theorem in the degree* a*, or provable in the degree* a*, in the fuzzy theory* T*.*

(ii) *If* $\mathcal{C}^{sem}(\mathrm{LAx}\cup\mathrm{SAx})(A) = a$ *then we write* $T \models_a A$ *and say that* A *is true in the degree* a *in the fuzzy theory* T*.*

(iii) *Let* $\mathcal{D}$ *be a truth valuation of formulas. Then it is a model of the fuzzy theory* T, $\mathcal{D} \models T$*, if* $\mathrm{SAx}(A) \leq \mathcal{D}(A)$ *holds for all formulas* $A \in F_J$*.*

On the basis of this definition, we may restate (4.7) as follows:

$$T \vdash_a A \quad \text{iff} \quad a = \bigvee\{\mathrm{Val}(w) \mid w \text{ is a proof of } A \text{ from } \mathrm{LAx} \cup \mathrm{SAx}\}. \tag{4.12}$$

It follows from (4.12) that finding a proof of a formula A gives us information only about the lower bound of the degree in which it is a theorem in T.

The item (i) of Definition 4.10 should be more subtly distinguished as follows. Given a fuzzy theory T and a formula A and let $T \vdash_a A$. In the special case there may exist a proof w_A of A such that $\mathrm{Val}_T(w_A) = a$. Then we say that the formula A is *effectively provable in the degree* a in the fuzzy theory T. This may especially be the case of the tautologies for which $T \vdash A$. Note that "to be effectively provable in the degree a" is a stronger property than "to be provable (theorem) in the degree a" since the latter assures only the existence of a sequence of proofs the supremum of which is equal to a.

Due to the assumption made on the fuzzy set of logical axioms LAx, we conclude that $\mathrm{LAx}(A) \leq \mathcal{D}(A)$ holds for every interpretation $\mathcal{D}$ of formulas. Then

$$T \models_a A \quad \text{iff} \quad a = \bigwedge\{\mathcal{D}(A) \mid \mathcal{D} \models T\}. \tag{4.13}$$

Occasionally, SAx may be extended by some fuzzy set of formulas $\Gamma \underset{\sim}{\subset} F_J$. Then by $T \cup \Gamma$ we understand an extended fuzzy theory

$$T \cup \Gamma = \langle \mathrm{LAx}, \mathrm{SAx} \cup \Gamma, R\rangle.$$

Soundness and completeness. Recall that the classical syntactic and semantic consequence operations are two different characterizations of true formulas. An important achievement of classical logic is the theorem stating that, even though each characterization is based on entirely different assumptions, both of them lead to the same result. Is such a conclusion true also in fuzzy logic?

We have introduced the concept of evaluated syntax and semantics in the graded formal logical systems. This means that each formula is assigned a syntactic evaluation on the level of syntax and also a truth value on the level of semantics, both belonging to the same set L. Hence, we can compare them.

Note, that the evaluated syntax is a natural generalization of the classical one. We have noted that syntactic consequence operation is another characterization of true formulas. In the classical syntax, we use sound inference rules which make us sure to derive only true conclusion from true assumptions. Thus, in the classical syntactic derivation we start from axioms assumed to be true formulas and using sound inference rules we derive again true formulas. Hence, the syntax in classical logic is virtually evaluated, too. However, the evaluation is always equal to $\mathbf{1}$ (true) since $\mathbf{0}$ (false) as the other possibility is uninteresting. Consequently, we do not need to stress explicitly the former.

We say that the graded formal logical system is *sound* if the fuzzy set of the syntactic consequences of any fuzzy set of formulas is a fuzzy subset of its semantic consequences, i.e. if

$$\mathcal{C}^{syn}(X)(A) \leq \mathcal{C}^{sem}(X)(A) \tag{4.14}$$

holds for every $X \underset{\sim}{\subseteq} F_J$ and a formula $A \in F_J$. Using the concept of the formal fuzzy theory, we may write (4.14) equivalently as

$$T \vdash_a A \quad \text{and} \quad T \models_b A \quad \text{implies} \quad a \leq b \tag{4.15}$$

holds for every theory T and a formula $A \in F_J$.

The graded formal logical system is *complete* if

$$\mathcal{C}^{syn}(X)(A) = \mathcal{C}^{sem}(X)(A) \tag{4.16}$$

holds for every $X \underset{\sim}{\subseteq} F_J$ and a formula $A \in F_J$. Equivalently, the system is complete if

$$T \vdash_a A \quad \text{iff} \quad T \models_a A \tag{4.17}$$

holds for every theory T and a formula $A \in F_J$. Note that due to the definition, we always have $T \vdash_a A$ and $T \models_b A$ for some $a, b \in L$, possibly equal to $\mathbf{0}$.

Note that completeness of the graded logical system is not so straightforward as that of the classical one. Unlike the latter, where the syntactic consequence operation means finding some proof, in the former we must find a sequence (possibly infinite) of all proofs and evaluate its supremum. Similarly, the semantic consequence operations means finding the infimum of all the possible truth evaluations. Hence, completeness here means a kind of coincidence of different truth characterizations "in a limit".

We will show in this chapter that FLn is complete. However, we are going to demonstrate that this conclusion does not hold in full generality.

Bounds for completeness of the graded formal logical systems with implication. A connective of significant importance in every formal logical systems is that of implication $\Rightarrow$. Let us, therefore, assume that the language J contains it, too. The question studied in this subsection is, whether it is possible to construct a graded formal logical system in such a way that it is complete, i.e. that the equality (4.16) holds true. The following theorem, originally proved in [103], gives a surprisingly unambiguous answer to it.

THEOREM 4.2
Given a graded formal logical system with the implication connective $\Rightarrow$ based on a complete residuated lattice (4.1). If the interpretation $\rightarrow$ of $\Rightarrow$ does not fulfil the equations

$$\bigvee_{i\in I}(a \rightarrow b_i) = a \rightarrow (\bigvee_{i\in I} b_i) \tag{4.18}$$

$$\bigvee_{i\in I}(a_i \rightarrow b) = (\bigwedge_{i\in I} a_i) \rightarrow b \tag{4.19}$$

for arbitrary subset of L then such a system cannot be complete.

PROOF: Let us assume that equality (4.18) does not hold for the set $E = \{b_i \mid i \in I_0\}$ and denote $e_1 = \bigvee_{i\in I_0}(a \rightarrow b_i)$ and $e_2 = a \rightarrow (\bigvee_{i\in I_0} b_i)$ for some $a \in L$. Then it follows from Lemma 2.9(g) on page 29 that $e_1 < e_2$. Let $B \in F_J$ and T be a fuzzy theory given by the following fuzzy set of special axioms

$$\mathrm{SAx}(A) = \begin{cases} \mathbf{1} & \text{if } A := \mathbf{b}_i \Rightarrow B, b_i \in E, \\ \mathbf{0} & \text{otherwise.} \end{cases}$$

Now let $A := \mathbf{a} \Rightarrow B$ for the a above. Then

$$\mathcal{C}^{sem}(\mathrm{LAx} \cup \mathrm{SAx})(A) = \bigwedge\{\mathcal{D}(A) \mid \mathcal{D} \models T\} = a \rightarrow \bigwedge_{\mathcal{D}\models T} \mathcal{D}(B) =$$
$$= a \rightarrow \bigvee_{i\in I_0} b_i = e_2$$

by Lemma 2.9(d) on page 29 and due to the fact that $\bigvee_{i\in I_0} b_i \leq \mathcal{D}(B)$ holds for every model $\mathcal{D} \models T$.

Since $\mathcal{C}^{syn}$ is compact, the equality (4.16) implies that also $\mathcal{C}^{sem}$ must be compact. Let $G \subset F_J$ be a finite set. Then

$$\mathcal{C}^{syn}((\mathrm{LAx} \cup \mathrm{SAx})|G)(A) = \bigwedge\{\mathcal{D}(A) \mid ((\mathrm{LAx} \cup \mathrm{SAx})|G)(A) \leq \mathcal{D}(A), \mathcal{D} \models T\}$$
$$= a \rightarrow e_G$$

for some $e_G \leq \bigvee_{i\in I_0} b_i$. Using (4.8) we then obtain

$$\mathcal{C}^{syn}(\mathrm{LAx} \cup \mathrm{SAx})(A) = \bigvee\{\mathcal{C}^{syn}((\mathrm{LAx} \cup \mathrm{SAx})|G)(A) \mid G \subset F_J, G \text{ finite}\} =$$
$$= \bigvee\{a \rightarrow e_G \mid G \subset F_J, G \text{ finite}\} \leq e_1 < e_2.$$

Consequently, the equality (4.16) is not fulfilled.

If the second equality (4.19) is not fulfilled then we put

$$\mathrm{SAx}(A) = \begin{cases} \mathbf{1} & \text{if } A := B \Rightarrow \mathfrak{b}_i, i \in I_0, \\ \mathbf{0} & \text{otherwise} \end{cases}$$

and proceed analogously as above. □

When joining this theorem with Lemma 2.9(d), (e) on page 29, we see that the interpretation of the implication in a complete graded formal logical system must preserve all infima as well as suprema in the second variable, and exchange suprema for infima and vice-versa in the first one. Theorem 4.2 leads to significant consequences for the structure of truth values. This will be discussed in the next section.

4.2 Truth Values

4.2.1 Intuition on truth values and operations with them

We have given a lot of arguments in favour of the use of grades for the description of vagueness. A question now arises, what should be the structure of truth values in fuzzy logic. This question is analyzed in detail in this section.

Let us denote by T the set of truth values and by $\mathcal{T}$ a structure defined on it. We will give reasons for the assumption that $\mathcal{T}$ should form an infinite residuated lattice. Furthermore, we demonstrate that the necessary consequence of several other requirements on it is that $\mathcal{T}$ should be even the Łukasiewicz algebra $\mathcal{L}_{\text{Ł}}$.

Fundamental assumptions on truth values. Similarly as classical logic, fuzzy logic is also based on the Aristotelian logic. Therefore, $\mathcal{T}$ must be a generalization of classical two-valued Boolean algebra for classical logic $\mathcal{L}_B$ (see Example 2.1). Since both full truth as well as full falsity must obviously be distinguished, $\mathcal{T}$ must have both minimal as well as maximal elements, $\mathbf{0}, \mathbf{1} \in T$.

An important question is whether the truth values should be comparable. Consider the statement "a man having x years is old". Is it sensible to say that it is more true for $x = 60$ than for $x = 40$? Apparently yes. Therefore, we assume that the truth values are ordered. At the same time it seems natural that each two truth values can be compared, i.e. $\mathcal{T}$ should be linearly ordered.

The following objection towards the linear ordering may be met. Let us consider the concept of colour. Every colour as perceived by people has three characteristics, namely hue, saturation and brightness. It is difficult to say that a colour with a hue corresponding to 770 nm and given brightness is more (less) red than a colour with the same hue but different brightness. This would suggest that there should exist non-comparable truth values. However, it is possible to solve this problem by introducing more dimensional truth values where the truth values in each dimension are taken from a chain. Consequently, for more complicated deliberation about the vagueness phenomenon it is still appropriate to take a chain of truth values as basic[1]. Therefore, we will assume in the sequel that $\mathcal{T}$ is a chain.

In Chapter 1, page 4 we have stressed that one of important features of the vagueness phenomenon is its continuity. Thus, the graded approach must be able to render a continuous transition from truth to falsity. Consequently, the cardinality of T should be that of continuum. Without loss of generality we may assume that $T = [0, 1]$.

REMARK 4.1
The interval $[0, 1]$ is used as a natural measuring scale for various purposes. One of them is also measuring probabilities in it. However, we should by no means mistake probabilities with truth values.

Fundamental assumptions on truth functions. Our reasoning, which proceeds in natural language, is abstracted by logic in such a way that intricate relations inside the meanings of words are transformed into the assignment of truth values, and into operations with them. To obtain more complex propositions, logic introduces special logical connectives. In classical logic, we distinguish conjunction, disjunction, implication and negation. They represent simplification and abstraction of specific natural language phenomena. Namely, *conjunction* and *disjunction* correspond to the so called *coordination* (we distinguish close and free one), *implication* corresponds to *conditional sentences* and logical *negation* is an abstraction of the quite complicated phenomenon of *linguistic negation.* We will not be more specific, because it is not our purpose to go here into the general linguistic theory. For our purpose, it will be sufficient to take into account, that the mentioned linguistic phenomena are complex but can be decomposed into simpler ones, which are, however, vague in essence.

Classical logic made two principal assumptions. The first one is truth functionality, i.e. the truth of the compound statement is a function of the truths of its constituents. The second one is neglecting vagueness. Both assumptions lead to the well known truth tables (cf. Example 2.1 on page 17).

In fuzzy logic we keep the truth functionality but take vagueness as the subject of its study. Remember that the latter is continuous. Therefore, continuity of a vague compound proposition must not be harmed when passing to it from its (also vague) constituents. Consequently, the truth functions on T, which will be assigned to the logical connectives, are assumed to be continuous, too.

Conjunction, disjunction and negation. In logic, *conjunction* corresponds to a specific token of the coordination in natural language, which relates linguistic units inside a compound one usually using the connective 'and'. It comes out that the latter should be a binary connective. The linguistic units can be simple words (such as "young and stout"), as well as sentences. Since a deeper linguistic analysis would lead to great complexity, we will take these

[1]This reminds the role of chain in the subdirect representation theorem of MV-algebras.

units as propositions without their detailed specification. We will now analyze conjunction from the point of view of the truth assignment

Let P, Q be some (vague) propositions. In simple situations, such as "Lucy is young and Lucy is clever" we feel that the truth of "P and Q" cannot be different from that of "Q and P". Therefore, we will assume that conjunction is *commutative*[2].

If the truth of P' is greater than the truth of P then the truth of "P' and Q" should be higher that that of "P and Q". Consequently, conjunction is *monotone*. Similarly we come to the conclusion that "(P and Q) and R" is the same as "P and (Q and R)" and so conjunction is also *associative*.

The fourth characteristic property of the conjunction follows from the necessity that "P and Q" can be true, at least a little, only if both P as well as Q are at least a little true, and the truth of the former cannot be greater than that of either of P a Q. Hence, we come to the *boundary condition* stating that if, e.g. P is fully true then "P and Q" can be true at most as much as is Q true.

One may see that these properties are exactly those occurring in Definition 2.19 on page 42 of triangular norms. Hence, it seems natural to *interpret logical conjunction using a t-norm* $\mathbf{t}$.

Quite similar arguments we find when discussing *disjunction*, which raises as a token of the close coordination when the connective 'or' is used. We find that commutativity, associativity and monotonicity should be the same as in the case of conjunction. The boundary condition, however, is modified as follows.

Two properties are joined by 'or' into a compound property so that elements having it have *at least one* of its constituents. Therefore, "P or Q" can be true, at least a little, *even* when only one of P, or Q is a little true (the second one can be false). Comparing this reasoning with Definition 2.20 on page 43 we conclude, that good *interpretation of the logical disjunction in fuzzy logic is a t-conorm* $\mathbf{s}$.

Negation in natural language is a quite complicated phenomenon. When dealing with it, we must especially take care of the so called *topic-focus articulation*[3] and *presupposition*[4]. In logic, negation is again simplified to an operation on truth values assigned to propositions.

Negation of a proposition means reversing its truth value. If a proposition represents a vague property then its negation represents again a vague prop-

---

[2]The close coordination in natural languages, where also joining of words by the connective 'and' belongs, is more subtle. If we want to stress an importance of one proposition over the second one then the conjunction becomes non-commutative. For example, if the youth is for some reasons more important than cleverness then the sentence "Lucy is young and clever" has other meaning than "Lucy is clever and young". Of course, it is very difficult to capture such finenesses in logic and therefore, neither we will consider them in fuzzy logic.

[3]Each sentence can be divided into two parts, namely the topic — what is spoken about, and the focus — what is the new information. Of course, there are many possibilities how to do it, and the division may not be unambiguous. This phenomenon is one of the sources of extreme information power of natural language.

[4]Roughly speaking, presupposition are properties supposed to hold in the topic uttered by the sentence.

erty. Therefore, negation is also supposed to be a continuous operation. Also strictness (cf. Definition 2.26 on page 50) seems to be an obvious requirement. A quite debatable is the law of double negation. For example, in constructive or intuitionistic mathematics this law has been omitted. The price payed for this is the smaller expressive power and much more difficult proving. Because of the lack of substantial reasons for rejection, we assume the law of double negation to hold also in fuzzy logic. Consequently, only strong negation is really convincing for us.

Implication. The most important connective in logical systems is implication since it is a basis of the principal inference rule of the human mind studied already by Aristoteles, namely that of modus ponens. Classical logic works with the so called material implication. However, there is a lot of controversy about it since it admits propositions such as "if $2+2=5$ then Earth circulates around Sun" to be valid. This is caused by the fact that classical material implication operates solely with truth values and does not care about the meaning of the propositions occurring in the given conditional sentence.

In the formal system of FLn, we have no other choice than to copy the properties of classical material implication since neither here we have information about the meaning of the propositions. We will lay two main requirements on the implication in fuzzy logic: it must be a generalization of classical one, and it must keep the rule of modus ponens.

To be a generalization of classical implication means that it should keep its main properties and reduce to it when confining only to two truth values $\{\mathbf{0}, \mathbf{1}\}$. In the literature, various sets of requirements have been imposed on the implication. For example, G. Klir and Bo Yuan in the book [57] give a list of 9 axioms, which are derived from reasoning on the possible properties of implication. The first seven are obvious requirements on its behaviour in both arguments, namely antitonicity in the first and isotonicity in the second argument, boundary condition ($a \to b = \mathbf{1}$ iff $a \leq b$), etc. The last two are the contraposition (classical axiom (CL2)) and continuity of $\to$ as a function on $[0,1]^2$. Since classical axioms (CL1)–(CL3) in Definition 3.2 represent the essence of the basic properties of implication, we focus on them and on the modus ponens in the sequel.

Modus ponens can be characterized by the following situation. Let an element x have a property P and we know that "if x has P then an element y has a property Q. We conclude that y has Q. It seems natural that the truth of Q cannot be smaller than the truth of the compound statement "P and (if P then Q)". The problem is how to estimate the truth of "if P then Q". Let γ denote this estimation. Then

$$\text{truth}(P \text{ and } \gamma) \leq \text{truth}(Q) \tag{4.20}$$

follows

$$\gamma \leq \text{truth}(\text{if } P \text{ then } Q) \tag{4.21}$$

since γ is the estimation of truth of the implication in concern. But also vice-versa, if (4.21) holds, i.e. γ is some estimation of truth of the implication then this must keep modus-ponens, i.e. the inequality (4.20). Joining both speculations and using the symbols a, b, c for the truth values, we come to the conclusion that the adjunction property, which states that

$$a \mathbf{t} x \leq b \quad \text{iff} \quad x \leq a \to b, \tag{4.22}$$

should hold for the implication $\to$ (cf. (4.22) and (2.15) on page 24) where $\mathbf{t}$, as assumed above, interprets the conjunction 'and' in (4.20).

The most important properties of implication, which follow from adjunction are summarized in Lemma 2.5 on page 25. It can be verified that they are at the same time properties required in majority of the above mentioned axioms from [57].

If we follow formal properties of classical material implication then we obtain two basic possibilities.

DEFINITION 4.11

(a) *The operation* $\to$: $[0,1]^2 \longrightarrow [0,1]$ *is an* s*-implication if there is a t-conorm* $\mathbf{s}$ *and negation* $\mathbf{n}$ *such that*

$$a \to b = \mathbf{n}(a)\, \mathbf{s}\, b.$$

(b) *The operation* $\to$: $[0,1]^2 \longrightarrow [0,1]$ *is an R-implication if there is a t-norm such that* $\to$ *is a residuation with respect to* $\mathbf{t}$.

Obviously, both kinds of implication reduce to the classical one if we confine only to $T = \{\mathbf{0}, \mathbf{1}\}$.

On the basis of Lemma 2.5(g) on page 25, Theorem 2.11 on page 46 and Lemma 2.21 on page 46, the following lemma disqualifies most of s-implications from our further consideration.

LEMMA 4.2

(a) *Let* $\mathbf{t}$ *be a continuous Archimedean t-norm and* $\mathbf{s}$ *the corresponding t-conorm. Let* $\to_{\mathbf{s}}$ *be an* s*-implication. If it is true that* $a \leq b$ *implies* $a \to_{\mathbf{s}} b = 1$ *for all* $a, b \in [0,1]$ *then* $\mathbf{t}$ *cannot be strict.*

(b) *There are* $a \leq b$ *such that* $\neg a \vee b \neq 1$.

PROOF: (a) It follows from Theorem 2.11 that

$$a\, \mathbf{s}\, b = g^{-1}(\min(g(1), g(a) + g(b)))$$

for a continuous increasing function $g : [0,1] \longrightarrow [0,\infty]$. Due to Lemma 2.21 on page 46, if $\mathbf{t}$ is strict then $g(1) = \infty$. Let $0 < a \leq b < 1$ and $a \to_{\mathbf{s}} b = 1$. Then $g(1) = \min(g(1), g(1-a) + g(b))$, i.e. $g(1-a) + g(b) = \infty$, which is not possible.

(b) is obvious. □

We have already given reasons for the conjunction, disjunction and negation to be continuous operations. Recall our discussion on page 48 where we concluded that all continuous t-norms are isomorphic either to Łukasiewicz t-norm

'$\otimes$', product '$\cdot$', infimum '$\wedge$', or are a "mixture" of all of them. Therefore, it makes sense to confine only to these three operations to serve as the interpretation of the basic connectives in fuzzy logic (note that we did not exclude other t-norms as possible connectives). But with respect to the previous theorem and the necessary consequences of the adjunction (cf. Lemma 2.5) we conclude that only $\oplus$-implication is acceptable for the interpretation of the logical (fuzzy) implication. And it is easy to verify that this is equal to R-implication based on $\otimes$. Consequently, we may exclude **s**-implications from further considerations. The only candidates for the implication in fuzzy logic are therefore R-implications (i.e. residuations with respect to t-norms), and namely those from Proposition 2.1 — Łukasiewicz, Gödel and Goguen ones.

Let us now check, which of the classical logical axioms (CL1)–(CL3) from Definition 3.2 on page 63 can be tautologies, provided that we interpret the implication by some of these three implications.

PROPOSITION 4.1
Let $a, b, c \in [0, 1]$.

(CL1): The equality $a \to (b \to a) = 1$ holds for Łukasiewicz, Gödel, as well as Goguen implication.

(CL2): The equality $(\neg b \to \neg a) \to (a \to b) = 1$ holds only for $\to = \to_L$.

(CL3): The equality $(a \to (b \to c)) \to ((a \to b) \to (a \to c)) = 1$ holds only for $\to = \to_G$.

We see that Goguen implication satisfies only (CL1), Gödel one satisfies (CL1) and (CL3), and Łukasiewicz one satisfies (CL1) and (CL2). This implication does not satisfy the transitivity (CL3) but, as we will see later, there is a weaker form of (CL3), which is satisfied by it.

Local character of the connectives in fuzzy logic. On many places in this book, we took motivation for basic properties of fuzzy logic from natural language since this is the principal bearer of the vagueness phenomenon. Apparently, the phenomena of natural language are much more complicated to be well modelled by only few functions forming a simple algebra, such as the MV-one. At the same time, it is argued in [4] that fuzzy logic should take into account various kinds of connectives since each connective can be used only for certain class of propositions of natural language. We say that such a connective is valid only *locally*. Since in fuzzy logic, there is infinite number of possible operations at disposal, we have a potential to find a suitable operation for each local situation. The explication power of fuzzy logic can thus be extended to make its applications in the practice more realistic.

4.2.2 Truth values as algebras

In Sections 2.1–2.3, various algebraic structures selected from the point of view of their properties potentially appropriate to fit the intuitive requirements put on the structure of truth values in fuzzy logic have been presented. We have started with Boolean algebras. Many-valued logic based on non-trivial Boolean algebras mimics all the properties of classical logic, namely the law of the excluded middle, idempotency, distributivity (leading to the classical normal forms) and thus does not bring significant generalization necessary for fulfilling of the agenda of fuzzy logic outlined in Section 1.4. Hence, Boolean algebras, though very important for classical and categorical logic, are not convenient as structures of truth values for fuzzy logic. The requirements on the structure of truth values discussed in the previous section lead to the following conclusions.

The most general acceptable structure $\mathcal{T}$ of truth values, which meets most requirements for implication is that of *residuated lattice.* If no more specific requirements are imposed then we will assume this.

The second, more specific requirement is that the basic connectives of conjunction and disjunction should be interpreted by continuous t-norms and t-conorms, respectively. Consequently, due to Corollary 2.4 on page 49, $\mathcal{T}$ should be even a *BL-algebra.*

The third requirement is connected with the negation, namely with the classical axiom (CL2). This axiom reflects the intuitive property of implication widely used in human reasoning, even in the case we deal with vague properties. Let us now consider the algebraic counterpart of (CL2) and the case when $a = 1$. Then we immediately obtain $(\neg b \to 0) = (1 \to a) = a$, i.e. (CL2) is equivalent with the law of double negation. However, when accepting that $\mathcal{T}$ is already a BL-algebra then by Theorem 2.2(d) on page 33, it is necessarily an *MV-algebra.*

4.2.3 *Why Łukasiewicz algebra*

We will try to give reasons for the assumption that truth values should take the structure of Łukasiewicz algebra $\mathcal{L}_{\text{Ł}}$. Till now, we have not mentioned the interpretation of quantifiers. By Definition 3.12 on page 70, the universal quantifier in classical logic is interpreted using the min-operation and the existential one using the max-operation. A direct generalization leads to the respective interpretation using the operations of $\wedge$ and $\vee$ over any subset of T. Consequently, the truth values $\mathcal{T}$ should form a complete lattice[5]. However, the following lemma holds true for linearly ordered residuated lattices.

LEMMA 4.3
Let $\mathcal{L}$ be a complete residuated chain and I and arbitrary index set. Let a topology τ be given by the open basis $B = \{\{x \in L \mid a < x < b\} \mid a, b \in L\}$. Then the residuation operation $\to$ is continuous with respect to τ iff it fulfils the equalities (4.18), (4.19) and (d), (e) from Lemma 2.9 on page 29.

---

[5] Strictly speaking, this assumption may be weakened when using the concept of "safe models" introduced in [41]

PROOF: All four conditions are equivalent to the upper as well as lower continuity of $\to$ in both arguments. Therefore, it is continuous. □

Let us remember that by Lemma 2.12(d),(e) on page 34, the equalities (4.18) and (4.19) are fulfilled in MV-algebras. However, by Theorem 4.2 on page 105 they must be fulfilled in the complete graded formal logical system with implication. Consequently, we can again conclude that the structure of truth values $\mathcal{T}$ should form an MV-algebra. But recall our previous reasoning on the implication, take $[0,1]$ as the set of truth values and add the requirement that the implication should be continuous due to Lemma 4.3. Then the necessary conclusion is that the most plausible structure of truth values $\mathcal{T}$ is that of Łukasiewicz algebra $\mathcal{L}_L$. A distinguished role of the latter is supported also formally by the representation Theorem 2.7 on page 40.

On the other hand, it may be interesting for some applications (e.g. in expert systems) to use also the other two kinds of implications, namely Gödel and Goguen ones, so that the completeness property would be in some sense preserved. One possibility is to introduce an infinite inference rule. Some other algebraic means are now thoroughly investigated (see [41, 25]).

4.2.4 Enriching structure of truth values

Recall that the local character of logical connectives in fuzzy logic requires more interpretations of them dependingly on the given specific situation. From the formal point of view, we enrich the structure of truth values $\mathcal{T}$ being in the most general case a residuated lattice, by additional operations. A question arises, what properties should be fulfilled by them. One of the crucial ones is preservation of the logical equivalence. We thus come to the following definition.

DEFINITION 4.12
Let $\Box : L^n \longrightarrow L$ be an n-ary operation. We say that it is logically fitting if it satisfies the following condition: There are natural numbers $k_1 > 0, \ldots, k_n > 0$ such that

$$(a_1 \leftrightarrow b_1)^{k_1} \otimes \cdots \otimes (a_n \leftrightarrow b_n)^{k_n} \leq \Box(a_1, \ldots, a_n) \leftrightarrow \Box(b_1, \ldots, b_n) \qquad (4.23)$$

holds for all $a_1, \ldots, a_n, b_1, \ldots, b_n \in L$, where the power is taken with respect to the operation $\otimes$.

The following theorem demonstrates that all the considered operations are logically fitting.

THEOREM 4.3
(a) *Let $\mathcal{L}$ be a residuated lattice. Then each basic operation $\vee, \wedge, \otimes, \to$ is logically fitting.*

(b) *Any composite operation obtained from logically fitting operations is logically fitting.*

(c) *An operation $\Box : [0,1]^n \longrightarrow [0,1]$ in Łukasiewicz algebra $\mathcal{L}_L$ is logically fitting iff it is Lipschitz continuous.*

PROOF: (a) is a consequence of Lemma 2.7 on page 27.

(b) follows from the monotonicity of $\otimes$ after rewriting (4.23) for all the members of the composite operation.

(c) For simplicity, we assume $n = 2$. Then (4.23) holds iff there is some $k \in \mathbb{N}$ such that

$$((1 - |a_1 - b_1|) \otimes (1 - |a_2 - b_2|))^k \leq 1 - |\Box(a_1, a_2) - \Box(b_1, b_2)|$$

for all $a_1, a_2, b_1, b_2 \in [0, 1]$, i.e.

$$|\Box(a_1, a_2) - \Box(b_1, b_2)| \leq k(|a_1 - b_1| + |a_2 - b_2|),$$

which is the condition for the Lipschitz continuity. □

DEFINITION 4.13
Let $\mathcal{L}$ be a residuated lattice 2.14. Let $\Box_i$, $i = 1, \ldots, k$ be a set of additional n_i-ary operations on L being logically fitting. The enriched residuated lattice is an algebra

$$\mathcal{L}_e = \langle L, \vee, \wedge, \otimes, \rightarrow, \Box_1, \ldots, \Box_k, \mathbf{0}, \mathbf{1} \rangle. \tag{4.24}$$

The following theorem demonstrates that introducing logically fitting operations have no influence on congruences, which play an important role in the algebraic proof of the completeness theorem in FLn.

THEOREM 4.4
A binary relation $\sim$ is a congruence in the enriched residuated lattice $\mathcal{L}_e$ iff it is a congruence with respect to the operations $\otimes$ and $\rightarrow$.

PROOF: The implication left-to-right is trivial.

Conversely, let F be a filter in $\mathcal{L}$ corresponding to the congruence $\sim$ due to Theorem 2.3 on page 37. Furthermore, let $a_i \sim b_i$, $i = 1, \ldots, n$. Then $(a_i \leftrightarrow b_i)^{k_i} \in F$ for all $k_i > 0$. Let $\Box$ be an n-ary logically fitting operation. Then $\Box(a_1, \ldots, a_n) \leftrightarrow \Box(b_1, \ldots, b_n) \in F$. We conclude that $\Box(a_1, \ldots, a_n) \sim \Box(b_1, \ldots, b_n)$ and thus, $\sim$ is a congruence also in $\mathcal{L}_e$. □

4.3 Predicate Fuzzy Logic of First-Order

This section is devoted to the predicate first-order fuzzy logic in narrow sense with graded syntax. The way of presentation is heavily influenced by the classical books on mathematical logic, especially by those of J. R. Shoenfield [121], E. Mendelson [76], J. D. van Dalen [17], and C. C. Chang and H. J. Keisler [15]. We will follow the general presentation from Section 4.1 by concrete specification of the language, deduction rules and semantics. Furthermore, we will employ the concept of a fuzzy theory and prove various properties of it. Finally, we will provide two independent proofs of the completeness of the predicate fuzzy logic (and hence, obviously, also of the propositional one).

If not stated otherwise, the set of truth values is assumed to form the Łukasiewicz algebra $\mathcal{L}_{\text{Ł}}$ with no designated truth values. Of course, special algebraic properties of 0 and 1 will be employed. On the other hand, various properties presented below hold either in more general MV-algebras, or even in the residuated lattices not being MV-algebras. We will not especially stress this and let the reader to distinguish these cases himself. Therefore we prefer to denote the set of truth values by L. However, if greater generality is not required, the reader, may always think that $L = [0, 1]$. Let us remark that first-order fuzzy logic is equivalent with classical first-order logic if $L = \{\mathbf{0}, \mathbf{1}\}$.

In the sequel, we will often use the extended order relation

$a >^* b$ iff either $b < \mathbf{1}$ and $a > b$, or $a = \mathbf{1}$ and $b = \mathbf{1}$.

The symbol $\mathbb{N}^+$ denotes the set of nonzero natural numbers $\mathbb{N} \setminus \{0\}$.

4.3.1 Syntax and semantics

Language. The *language* of the first-order fuzzy logic is the same as that of classical logic (see page 67), further extended by logical constants $\{\mathbf{a} \mid a \in L\}$.

The definition of *terms* in FLn coincides with Definition 3.9 on page 67. The definition of *formulas* is the same as Definition 3.10 on page 68 with the exception that logical constants $\mathbf{a}$ for all $a \in L$ are also atomic formulas (cf. the discussion on page 97).

A set of all terms of the language J is denoted by M_J and a set of all formulas by F_J. Furthermore, we will often use the symbol M_V to denote the set of all closed terms of the language J. Note that logical constants can be understood also as nullary predicates.

All the concepts of the scope of the quantifier, sets $FV(t), FV(A)$ of free variables, substitutible terms, open and closed formula are introduced in the same way as on page 68 and its sequel.

We will furthermore work with the following abbreviations of formulas.

$$\begin{aligned}
\neg A &:= A \Rightarrow \bot && \text{(negation)}\\
A \vee B &:= (B \Rightarrow A) \Rightarrow A && \text{(disjunction)}\\
A \wedge B &:= \neg((B \Rightarrow A) \Rightarrow \neg B) && \text{(conjunction)}\\
A \,\&\, B &:= \neg(A \Rightarrow \neg B) && \text{(Łukasiewicz conjunction)}\\
A \nabla B &:= \neg(\neg A \,\&\, \neg B) && \text{(Łukasiewicz disjunction)}\\
A \Leftrightarrow B &:= (A \Rightarrow B) \wedge (B \Rightarrow A) && \text{(equivalence)}\\
A^n &:= \underbrace{A \,\&\, A \,\&\, \cdots \,\&\, A}_{n-\text{times}} && \text{(n-fold conjunction)}\\
nA &:= \underbrace{A \nabla A \nabla \cdots \nabla A}_{n-\text{times}} && \text{(n-fold disjunction)}\\
(\exists x)A &:= \neg(\forall x)\neg A && \text{(existential quantifier)}
\end{aligned}$$

Inference rules. Recall from Definition 4.2 on page 98 that inference rules in the graded formal system manipulate with evaluated formulas. Their set R in FLn consists of three rules defined below. Later on, we will extend R

by additional rules, which, however, are not necessary for the proof of the fundamental properties of FLn.

(i) The inference rule of *modus ponens* is the scheme

$$r_{MP} : \frac{{}^{a}/_{A},\ {}^{b}/_{A \Rightarrow B}}{{}^{a \otimes b}/_{B}}. \tag{4.25}$$

(ii) The inference rule of *generalization* is the scheme

$$r_{G} : \frac{{}^{a}/_{A}}{{}^{a}/_{(\forall x)A}}. \tag{4.26}$$

(iii) The inference rule of *logical constant introduction* is the scheme

$$r_{LC} : \frac{{}^{a}/_{A}}{{}^{a \to a}/_{\mathbf{a} \Rightarrow A}} \tag{4.27}$$

(note that the evaluation operation of r_{LC} is $r_{LC}^{evl}(x) = a \to x$). It is easy to verify that these rules fulfil the conditions of Definition 4.2 on page 98 and are sound in the sense of Definition 4.6 on page 101. Obviously, the operation $\otimes$ in the rule of Modus ponens cannot be replaced by $\wedge$.

Given an evaluated formula ${}^{a}/_{A}$, the rule r_{LC} enables to represent the evaluation a, which otherwise comes from the outside and does not belong to the language J, inside the latter. Consequently, we may identify the evaluated formula with the formula $\mathbf{a} \Rightarrow A$[6].

Logical axioms. Recall that we deal with fuzzy sets of axioms in FLn. Concerning the fuzzy set LAx of logical axioms, various systems have been proposed. J. Pavelka in [103] deals with more than thirty of them adding a new axiom whenever he needed it. This number has been significantly reduced in [88]. As pointed out by P. Hájek in [38] and S. Gottwald in [34], it is possible to reduce the fuzzy set of propositional logical axioms to that of A. Rose and J. B. Rosser [114]. Hence, we come to the following definition.

DEFINITION 4.14
Given the following schemes of formulas.

(R1) $A \Rightarrow (B \Rightarrow A)$.

(R2) $(A \Rightarrow B) \Rightarrow ((B \Rightarrow C) \Rightarrow (A \Rightarrow C))$.

(R3) $(\neg B \Rightarrow \neg A) \Rightarrow (A \Rightarrow B)$.

(R4) $((A \Rightarrow B) \Rightarrow B) \Rightarrow ((B \Rightarrow A) \Rightarrow A)$.

[6] This identification makes it possible to reinterpret the evaluated syntax inside the non-evaluated one — see [41] for the details.

(B1) $(\mathbf{a} \Rightarrow \mathbf{b}) \Leftrightarrow (\overline{\mathbf{a} \to \mathbf{b}})$
where $\overline{\mathbf{a} \to \mathbf{b}}$ *denotes the logical constant (atomic formula) for the truth value* $a \to b$ *when* a *and* b *are given.*

(T1) $(\forall x)A \Rightarrow A_x[t]$
for any substitutible term t.

(T2) $(\forall x)(A \Rightarrow B) \Rightarrow (A \Rightarrow (\forall x)B)$
provided that x *is not free in* A.

The fuzzy set LAx *of logical axioms is specified as follows:*

$$\mathrm{LAx}(A) = \begin{cases} a \in L & \text{if } A := \mathbf{a}, \\ \mathbf{1} & \text{if } A \text{ has some of the forms (R1)–(R4), (T1)–(T3),} \\ \mathbf{0} & \text{otherwise.} \end{cases}$$

The formulas of the form (R1)–(R4) are those originally proposed by A. Rose and J. B. Roser. The axiom (B1) is called *book-keeping*[7]. It assures correct behaviour of truth constants with respect to the connectives. The axiom (T1) is the *substitution axiom*. Both (T1) and (T2) coincide with the corresponding classical logical axioms (CL4) and (CL5), respectively (see page 68).

REMARK 4.2
It is easy to prove that in (B1), only implication is necessary and the opposite implication is a consequence of the above defined inference rules and axioms. However, since this brings no interesting simplification, we directly assumed equivalence in (B1).

In the sequel, we will denote by $T = \langle \mathrm{LAx}, \mathrm{SAx}, R\rangle$ a fuzzy theory in the sense of Definition 4.9 on page 102 where LAx is the above fuzzy set of logical axioms and R consists of the inference rules (4.25)–(4.27). If there is no danger of misunderstanding then we will omit the adjective "fuzzy".

Interpretation of formulas. The following definition is a straightforward generalization of Definition 3.11 on page 69.

DEFINITION 4.15
A structure for the language J *of FLn is*

$$\mathcal{D} = \langle D, P_D, \dots, f_D, \dots, u_1, \dots\rangle$$

where D *is a set,* $P_D \underset{\sim}{\subset} D$, ... *are* n*-ary*[8] *fuzzy relations assigned to each* n*-ary predicate symbol* $P, \dots$, *and* f_D *are ordinary (crisp)* n*-ary functions on* D *assigned to each* n*-ary functional symbol* f. *Finally,* $u_1, \dots \in D$ *are*

[7] This name has been proposed by P. Hájek.
[8] Of course, n's are, in general, different for different P_D's. In accordance with our agreement, we do not explicitly write this.

designated elements which are assigned to each constant $\mathbf{u}_1, \dots$ *of the language* J*, respectively.*

Note that functional symbols are interpreted by ordinary classical functions though various concepts of fuzzy functions can be found (cf. [22, 57, 85, 87]). The reason is, that introducing symbols for fuzzy functions in the language J would make all the definitions unnecessarily complicated. On the other hand, fuzzy functions can be easily introduced as new predicates determined by some special axioms.

A *truth valuation* of formulas in FLn is realized by means of their interpretation in some structure $\mathcal{D}$ in a way analogous to classical logic. For the given structure, it is a function $F_J \longrightarrow \mathcal{L}$ defined below.

Let $\mathcal{D}$ be a structure for the language J. In the same way as in classical logic, we extend J to the language $J(\mathcal{D})$ by new constants being names for all the elements from D (cf. page 70), i.e.constants will be denoted by the corresponding bold-face letter, namely $J(\mathcal{D}) = J \cup \{\mathbf{d} \mid d \in D\}$.

DEFINITION 4.16

(i) *Interpretation of closed terms is the same as in Definition 3.12 on page 70.*

(ii) *Interpretation of closed formulas: let* $t_1, \dots, t_n$ *be closed terms. Then*

$$\begin{aligned}
\mathcal{D}(\mathbf{a}) &= a, \quad \textit{for all} \quad a \in L, \\
\mathcal{D}(P(t_1, \dots, t_n)) &= P_D(\mathcal{D}(t_1), \dots, \mathcal{D}(t_n)), \\
\mathcal{D}(A \Rightarrow B) &= \mathcal{D}(A) \rightarrow \mathcal{D}(B), \\
\mathcal{D}((\forall x)A) &= \bigwedge\{\mathcal{D}(A_x[\mathbf{d}]) \mid d \in D\},
\end{aligned}$$

and in the case of the derived connectives,

$$\begin{aligned}
\mathcal{D}(A \wedge B) &= \mathcal{D}(A) \wedge \mathcal{D}(B), \\
\mathcal{D}(A \,\&\, B) &= \mathcal{D}(A) \otimes \mathcal{D}(B), \\
\mathcal{D}(A \vee B) &= \mathcal{D}(A) \vee \mathcal{D}(B), \\
\mathcal{D}(A \nabla B) &= \mathcal{D}(A) \oplus \mathcal{D}(B), \\
\mathcal{D}(A \Leftrightarrow B) &= \mathcal{D}(A) \leftrightarrow \mathcal{D}(B), \\
\mathcal{D}((\exists x)A) &= \bigvee\{\mathcal{D}(A_x[\mathbf{d}]) \mid d \in D\}.
\end{aligned}$$

Interpretation of a general formula A is analogous to the classical definition on page 70, i.e. using an evaluation $e : FV(A) \longrightarrow D$.

DEFINITION 4.17

(i) *A formula* $A(x_1, \dots, x_n)$ *is satisfied in* $\mathcal{D}$ *by the evaluation* e, $e(x_1) = d_1$, $\dots$, $e(x_n) = d_n$ *in the degree* a *if* $\mathcal{D}(A_{x_1,\dots,x_n}[\mathbf{d}_1, \dots, \mathbf{d}_n]) = a$.

(ii) *A formula* A *is* true *in* $\mathcal{D}$ *in the degree* a *if*

$$a = \mathcal{D}(A) = \bigwedge\{\mathcal{D}(A_{x_1,\dots,x_n}[\mathbf{d}_1, \dots, \mathbf{d}_n]) \mid \textit{for all evaluations } e, \\ e(x_1) = d_1, \dots, e(x_n) = d_n\}.$$

Canonical structure for the language J. Semantics of the given logical system requires construction of a structure, in which we can interpret formulas (provided it exists). The basic idea behind the concept of canonical structure is to construct it from the syntactic material of the given logical system. In fuzzy logic we, moreover, need an initial evaluation of the closed atomic formulas.

We suppose that J contains at least two different constants and let M_V be the set of all the closed terms. Furthermore, let $\mathrm{AF}_J \subset F_J$ be a set of all the closed atomic formulas from F_J not being logical constants.

DEFINITION 4.18
Let $C : \mathrm{AF}_J \longrightarrow L$ be an evaluation of the closed atomic formulas. The structure

$$\mathcal{D}_0 = \langle D_0, P_{D_0}, \dots, f_{D_0}, \dots, u, \dots \rangle$$

where

$$\begin{aligned} D_0 &= M_V, \\ \mathcal{D}_0(t) &= t, \qquad t \in M_V, \\ f_{D_0}(t_1, \dots, t_n) &= f(t_1, \dots, t_n), \\ \mathcal{D}_0(\mathbf{a}) &= a, \qquad a \in L, \\ P_{D_0}(t_1, \dots, t_n) &= C(P(t_1, \dots, t_n)) \end{aligned}$$

is the canonical structure for the language J.

Given a fuzzy theory T with constants. We may put $C(P(t_1, \dots, t_n)) = a$ iff $T \vdash_a P(t_1, \dots, t_n)$. If, furthermore, the language of T contains equality then we can introduce the equivalence relation between terms $t \sim s$ iff $T \vdash t = s$, $t, s \in M_V$ and construct $D_0 = M_V | \sim$ in the same way as is done in classical logic. Then we obtain a canonical structure for the language $J(T)$. The proof that this is also a model of T is the principal goal of the the completeness theorem below.

Validity and tautological consequence. Tautologies (cf. Definition 4.8 on page 102) play central role in every logical system. The following lemma gives simple rules how to verify tautologies in fuzzy logic.

LEMMA 4.4
Let A, B be formulas in the language J.

(a) $\models A \Rightarrow B$ iff $\mathcal{D}(A) \le \mathcal{D}(B)$ holds for every structure $\mathcal{D}$ for the language J.

(b) $\models A \Leftrightarrow B$ iff $\mathcal{D}(A) = \mathcal{D}(B)$ holds for every structure $\mathcal{D}$ for the language J.

PROOF: This follows from (4.10) and (4.11) using the fact that $\mathcal{D}(A \Rightarrow B) = \mathbf{1}$ iff $\mathcal{D}(A) \le \mathcal{D}(B)$ (cf. Lemmas 2.5 on page 25 and 2.7 on page 27). □

EXAMPLE 4.2
Using this lemma it is easy to verify that all formulas (R1)–(R4), (B1) and (T1)–(T2) are (**1**-)tautologies. Let us do it for (R3).

We will denote $a = \mathcal{D}(A)$ (and similarly for B) where $\mathcal{D}$ is an arbitrary structure. Then (R3) is equivalent with $\neg b \rightarrow \neg a \leq a \rightarrow b$ which follows from Lemma 2.11(d) on page 33.

Analogously, we obtain (T1) from the properties of infimum since

$$\bigwedge_{d \in D} \mathcal{D}(A_x[\mathbf{d}]) \leq \mathcal{D}(A_x[t])$$

and t is a term representing some element $d \in D$. □

We may also generalize the concept of the tautological consequence as follows.

DEFINITION 4.19
A formula B is a tautological consequence of the formulas $A_1, \ldots, A_n$ in the degree b, if

$$\models_b A_1 \Rightarrow (\cdots \Rightarrow (A_n \Rightarrow B) \cdots) \tag{4.28}$$

(cf. Definition 4.8).

It follows from Lemma 4.4 that B is a tautological consequence of $A_1, \ldots, A_n$ in the degree **1** iff $\mathcal{D}(A_1) \otimes \cdots \otimes \mathcal{D}(A_n) \leq \mathcal{D}(B)$ holds in every structure $\mathcal{D}$.

The following theorem is a consequence of Definitions 4.5 and 4.7 (on page 100 and its sequel) of the syntactic as well as semantic consequence operations. Due to it, we may derive formulas whose syntactic evaluation is at most as big as their semantic interpretation.

THEOREM 4.5 (VALIDITY THEOREM)
Let T be a fuzzy theory. If $T \vdash_a A$ and $T \models_b A$ then $a \leq b$ holds for every formula A.

As a special case, it follows from the definition that $T \vdash_a \mathbf{a}$ and $T \models_a \mathbf{a}$ holds for every logical constant $\mathbf{a}$, $a \in L$, in every fuzzy theory T.

4.3.2 Basic properties of fuzzy theories

Provability in fuzzy theories. In this subsection, we present some fundamental properties of fuzzy theories which are mostly valid in the pure *fuzzy predicate calculus*, i.e. a fuzzy theory with SAx = ∅. Let us remark that most of the provable formulas (formal theorems) presented below are proved using proofs in the degree **1** and thus, they are at the same time theorems of the Łukasiewicz predicate logic (cf. [34, 41]).

Recall from Section 4.1 that due to our assumption about equal importance of all the truth degrees, every formula is a theorem (provable) in some degree, possibly equal to **0**. Remind also the difference between the (general) provability degree and the effective provability degree discussed on page 103. To find the effective provability degree (if it exists) is not an easy task. Therefore, the following procedure will generally be kept in the proofs of most of subsequent lemmas and theorems.

Suppose we have learned that $T \vdash_a A$. Then we may consider some proof w_A of the formula A with the value $\mathrm{Val}(w_A) = a' \leq a$. Furthermore, let a proof w_B of some formula B be derived, which contains the formula A and which has the value $\mathrm{Val}(w_B) = f(a', \dots)$ (f is some operation). Using (4.12) we conclude that $T \vdash_b B$ where

$$b \geq \bigvee \{f(a', \dots) \mid \text{all proofs } w_A\}.$$

If f preserves suprema then we, moreover, know that $b \geq f(\bigvee a', \dots) = f(a, \dots)$. The latter case is fulfilled, for example, by the product $\otimes$.

When demonstrating the provability, we will explicitly write the evaluated proofs, which will consist of the sequence of evaluated formulas, each of which is commented by its origin (i.e. that it is an axiom, result of some inference rule, or the result of some previous proof). However, since the proofs may be quite long with quite technical big parts where we usually apply some inference rule, we will shorten them by replacing the explicit formulation of these parts only by dots. Thus, if we write something like "r_{MP}" in the comment of the evaluated formula following the dots, this will usually mean application of the the rule of modus ponens on an instance of the axiom (R2) as well as some formulas preceding it.

Provable propositional tautologies. In the lemmas below, we denote by T the fuzzy predicate calculus.

LEMMA 4.5
The following are schemes of effectively provable formal propositional theorems in T, i.e. for each theorem there exists a proof with the degree equal to **1**.

(P1) $T \vdash A \Rightarrow A$,

(P2) $T \vdash A \Rightarrow \top$,

(P3) $T \vdash \neg\neg A \Rightarrow A$, $\quad T \vdash A \Rightarrow \neg\neg A$,

(P4) $T \vdash (A \wedge B) \Rightarrow A$, $\quad T \vdash (A \wedge B) \Rightarrow B$,

(P5) $T \vdash (A \,\&\, B) \Rightarrow A$, $\quad T \vdash (A \,\&\, B) \Rightarrow B$,

(P6) $T \vdash A \Rightarrow (A \vee B)$, $\quad T \vdash B \Rightarrow (A \vee B)$,

(P7) $T \vdash (A \,\&\, \neg A) \Rightarrow B$,

(P8) $T \vdash (A \,\&\, B) \Leftrightarrow (B \,\&\, A)$,

(P9) $T \vdash (A \wedge B) \Rightarrow (B \wedge A)$,

(P10) $T \vdash (C \Rightarrow A) \Rightarrow ((C \Rightarrow B) \Rightarrow (C \Rightarrow (A \wedge B)))$,

(P11) $T \vdash (A \Rightarrow C) \Rightarrow ((B \Rightarrow C) \Rightarrow (A \vee B) \Rightarrow C)$,

(P12) $T \vdash ((A \,\&\, B) \Rightarrow C) \Leftrightarrow (A \Rightarrow (B \Rightarrow C))$,

(P13) $T \vdash A \Rightarrow (B \Rightarrow (A \,\&\, B))$,

(P14) $T \vdash ((A \Rightarrow B) \Rightarrow C) \Rightarrow (A \Rightarrow (B \Rightarrow C))$,

(P15) $T \vdash \neg(A \Rightarrow B) \Leftrightarrow (A \,\&\, \neg B)$,

(P16) $T \vdash (A \Rightarrow B) \Leftrightarrow (\neg A \nabla B)$,

(P17) $T \vdash (A \Rightarrow B) \vee (B \Rightarrow A)$,

(P18) $T \vdash (A \vee B)^n \Rightarrow (A^n \vee B^n), \quad n \in \mathbb{N}^+$,

(P19) $T \vdash (A \wedge B) \Leftrightarrow \neg(\neg A \vee \neg B)$,

(P20) $T \vdash (A \Rightarrow (B \Rightarrow C)) \Leftrightarrow (B \Rightarrow (A \Rightarrow C))$,

(P21) $T \vdash (A^m \Rightarrow (B \Rightarrow C)) \Rightarrow ((A^n \Rightarrow B) \Rightarrow (A^{m+n} \Rightarrow C)), \quad m, n \in \mathbb{N}^+$,

(P22) $T \vdash (A \,\&\, B) \Rightarrow (A^2 \vee B^2)$,

(P23) $T \vdash (A \,\&\, B)^n \Rightarrow (A^n \Rightarrow B^n),\ n \in \mathbb{N}^+$.

PROOF: All the schemes are proved from the logical axioms using r_{MP}. The proofs should be done systematically starting from simpler schemes, which may then be used when proving the other ones. We will demonstrate the proof of few of them.

(P6): Using the definition $(A \vee B) := (B \Rightarrow A) \Rightarrow A)$ we have the proof

$$\mathbf{1}/A \Rightarrow ((B \Rightarrow A) \Rightarrow A \ \{\textit{instance of (R1)}\}.$$

The second form can be obtained from this and (R4) using modus ponens.

(P20): Realizing that $(B \vee C) := (B \Rightarrow C) \Rightarrow C)$ we have the proof

$$\mathbf{1}/((B \Rightarrow (B \vee C)) \Rightarrow (((B \vee C) \Rightarrow (A \Rightarrow C)) \Rightarrow (B \Rightarrow (A \Rightarrow C))) \ \{(R2)\},$$
$$\mathbf{1}/((A \Rightarrow (B \Rightarrow C)) \Rightarrow (((B \Rightarrow C) \Rightarrow C) \Rightarrow (A \Rightarrow C))) \ \{(R2)\},$$
$$\ldots,\ \mathbf{1}/(A \Rightarrow (B \Rightarrow C)) \Rightarrow (B \Rightarrow (A \Rightarrow C)) \ \{(R2), r_{MP}\}.$$

(P1):

$$\mathbf{1}/(A \Rightarrow (\top \Rightarrow A)) \Rightarrow (\top \Rightarrow (A \Rightarrow A)) \ \{(P20)\},\ \mathbf{1}/\top \ \{\textit{log. axiom}\},$$
$$\mathbf{1}/(A \Rightarrow (\top \Rightarrow A)) \ \{(R1)\} \ldots \ \mathbf{1}/(A \Rightarrow A) \ \{(R2), r_{MP}\}$$

(P5): Recall that $A \,\&\, B := \neg(B \Rightarrow \neg A)$. Then

$$\mathbf{1}/((\neg A \Rightarrow (A \Rightarrow \bot)) \Rightarrow (A \Rightarrow (\neg A \Rightarrow \bot))) \ \{P(20)\},$$
$$\mathbf{1}/((B \Rightarrow \neg A) \Rightarrow \neg\neg(B \Rightarrow \neg A)) \ \{(P3)\}, \ldots,$$
$$\mathbf{1}/(\neg A \Rightarrow (\neg\neg(B \Rightarrow \neg A)) \ \{r_{MP}\},$$
$$\mathbf{1}/((\neg A \Rightarrow \neg\neg(B \Rightarrow \neg A)) \Rightarrow (\neg(B \Rightarrow \neg A) \Rightarrow A))) \ \{(R3)\},$$
$$\ldots,\ \mathbf{1}/(A \,\&\, B) \Rightarrow A) \ \{(R2), r_{MP}\}$$

(P22): We start with (P6) and the instance of (R2):

$\mathbf{1}/((A \Rightarrow (A \vee B)) \Rightarrow ((A \vee B) \Rightarrow (B \Rightarrow (A \vee B) \,\&\, B) \Rightarrow (A \Rightarrow (B \Rightarrow ((A \vee B) \,\&\, B)))$ {(R2)}, ..., $\mathbf{1}/((A \,\&\, B) \Rightarrow ((A \vee B) \,\&\, B))$ $\{(P12), r_{MP}\}$,
$\mathbf{1}/(((A \vee B) \,\&\, B) \Rightarrow (A \vee B)^2)$ $\{r_{MP}\}$, ...,
$\mathbf{1}/((A \,\&\, B) \Rightarrow (A^2 \vee B^2))$ $\{(R2), (P18), r_{MP}\}$.

□

It can be verified using Lemma 4.4 that all these schemes of formulas are tautologies[9].

The following lemma demonstrates that Łukasiewicz conjunction is isotonic in both arguments, and implication is isotonic in the second and antitonic in the first arguments with respect to the provability in the degree **1**.

LEMMA 4.6
Let T be a fuzzy theory, $T \vdash A \Rightarrow B$ and C be a formula. Then the following formulas holds true.

(a) $T \vdash (A \,\&\, C) \Rightarrow (B \,\&\, C)$,

(b) $T \vdash (B \Rightarrow C) \Rightarrow (A \Rightarrow C)$,

(c) $T \vdash (C \Rightarrow A) \Rightarrow (C \Rightarrow B)$.

(d) If $T \vdash A \Rightarrow B$ *and* $T \vdash C \Rightarrow D$ *then* $T \vdash (A \,\&\, C) \Rightarrow (B \,\&\, D)$.

PROOF: We will demonstrate (a). We construct the proof

$$w := \ a/A \Rightarrow B \ \{\textit{some proof } w_{A \Rightarrow B}\},$$
$$\mathbf{1}/((A \Rightarrow B) \Rightarrow ((B \Rightarrow (C \Rightarrow (B \,\&\, C))) \Rightarrow (A \Rightarrow (C \Rightarrow (B \,\&\, C)))$$
$$\{\text{instance of (R2)}\},\ \mathbf{1}/(B \Rightarrow (C \Rightarrow (B \,\&\, C)) \ \{\textit{instance of (P13)}\},$$
$$\mathbf{1}/((A \Rightarrow (C \Rightarrow (B \,\&\, C)) \Rightarrow ((A \,\&\, C) \Rightarrow (B \,\&\, C))) \ \{\textit{instance of (P13)}\},$$
$$\ldots,\ a/((A \,\&\, C) \Rightarrow (B \,\&\, C)) \ \{r_{MP}\},$$

which follows (a) since $\bigvee\{a \mid \text{all proofs } w_{A \Rightarrow B}\} = \mathbf{1}$. □

The following simple but useful lemma is obvious.

LEMMA 4.7
Let T_1, T_2 be fuzzy theories and A_1, A_2 formulas. If for every $a, b, c, d \in L$, $T_1 \vdash_a A_1$ implies $T_2 \vdash_b A_2$ where $a \le b$, and at the same time $T_2 \vdash_c A_2$ implies $T_1 \vdash_d A_1$ where $c \le d$ then

$$T_1 \vdash_a A \quad \textit{iff} \quad T_2 \vdash_a A.$$

[9]Note that all them are at the same time tautologies of Łukasiewicz logic (cf. [34, 41]).

The following lemma replaces the so called lifting rules[10] introduced originally in [103].

LEMMA 4.8

(a) *To every proof* w_A *with* $\mathrm{Val}(A) = a$ *there is a proof* $w_{\mathbf{b} \Rightarrow A}$ *with the value* $\mathrm{Val}(w_{\mathbf{b} \Rightarrow A}) = b \to a$.

(b) *In every fuzzy theory,* $T \vdash_a A$ *implies* $T \vdash \mathbf{a} \Rightarrow A$ *for all formulas* A *and* $a \in L$.

PROOF: (a)

$$\begin{aligned} w_{\mathbf{b} \Rightarrow A} := {} & a/A\ \{w_A\},\ \ 1/\mathbf{a} \Rightarrow A\ \{r_{LC}\},\ \ b \to a/\overline{\mathbf{b} \Rightarrow \mathbf{a}}\ \{\textit{logical axiom}\}, \\ & \ldots,\ 1/(\mathbf{b} \Rightarrow \mathbf{a}) \Rightarrow ((\mathbf{a} \Rightarrow A) \Rightarrow (\mathbf{b} \Rightarrow A))\ \{(R2)\}, \ldots, \\ & b \to a/\mathbf{b} \Rightarrow A\ \{r_{MP}\}. \end{aligned}$$

(b) is is a direct consequence of (a) and the rule r_{LC}. □

LEMMA 4.9

Let T *be a fuzzy theory. Then the following holds true.*

(a) $T \vdash A \Leftrightarrow B$ *iff* $T \vdash A \Rightarrow B$ *and* $T \vdash B \Rightarrow A$.

(b) *Let* $T \vdash_a A$ *and* $T \vdash A \Leftrightarrow B$. *Then* $T \vdash_a B$.

(c) *If* $T \vdash_a A \Rightarrow B$ *and* $T \vdash_b B \Leftrightarrow B'$ *then* $T \vdash_c A \Rightarrow B'$ *where* $c \geq a \otimes b$.

(d) *If* $T \vdash_a A \Leftrightarrow A'$ *then* $T \vdash_c (A \Rightarrow B) \Leftrightarrow (A' \Rightarrow B)$ *where* $c \geq a$.

PROOF: Let us demonstrate (c) and (d). Write down the proof

$$\begin{aligned} w := {} & a'/A \Rightarrow B\ \{\textit{some proof}\ w_{A \Rightarrow B}\},\ \ b'/B \Leftrightarrow B'\ \{\textit{some proof}\ w_{B \Leftrightarrow B'}\}, \\ & 1/(B \Leftrightarrow B') \Rightarrow (B \Rightarrow B')\ \{\textit{log. axiom}\}, \ldots,\ \ a' \otimes b'/A \Rightarrow B'\ \{r_{MP}\}, \end{aligned}$$

which gives the proposition.

(d) Using the instance of the logical axiom (R2) $\vdash (A \Rightarrow A') \Rightarrow ((A' \Rightarrow B) \Rightarrow (A \Rightarrow B))$ we obtain the proofs w' of $(A' \Rightarrow B) \Rightarrow (A \Rightarrow B)$ and w'' of $(A \Rightarrow B) \Rightarrow (A' \Rightarrow B)$, both with the value $a' \leq a$. From them, we can find a proof of the formula $(A \Rightarrow B) \Leftrightarrow (A' \Rightarrow B)$ with the value a'. □

LEMMA 4.10

Let T *be a fuzzy predicate calculus. Then the following formula is effectively provable formal theorem in the degree* **1**.

$$T \vdash (A_1 \Rightarrow C) \Rightarrow (\cdots((A_n \Rightarrow C) \Rightarrow (A_1 \nabla \cdots \nabla A_n \Rightarrow nC)\cdots).$$

---

[10] $r_{Rb} : \dfrac{a/A}{b \to a/\mathbf{b} \Rightarrow A}$

PROOF: The formulas

$$T \vdash \neg C \,\&(A \Rightarrow C) \Rightarrow \neg A, \qquad T \vdash \neg A \,\&(\neg A \Rightarrow B) \Rightarrow B,$$
$$T \vdash ((A \Rightarrow C)\,\&(B \Rightarrow C)\,\&(\neg A \Rightarrow B)\,\&(C \Rightarrow \bot)) \Rightarrow C$$

are effectively provable formal theorems in the degree **1** where the latter is equivalent with

$$T \vdash (A \Rightarrow C) \Rightarrow ((B \Rightarrow C) \Rightarrow (A \nabla B \Rightarrow 2C)).$$

From this, we get the proposition of lemma using induction. □

Equivalence theorem. The following theorem is a generalization of the analogous classical equivalence one. Note that if all the provability degrees are equal to **1**, this theorem is equivalent to the corresponding classical one.

THEOREM 4.6 (EQUIVALENCE)
Let T be a fuzzy theory, A be a formula and $B_1, \dots, B_n$ some of its subformulas. Let $T \vdash_{a_i} B_i \Leftrightarrow B_i'$, $i = 1, \dots, n$. Then there are $m_1, \dots, m_n$ such that

$$T \vdash_b A \Leftrightarrow A', \qquad b \geq a_1^{m_1} \otimes \dots \otimes a_n^{m_n} \tag{4.29}$$

where A' is a formula which is a result of replacing of the formulas $B_1, \dots, B_n$ in A by $B_1', \dots, B_n'$.

PROOF: By induction on the length of the formula. If A is an atomic formula then its only subformula is A itself from which (4.29) trivially follows.

Let $A := B \Rightarrow C$ and

$$T \vdash_b B \Leftrightarrow B', \qquad b \geq a_1^{m_1} \otimes \dots \otimes a_n^{m_n},$$
$$T \vdash_c C \Leftrightarrow C', \qquad c \geq a_1^{m_1'} \otimes \dots \otimes a_k^{m_k'}$$

where B', C' are formulas in which the replacements have been done. By Lemma 4.9(d) we have

$$T \vdash_d (B \Rightarrow C) \Rightarrow (B' \Rightarrow C), \qquad d \geq b,$$

and using (R2), (P20) and the second assumption

$$T \vdash_{d'} (B' \Rightarrow C) \Rightarrow (B' \Rightarrow C'), \qquad d' \geq c,$$

i.e. $T \vdash_{e'} (B \Rightarrow C) \Rightarrow (B' \Rightarrow C')$, $e' \geq d \otimes d'$. Analogously we prove the opposite implication.

Let $A := (\forall x)B$. Then we proceed analogously as above using the rules of generalization r_G and modus ponens r_{MP}, and also the tautology (Q1) proved in Lemma 4.11 below. □

Provable predicate tautologies.

LEMMA 4.11
Let T be a fuzzy predicate calculus. The following are schemes of the effectively provable formal predicate theorems in T, i.e. for each theorem there exists a proof with the degree equal to **1**.

(Q1) $T \vdash (\forall x)(A \Rightarrow B) \Rightarrow ((\forall x)A \Rightarrow (\forall x)B)$,

(Q2) $T \vdash (\forall x)(A \,\&\, B) \Leftrightarrow ((\forall x)A \,\&\, B)$ *where x is not free in B,*

(Q3) $T \vdash (\forall x)(A \Rightarrow B) \Leftrightarrow (A \Rightarrow (\forall x)B)$ *where x is not free in A,*

(Q4) $T \vdash (\forall x)(A \Rightarrow B) \Leftrightarrow ((\exists x)A \Rightarrow B)$ *where x is not free in B,*

(Q5) $T \vdash (\exists x)(A \Rightarrow B) \Leftrightarrow (A \Rightarrow (\exists x)B)$ *where x is not free in A,*

(Q6) $T \vdash (\exists x)(A \Rightarrow B) \Leftrightarrow ((\forall x)A \Rightarrow B)$ *where x is not free in B,*

(Q7) $T \vdash (\forall x)A \Leftrightarrow \neg(\exists x)\neg A$,

(Q8) $T \vdash A_x[t] \Rightarrow (\exists x)A$ *for every term t.*

PROOF: (Q1): From the axioms (T1) and (R2) we obtain $T \vdash (A \Rightarrow B) \Rightarrow ((\forall x)A \Rightarrow B$. Then apply r_G, (T2) and (R2).

(Q2): (a) From definition of the disjunction connective and (P15) we have $T \vdash (\neg A \vee B) \Leftrightarrow (A \Rightarrow (A \,\&\, B))$. (b) Similarly, using (Q1) we obtain $T \vdash (\forall x)(A \vee B) \Rightarrow ((\forall x)A \vee B)$. (c) From the instance $(\forall x)(A \,\&\, B) \Rightarrow (A \,\&\, B)$ of (T1), axiom (R2), (P10), and (P4) we prove that $T \vdash (\forall x)(A \,\&\, B) \Leftrightarrow (B \wedge (\forall x)(A \,\&\, B))$.

From (a) and (T2), we conclude that $T \vdash (B \Rightarrow (\forall x)(A \,\&\, B)) \Leftrightarrow (\forall x)(\neg B \vee A)$. Then, using Theorem 4.6 and realizing that $T \vdash A \,\&(A \Rightarrow B) \Leftrightarrow (A \wedge B)$, we obtain $T \vdash (B \wedge (\forall x)(A \,\&\, B)) \Leftrightarrow (B \,\&(\forall x)(\neg B \vee A))$. Now using (b) and (c) and Lemma 4.6(a) we have the following sequence:

$$\begin{aligned}
&T \vdash (\forall x)(A \,\&\, B) \Rightarrow (B \,\&(\forall x)(\neg B \vee A)),\\
&T \vdash (\forall x)(A \,\&\, B) \Rightarrow (B \,\&(\neg B \vee (\forall x)A)),\\
&T \vdash (\forall x)(A \,\&\, B) \Rightarrow ((B \,\&\neg B) \vee (B \,\&(\forall x)A)),\\
&T \vdash (\forall x)(A \,\&\, B) \Rightarrow (B \,\&(\forall x)A).
\end{aligned}$$

Conversely, from the instance $(\forall x)A \Rightarrow A$ and Lemma 4.6(a) and using r_G and r_{MP} we get $T \vdash ((\forall x)A \,\&\, B) \Rightarrow (\forall x)(A \,\&\, B)$.

(Q3): One implication is axiom (T2). To find the proof of the converse implication, we start with the instance of the axiom (T2) $(\forall x)B \Rightarrow B$ and so $(A \Rightarrow (\forall x)B) \Rightarrow (A \Rightarrow B$ is provable by Lemma 4.6(c). The proof of formula (Q2) is then obtained using the rule r_G, axiom (T2), theorem (P10) and modus ponens.

(Q4): Starting from substitution (T1) and (R3) we get $T \vdash (\forall x)(A \Rightarrow B) \Rightarrow (\neg B \Rightarrow \neg A)$. Now, using r_G and (T2) we have $T \vdash (\forall x)(A \Rightarrow B) \Rightarrow (\neg B \Rightarrow (\forall x)\neg A)$. Finally, using (R3) we get $T \vdash (\forall x)(A \Rightarrow B) \Rightarrow ((\exists x)A \Rightarrow B)$. The converse implication is proved again using (R3), definition of the existential quantifier and (T2).

The other schemes are are proved analogously, using substitution axiom and (Q1)–(Q4). □

This lemma demonstrates that quantifiers in FLn preserve all the properties of those in classical logic. As can be verified using Lemma 4.4, all these schemes of formulas are tautologies of FLn[11].

Basic properties of the provability degrees.

LEMMA 4.12
Let T be a fuzzy theory.

(a) Let $T \vdash_a A$ where $a > 0$. Then there is a proof w_A of A such that $\mathrm{Val}(w_A) > \mathbf{0}$.

(b) $T \vdash_{\mathbf{0}} A \wedge B$ iff $T \vdash_a A$ and $T \vdash_b B$ and $a \wedge b = \mathbf{0}$.

(c) Let $T \vdash_a A$. Then $T \vdash_b A^m$ for every $m \geq 1$ where $b \leq a$.

PROOF: (a) If such a proof does not exist the $\mathrm{Val}(w_A) = \mathbf{0}$ holds for every proof w_A of A which gives $T \vdash_{\mathbf{0}} A$ — a contradiction.

(b) Let $T \vdash_c A \wedge B$. Using (P4) we obtain $c \leq a$ and $c \leq b$.

Vice versa, let $T \vdash_a A$ and $T \vdash_b B$. Then obviously $T \vdash \mathbf{a} \Rightarrow A$ as well as $T \vdash \mathbf{b} \Rightarrow B$, and thus $T \vdash \mathbf{d} \Rightarrow A \wedge B$ for every $d \leq a \wedge b$. Therefore $T \vdash_{a \wedge b} A \wedge B$.

(c) Let $b > a$. Using (P5) we obtain $T \vdash_c A$ where $c \geq b > a$ which is a contradiction with the assumption. □

LEMMA 4.13
Let T be a fuzzy theory. If $T \vdash_a A \Rightarrow B$ then $T \vdash_b (\exists x)A \Rightarrow (\exists x)B$ where $a \leq b$.

PROOF: To every proof $w_{A \Rightarrow B}$ we construct a proof of $(\exists x)A \Rightarrow (\exists x)B$ using the instance of (Q8) $\vdash B \Rightarrow (\exists x)B$, (R2), rule of generalization and a provable formula $\vdash (\forall x)(A \Rightarrow B) \Leftrightarrow ((\exists x)A \Rightarrow B)$ having the same value as $w_{A \Rightarrow B}$. □

Closure theorem. We demonstrate in this subsection that we may confine ourselves only to closed formulas. This fact will be used in the algebraic proof of the completeness theorem below. The following is proved using the substitution axiom.

LEMMA 4.14
Let T be a fuzzy theory and $t_1, \ldots, t_n$ be terms, which are substitutible into the formula $A(x_1, \ldots, x_n)$. Then $T \vdash_a A$ and $T \vdash_b A_{x_1,\ldots,x_n}[t_1, \ldots, t_n]$ implies $a \leq b$.

[11]At the same time, they are also predicate tautologies of Łukasiewicz logic.

PROOF: Let w_A be a proof of A, $\mathrm{Val}(w_A) = a'$. By n-times repetition of the rule r_G, substitution axiom (T1) and modus ponens, we obtain a proof w of $A_{x_1,\dots,x_n}[t_1,\dots,t_n]$ with the value $\mathrm{Val}(w) = a'$. From it follows that $\bigvee\{w \mid \text{all proofs } w_A\} = a$ which gives the lemma. □

COROLLARY 4.1
Let $y_1,\dots,y_n$ be variables not occurring in a formula $A(x_1,\dots,x_n)$. Then $T \vdash_a A$ iff $T \vdash_a A_{x_1,\dots,x_n}[y_1,\dots,y_n]$.

THEOREM 4.7 (CLOSURE THEOREM)
Let T be a fuzzy theory. Let $A \in F_{J(T)}$ and A' be its closure. Then $T \vdash_a A$ iff $T \vdash_a A'$.

PROOF: Let w_A be a proof of A. Using r_G, we obtain a proof $w_{A'}$ of the closure A' such that $\mathrm{Val}(w_A) = \mathrm{Val}(w_{A'})$ and, hence, $T \vdash_b A'$, $b \geq a$.

Vice-versa, starting from $T \vdash_c A'$, using substitution axiom (T1), we obtain $T \vdash_d A_{x_1,\dots,x_n}[y_1,\dots,y_n]$, $c \leq d$. Then apply Corollary 4.1 and Lemma 4.7. □

Provability of special predicate formulas and prenex form. Recall that variant of A' is a formula in which $(\forall x)B$ is replaced by the formula $(\forall y)B_x[y]$ where y is a variable, which is not free in B.

LEMMA 4.15
Let T be a fuzzy theory, A be a formula and A' its variant. Then $T \vdash A \Leftrightarrow A'$.

LEMMA 4.16
Let T be a fuzzy predicate calculus. Then for all $n \in \mathbb{N}^+$, the following formulas are effectively provable formal theorems.

(Q9) $T \vdash (\exists x)A^n \Leftrightarrow ((\exists x)A)^n$,

(Q10) $T \vdash (\exists y)(A_x[y] \Rightarrow (\forall x)A)^n$.

PROOF: (Q9): By iteration of Lemma 4.6(d) on page 123 and (Q8) we obtain $T \vdash (A \Rightarrow (\exists x)A)^n \Rightarrow (A^n \Rightarrow ((\exists x)A)^n)$. Then using r_{MP}, r_G and (Q4) we obtain $T \vdash (\exists x)A^n \Rightarrow ((\exists x)A)^n$.

Conversely, after the iterative application (P22) we get

$$T \vdash (A(x_1) \,\&\, \cdots \,\&\, A(x_n)) \Rightarrow (A^n(x_1) \vee \cdots \vee A^n(x_n))$$

where $x_1,\dots,x_n$ are distinct variables. Using (Q8), we get

$$T \vdash (A(x_1) \,\&\, \cdots \,\&\, A(x_n)) \Rightarrow ((\exists x_1)A^n(x_1)) \vee \cdots \vee (\exists x_n)A^n(x_n)).$$

Now, using Lemma 4.15 on variant and (P11), we have

$$T \vdash (A(x_1) \,\&\, \cdots \,\&\, A(x_n)) \Rightarrow (\exists z)A^n(z).$$

Finally, we use r_G, (P12) and (Q4) to obtain

$$T \vdash (\exists x_1)A(x_1) \Rightarrow (\cdots \Rightarrow (\exists x_n)A(x_n)) \Rightarrow (\exists z)A^n(z).$$

Using again (P12) and Lemma 4.15), we conclude that $T \vdash ((\exists x)A)^n \Rightarrow (\exists x)A^n$.

(Q10): From the instance $T \vdash (\forall x)A \Rightarrow A_x[y]$ of (T1) we obtain $T \vdash (\exists y)(A_x[y] \Rightarrow (\forall x)A$. Then apply (P5) and (Q9). □

This subsection will be closed by introducing of the *prenex form* of a formula, which has the meaning analogous to that of classical logic. Let $Q(x)$ denote a quantifier and $Q'(x)$ its opposite. The *prenex operations* are defined as follows.

(i) Replace A by its variant.

(ii) Replace $\neg Q(x)A$ by $Q'(x)\neg A$.

(iii) Replace $Q(x)A \vee B$ by $Q(x)(A \vee B)$ provided that x is not free in B.

(iv) Replace $Q(x)A \nabla B$ by $Q(x)(A \nabla B)$ provided that x is not free in B.

(v) Replace $Q(x)A \wedge B$ by $Q(x)(A \wedge B)$ provided that x is not free in B.

(vi) Replace $Q(x)A \,\&\, B$ by $Q(x)(A \,\&\, B)$ provided that x is not free in B.

(vii) Replace $Q'(x)A \Rightarrow B$ by $Q(x)(A \Rightarrow B)$ provided that x is not free in B.

(viii) Replace $A \Rightarrow Q(x)B$ by $Q(x)(A \Rightarrow B)$ provided that x is not free in A.

The following theorem on prenex form can be proved using the equivalence theorem.

THEOREM 4.8
Let T be a fuzzy theory, A be a formula and A' a formula obtained by some prenex operation. Then $T \vdash A \Leftrightarrow A'$.

4.3.3 Consistency of fuzzy theories

The question whether a theory we are dealing with is consistent is one of the crucial questions in any logical system. The reason is that in classical logic, a contradictory theory (i.e. that which is not consistent) turns into a useless theory in which everything is provable. Quite surprizing is that the same result holds also in fuzzy logic. Later on, we will see that even the attempt to introduce some kind of degrees of consistency mostly does not work.

Characterization of consistent theories.

DEFINITION 4.20
A fuzzy theory T is contradictory if there is a formula A and proofs w_A and $w_{\neg A}$ of A and $\neg A$, respectively, such that

$$\mathrm{Val}_T(w_A) \otimes \mathrm{Val}_T(w_{\neg A}) > \mathbf{0}.$$

It is consistent in the opposite case.

Obviously, if T is contradictory then $T \vdash_a A$, $T \vdash_b \neg A$ and $a \otimes b > \mathbf{0}$.

The proof of the following lemma can be done by contradiction when constructing a proof following from the assumption.

LEMMA 4.17
Let T be a consistent fuzzy theory and $T \vdash_a A$. Then $T \vdash_b \neg A$ where $b \leq \neg a$.

PROOF: Let $b > \neg a$. We will write the proof

$$w := \ {}^{a'}/A \ \{\textit{some proof}\ w_A\},\ {}^{b'}/\neg A\ \{w_{\neg A}\},\ldots,\ {}^{a' \otimes b'}/A \,\&\, \neg A\ \{r_{MP}\}$$

where $b = \bigvee\{b' \mid \text{all proofs } w_{\neg A}\}$ which follows that

$$\bigvee\{\mathrm{Val}(w) \mid \text{all proofs } w_A, w_{\neg A}\} = a \otimes b$$

and thus, $T \vdash_c A \,\&\, \neg A$ where $c \geq a \otimes b > 0$. Therefore, T is contradictory due to Lemma 4.12 (a) — a contradiction. □

The following theorem is crucial in this section. It states that a contradictory fuzzy theory collapses into a degenerated theory just as in classical logic.

THEOREM 4.9 (CONTRADICTION)
A fuzzy theory T is contradictory iff $T \vdash A$ holds for every formula $A \in F_{J(T)}$.

PROOF: Let us consider proofs w_A, $w_{\neg A}$, $\mathrm{Val}(w_A) = a$, $\mathrm{Val}(w_{\neg A}) = b$, $\mathbf{0} < a \otimes b < \mathbf{1}$ (the case $a \otimes b = \mathbf{1}$ is trivial). Write down the proof

$$\begin{aligned} w := \ & {}^{a}/A\ \{\textit{some proof}\ w_A\},\ {}^{b}/\neg A\ \{\textit{some proof}\ w_{\neg A}\},\\ & {}^{\mathbf{1}}/A \Rightarrow (\neg A \Rightarrow (A \,\&\, \neg A))\ \{\textit{instance of (P13)}\},\\ & \ldots,\ {}^{a \otimes b}/A \,\&\, \neg A\ \{r_{MP}\},\ {}^{\mathbf{1}}/(A \,\&\, \neg A) \Rightarrow \bot\ \{\textit{instance of (P7)}\},\\ & {}^{a \otimes b}/\bot\ \{r_{MP}\},\ldots,\ {}^{\mathbf{1}}/\overline{\mathbf{a} \otimes \mathbf{b}} \Rightarrow \bot\ \{r_{LC}\},\ldots,\ {}^{\mathbf{1}}/\overline{\mathbf{a} \otimes \mathbf{b} \to \mathbf{0}}\ \{r_{MP}\}, \end{aligned}$$

i.e. $T \vdash \mathbf{c}$ for some $c < \mathbf{1}$. Since c is nilpotent with respect to $\otimes$, let us take $n \in \mathbb{N}$ such that $c^n = \mathbf{0}$. Then using (P13) we obtain a proof $w_{\mathbf{c}^n}$ with the value $\mathrm{Val}(w_{\mathbf{c}^n}) = \mathbf{1}$. Now let $B \in F_{J(T)}$ be an arbitrary formula. Now, we may write the proof

$$\begin{aligned} w' := \ & {}^{\mathbf{1}}/\mathbf{c}^n\ \{w_{\mathbf{c}^n}\},\ {}^{\mathbf{0}}/\bot,\ {}^{\mathbf{0} \to \mathbf{0} = \mathbf{1}}/\mathbf{c}^n \Rightarrow \bot\ \{\textit{Lemma 4.8}\},\ {}^{\mathbf{1}}/\bot\ \{r_{MP}\},\\ & {}^{\mathrm{SAx}(B)}/B\ \{\textit{special axiom}\},\ {}^{\mathbf{0} \to \mathrm{SAx}(B) = \mathbf{1}}/\bot \Rightarrow B\ \{\textit{Lemma 4.8}\},\\ & {}^{\mathbf{1}}/B\ \{r_{MP}\}, \end{aligned}$$

i.e. $T \vdash B$. Note that $\mathrm{SAx}(B)$ can be also $\mathbf{0}$. The converse implication is obvious. □

Using the similar technique, we obtain the following corollary.

COROLLARY 4.2
Let T be a fuzzy theory. Then T is contradictory iff any of the following holds:

(a) *There is a formula* A *and a proof* $w_{A\,\&\,\neg A}$ *such that* $\mathrm{Val}_T(w_{A\,\&\,\neg A}) > \mathbf{0}$.

(b) *There are* $a < \mathbf{1}$ *and a proof* $w_{\mathfrak{a}}$ *such that* $\mathrm{Val}_T(w_{\mathfrak{a}}) > a$.

(c) *There is a formula* A *and* $a > \mathbf{0}$ *such that* $T \vdash_a A\,\&\,\neg A$.

This corollary will often be used as a criterion for checking whether the theory in concern is consistent.

The following lemma demonstrates that contradictory theories cannot have models. Its converse is a completeness theorem.

LEMMA 4.18
Let a fuzzy theory T *have a model* $\mathcal{D}$. *Then it is consistent.*

PROOF: Let A be a formula and $\mathcal{D}(A) = a$. Then $\mathcal{D}(\neg A) = \neg a$. By the Validity theorem, $T \vdash_b A$ and $T \vdash_c \neg A$ where $b \le a$ and $c \le \neg a$. Consequently, $\mathrm{Val}_T(w_A) \otimes \mathrm{Val}(w_{\neg A}) = \mathbf{0}$ for all proofs w_A and $w_{\neg A}$. □

Provability degrees of compound formulas. Given a consistent fuzzy theory, a natural question arises, what are the provability degrees compound formulas when starting from the degrees of theorem of their components. The answer contained in the following theorem shows that their structure is complicated and does not, in general, copy that of the truth degrees. The latter is fulfilled only by the complete theories studied later.

THEOREM 4.10
Let T *be a consistent fuzzy theory and* $T \vdash_a A$ *and* $T \vdash_b B$.

(a) $T \vdash_c A \Rightarrow B$ *implies* $c \le a \to b$.

(b) $T \vdash_c A\,\&\,B$ *implies* $c \ge a \otimes b$.

(c) *If* $T \vdash_a (\forall x)A$ *then* $a \le \bigwedge\{b \mid T \vdash_b A_x[t], t \in M_J(T)\}$.

(d) *If* $T \vdash_c A \wedge B$, $T \vdash_a A$ *and* $T \vdash_b B$ *then* $c = a \wedge b$.

PROOF: (a) The case $a \le b$ is trivial, so assume the opposite and let $c > a \to b$. Write down the proof

$$\begin{aligned} w := \;& a'/A\ \{\textit{some proof } w_A\},\quad c'/A \Rightarrow B\ \{\textit{some proof } w_{A\Rightarrow B}\},\\ & a' \otimes c'/B\ \{r_{MP}\} \end{aligned}$$

whence $T \vdash_d B$, $d \ge a \otimes c$. However, it follows from the assumption that $b < a \otimes c$ — a contradiction.

(b) follows from (P13) and (c) follows from Lemma 4.14.

(d) The if-part follows from (P4). Vice versa, let $T \vdash_a A$ and $T \vdash_b B$. Then obviously $T \vdash \mathbf{a} \Rightarrow A$ as well as $T \vdash \mathbf{b} \Rightarrow B$, and thus $T \vdash \mathbf{d} \Rightarrow A \wedge B$ for all $d \le a \wedge b$. Therefore $T \vdash_{a \wedge b} A \wedge B$. □

4.3.4 *Extension of fuzzy theories*

This subsection is devoted to the problem, what happens if new axioms are added to a (consistent) fuzzy theory. Of course, the crucial question is whether the obtained fuzzy theory remains consistent. We show that though many concepts are direct generalizations of those of classical logic, some propositions or their proofs are not as straightforward as it might be expected from the first sight.

Extending language and a fuzzy set of axioms.

DEFINITION 4.21
A language J' is an extension of the language J if $J \subseteq J'$. A fuzzy theory T' is an extension of the fuzzy theory T if $J(T) \subseteq J'(T')$ and $T \vdash_a A$ and $T' \vdash_b A$ implies $a \leq b$ for every formula $A \in F_{J(T)}$.

The extension T' is a conservative extension of T if $T' \vdash_b A$ and $T \vdash_a A$ implies $a = b$ for every formula $A \in F_{J(T)}$. The extension T' is a simple extension of T if $J'(T') = J(T)$.

Let $T_1 = \langle \mathrm{LAx}_1, \mathrm{SAx}_1, R\rangle$, $T_2 = \langle \mathrm{LAx}_2, \mathrm{SAx}_2, R\rangle$ be fuzzy theories and $J(T_1) \subseteq J(T_2)$. If $\mathrm{SAx}_1 \subseteq \mathrm{SAx}_2$ then T_2 is an extension of T_1, where we understand that $\mathrm{SAx}_1(A) = \mathbf{0}$ for all $A \in F_{J(T_2)} \setminus F_{J(T_1)}$.

Let T_1 be a fuzzy theory and $\Gamma \underset{\sim}{\subset} F_{J(T_1)}$ a fuzzy set of formulas. By $T_2 = T_1 \cup \Gamma$ we denote a fuzzy theory whose special axioms are $\mathrm{SAx}_2 = \mathrm{SAx}_1 \cup \Gamma$.

The following natural lemma demonstrates that we may safely add a theorem of a fuzzy theory among its special axioms.

LEMMA 4.19
Let T be a fuzzy theory and $T \vdash_a A$. Then $T' = T \cup \{\, ^a/A\}$ is a simple conservative extension of T.

PROOF: Let $T \vdash_b B$. Using the induction on the length of proof we will demonstrate that $\mathrm{Val}_{T'}(w'_B) \leq b$ holds for every proof w'_B of B in T'. Then we will obtain $b \leq \bigvee_{w'_B} \mathrm{Val}_{T'}(w'_B) \leq b = \bigvee_{w_B} \mathrm{Val}_T(w_B)$ by the definition of provability and, at the same, by the fact that T' is extension of T.

(a) Let B be an axiom (logical or special). First, we assume that $B := A$. Then $w'_B := \; ^a/A$ $\{special\ axiom\}$. But since $T \vdash_a A$, we have $a = \mathrm{Val}_{T'}(w'_B) \leq \bigvee_{w_A} \mathrm{Val}_T(w_A) = a$.

If B is another special axiom then $w'_B := \; ^{b'}/B$ $\{axiom\}$ and we have $b' = \mathrm{Val}_{T'}(w'_B) \leq \bigvee_{w_B} \mathrm{Val}_T(w_B)$ by the definition of provability and the fact that B is also the axiom of T.

(b) Let

$$
\begin{aligned}
w'_B := \; & ^c/C \; \{some\ proof\ w_C\}, \; ^d/C \Rightarrow B \; \{some\ proof\ w_{C\Rightarrow B}\}, \\
& ^{c \otimes d}/B \; \{modus\ ponens\}.
\end{aligned}
$$

By the inductive assumption, $c \leq \bigvee_{w_C} \mathrm{Val}_T(w_C)$, $d \leq \bigvee_{w_{C\Rightarrow B}} \mathrm{Val}_T(w_{C\Rightarrow B})$ and so

$$\mathrm{Val}_{T'}(w'_B) = c \otimes d \leq \bigvee_{w_C, w_{C\Rightarrow B}} (\mathrm{Val}_T(w_C) \otimes \mathrm{Val}_T(w_{C\Rightarrow B})) \leq \bigvee_{w_B} \mathrm{Val}_T(w_B) = b.$$

The case when B is obtained using r_G follows from the semicontinuity of the semantic part of the rule and the inductive assumption. Similarly for r_{LC}. □

The following simple lemma demonstrates that closed and open axioms lead to the same consistency behaviour of fuzzy theories.

LEMMA 4.20
Let T be a fuzzy theory, A be a formula and A' be its closure. Then the fuzzy theory $T' = T \cup \{\, ^a\!/A'\}$ is contradictory iff $T'' = T \cup \{\, ^a\!/A\}$ is contradictory.

PROOF: Let T' be contradictory and $w_{C \,\&\, \neg C}$ be a proof containing a special axiom $^a\!/A'$ such that $\mathrm{Val}(w_{C \,\&\, \neg C}) > \mathbf{0}$. When extending the former by $^a\!/A$, $^a\!/(\forall x)A, \ldots$, $^a\!/A'$ then $^a\!/A$ takes the role of special axiom instead of $^a\!/A'$ and thus, we obtain a proof of $C \,\&\, \neg C$ in T'' with the value greater than $\mathbf{0}$. Consequently, T'' is contradictory. The proof of the converse is similar. □

The following is obvious.

LEMMA 4.21
Let T' be an extension of a fuzzy theory T.

(a) If $\mathcal{D} \models T'$ then $\mathcal{D} \models T$.

(b) If T' is consistent then T is also consistent.

LEMMA 4.22
Let T' be an extension of a fuzzy theory T and T' has a model $\mathcal{D}'$. Then its restriction $\mathcal{D}$ into $J(T)$ is a model of T.

PROOF: The structure $\mathcal{D}$ for $J(T)$ is obtained from $\mathcal{D}'$ by omitting some fuzzy relations, functions and constants corresponding to predicate, functional and constant symbols, respectively from $J(T')$ which do not occur in $J(T)$. Furthermore, we put $D = D'$. By induction on the length of formula we then prove that $\mathcal{D}(A) = \mathcal{D}'(A)$ holds for every $A \in F_{J(T)}$. □

Just as in the classical case, we say that two fuzzy theories T_1, T_2 are *equivalent* if one is an extension of the other one. This implies that T_1, T_2 are equivalent iff for every formula $A \in F_J$ we have $T_1 \vdash_a A$ iff $T_2 \vdash_a A$. However, the classical property that fuzzy theories are equivalent iff they have the same models follows only from the completeness theorem proved below. Let us note that there is also a weaker concept of equivalent fuzzy theories which will be mentioned in Section 4.3.6.

Chains of fuzzy theories. In this subsection, we prove one technical lemma which will be used below.

LEMMA 4.23
Let T be a consistent fuzzy theory and $\{E_i \underset{\sim}{\subset} F_{J(T)} \mid i < \lambda\}$, where λ is some ordinal, be a chain of fuzzy sets of formulas, $E_i \subseteq E_{i+1}$. Furthermore, put $T_0 = T$ and let $T_{i+1} = T_i \cup E_{i+1}$ be consistent for all $i < \lambda - 1$. Then

$$\overline{T} = T \cup \bigcup_{i<\lambda} E_i$$

is a consistent extension of T.

PROOF: Let $A \in F_{J(T)}$. By induction on the length of the proof and the semicontinuity of r^{evl} in each variable we prove that

$$\mathrm{Val}_{\overline{T}}(w_A) = \bigvee\{\mathrm{Val}_{T_i}(w_A) \mid i < \lambda\} \tag{4.30}$$

for every proof w_A of the formula A. From the assumption, it follows that $\mathrm{Val}_{T_i}(w_A) \otimes \mathrm{Val}_{T_i}(w_{\neg A}) = \mathbf{0}$, which together with (4.30) implies that $\overline{T}$ is consistent. □

Adding new constants. We will now prove several theorems which are analogues of the classical ones in FLn. We start with extension by constants which, similarly as in classical logic, is also conservative.

THEOREM 4.11 (CONSTANTS)
Let T be a fuzzy theory and V a set of constants not contained in $J(T)$. We put $\bar{J} = J(T) \cup V$ and define a fuzzy theory $\overline{T}$ by $\overline{\mathrm{SAx}}(A) = \mathrm{SAx}(A)$ for $A \in F_{J(T)}$ and $\overline{\mathrm{SAx}}(A) = \mathbf{0}$ otherwise. Then

$$\overline{T} \vdash_a A_{x_1,\dots,x_n}[\mathbf{v}_1,\dots,\mathbf{v}_n] \quad \textit{iff} \quad T \vdash_a A$$

holds for every formula $A \in F_{J(T)}$ and $\mathbf{v}_1,\dots,\mathbf{v}_n \in V$.

PROOF: Let $\overline{T} \vdash_a A_{x_1,\dots,x_n}[\mathbf{v}_1,\dots,\mathbf{v}_n]$. Since $\mathbf{v}_1,\dots,\mathbf{v}_n$ do not belong into $J(T)$, we can replace them in every proof by some variables $y_1,\dots,y_n$ not occurring in A. Consequently, $T \vdash_b A_{x_1,\dots,x_n}[y_1,\dots,y_n]$, $a \le b$ and since $A_{x_1,\dots,x_n}[y_1,\dots,y_n] \in F_{J(T)}$, we obtain $T \vdash_b A$ by Corollary 4.1.

Conversely, from $T \vdash_c A$ we obtain $\overline{T} \vdash_d A_{x_1,\dots,x_n}[\mathbf{v}_1,\dots,\mathbf{v}_n]$ $c \le d$ by Lemma 4.14. Then apply Lemma 4.7. □

Apparently, $\overline{T}$ is a conservative extension of T.

Deduction theorem. Deduction theorem plays a crucial role in the study of the extension of logical theories. Unlike classical logic, in fuzzy logic is the deduction theorem weaker due to the non-idempotency of the Łukasiewicz conjunction.

LEMMA 4.24
Let T be a fuzzy theory, A be a closed formula and $T' = T \cup \{ ^{1}/A\}$.

(a) *If $T \vdash_a A^n \Rightarrow B$ and $T' \vdash_b B$ for some n then $a \leq b$.*

(b) *To every proof w_B in T' there are n and a proof $w_{A^n \Rightarrow B}$ in T such that*

$$\mathrm{Val}_{T'}(w_B) = \mathrm{Val}_T(w_{A^n \Rightarrow B}).$$

PROOF: (a) follows from (P13) and modus ponens.

(b) is proved by induction on the length of w_B. The case $B := A$ is trivial. Let $w_B := \ ^{b}/B\ \{axiom\}$. Then

$$w_{A^n \Rightarrow B} := \ ^{b}/B\ \{axiom\},\ \ ^{1}/B \Rightarrow (A^n \Rightarrow B)\ \{log.\ axiom\ (R1)\}$$
$$^{b}/A^n \Rightarrow B\ \{r_{MP}\}.$$

Let B be obtained using the proof

$$w_B := \ ^{c}/C\ \{some\ proof\ w_C\},\ \ ^{d}/C \Rightarrow B\ \{some\ proof\ w_{C \Rightarrow B}\},$$
$$^{c \otimes d}/B\ \{r_{MP}\}.$$

Then by the inductive assumption

$$w_{A^n \Rightarrow B} := \ ^{c}/A^m \Rightarrow C\ \{some\ proof\ w_{A^m \Rightarrow C}\},$$
$$^{d}/A^p \Rightarrow (C \Rightarrow B)\ \{some\ proof\ w_{A^p \Rightarrow (C \Rightarrow B)}\},$$
$$^{1}/(A^p \Rightarrow (C \Rightarrow B)) \Rightarrow ((A^m \Rightarrow C) \Rightarrow (A^{m+p} \Rightarrow B))\ \{instance\ of\ (P21)\},$$
$$\ldots,\ \ ^{c \otimes d}/A^n \Rightarrow B\ \{r_{MP}\}$$

where $n = m + p$.

Let $B := (\forall x)C$ and $w_B := \ ^{c}/C\ \{some\ proof\ w_C\},\ \ ^{c}/(\forall x)C\ \{r_G\}$. Then by the inductive assumption

$$w_{A^n \Rightarrow B} := \ ^{c}/A^n \Rightarrow C\ \{some\ proof\ w_{A^n \Rightarrow C}\},\ \ ^{c}/(\forall x)(A^n \Rightarrow C)\ \{r_G\},$$
$$\ldots,\ \ ^{c}/A^n \Rightarrow (\forall x)C\ \{r_{MP}\}$$

using the axiom (T2).

If $B := \mathbf{c} \Rightarrow C$ and $w_B := \ ^{c}/C\ \{some\ proof\ w_C\},\ \ ^{1}/\mathbf{c} \Rightarrow C\ \{r_{LC}\}$ then we use the inductive assumption and the tautology (P20). □

COROLLARY 4.3
Let A be a closed formula and $T' = T \cup \{ ^{a}/A\}$. Then to every proof w'_B in T' there are an n and a proof $w_{A^n \Rightarrow B}$ in T such that $\mathrm{Val}_{T'}(w') \leq \mathrm{Val}_T(w)$.

As seen, the part (b) of the previous lemma is weaker then the opposite implication of the part (a) and thus, we do not obtain the strong enough analogy of the classical deduction theorem. The problem consists in the fact that the part (b) assures existence of some finite n, which is used in the formula $A^n \Rightarrow B$, to every proof of the latter. However, since there may exist infinite number of proofs, we cannot be sure that n cannot arbitrarily increase. The following lemmas are necessary for the proof that this is indeed impossible.

LEMMA 4.25
If a fuzzy theory $T' = T \cup \{ {}^{a}/A\}$ is contradictory then to every formula B there is m such that $T \vdash A^m \Rightarrow B$.

PROOF: This can be proved using Theorem 4.9 on page 130 and Lemma 4.24(b). □

LEMMA 4.26
Let $T \vdash_a A$ and $b > a$. Then

$$T' = T \cup \{ {}^{1}/A \Rightarrow \mathbf{b}\}$$

is a consistent extension of T.

PROOF: Let T' be contradictory and $c < 1$. By Lemma 4.25 there is m such that

$$T \vdash (A \Rightarrow \mathbf{b})^m \Rightarrow \mathbf{c}. \tag{4.31}$$

At the same time, using (P17) and (P18) we have

$$T \vdash (A \Rightarrow \mathbf{b})^m \vee (\mathbf{b} \Rightarrow A)^m,$$

which gives $T \vdash (A \Rightarrow \mathbf{b})^m \vee (\mathbf{b} \Rightarrow A)$ using (P5), (P6) and (P11). Using (4.31), (R2) and (P6) we obtain

$$T \vdash (A \Rightarrow \mathbf{b})^m \Rightarrow (\mathbf{c} \vee (\mathbf{b} \Rightarrow A)).$$

Furthermore, $T \vdash (\mathbf{b} \Rightarrow A) \Rightarrow (\mathbf{c} \vee (\mathbf{b} \Rightarrow A))$ holds from which and (P11) it follows that $T \vdash (\mathbf{b} \Rightarrow A) \vee \mathbf{c}$. Since $c < 1$ we can find n such that $c^n = 0$. Then $T \vdash (\mathbf{b} \Rightarrow A)^n \vee \mathbf{c}^n$. Now, using $T \vdash \mathbf{c}^n \Leftrightarrow \mathbf{0}$ and $\vdash (\mathbf{b} \Rightarrow A)^n \Rightarrow (\mathbf{b} \Rightarrow A)$ we obtain $T \vdash \mathbf{b} \Rightarrow A$.

Let us now write down the proof

$$w := \;\; {}^{b}/\mathbf{b}\ \{\textit{logical axiom}\},\;\; {}^{u}/\mathbf{b} \Rightarrow A\ \{\textit{some proof } w_{\mathbf{b}\Rightarrow A}\},\;\; {}^{b \otimes u}/A\ \{r_{MP}\}$$

where $\bigvee_{w_{\mathbf{b}\Rightarrow A}} u = \mathbf{1}$. Then $T \vdash_{b'} A$, $b' \geq b > a$, which is a contradiction with $T \vdash_a A$. □

LEMMA 4.27
Let T be a fuzzy theory and $T \vdash_a A$ where $a < 1$.

(a) $T \vdash_b A^m$ and $b \leq a$ holds for every $m > 1$.

(b) If $p \geq q$ then $T \vdash A^p \Rightarrow A^q$.

(c) There is m and $b \leq a$ such that $T \vdash_b A^n$ for all $n \geq m$.

PROOF: (a) Let $b > a$ and let w_{A^m} be a proof with the value $\mathrm{Val}_T(w_{A^m}) = b' \leq b$. Then using (P5) and Theorem 4.1 we conclude that $T \vdash_c A$ where $c \geq b > a$ — a contradiction. (b) follows from (a).

(c) Note the proposition obviously holds for contradictory T. Therefore, we will suppose that T is consistent.

Let the proposition does not hold. Then to every p there is q, $q > p$ such that

$$T \vdash_{c_p} A^p, \quad T \vdash_{c_q} A^q \quad \text{and} \quad 0 < c_q < c_p. \tag{4.32}$$

By Lemma 4.26, the theory $T' = T \cup \{\, ^{\mathbf{1}}/A^q \Rightarrow \mathbf{c}_p\}$ is consistent. Then $T \vdash (A^q \Rightarrow \mathbf{c}_p)^n$ for every n and using (P23) we obtain $T \vdash A^{qn} \Rightarrow \mathbf{c}_p^n$. Let us now take n such that $c_p^n = \mathbf{0}$. On the basis of the assumption we can find $m \geq qn$ such that $T' \vdash_{c'_m} A^m$ and $c'_m > \mathbf{0}$. By (b) we have $T' \vdash A^m \Rightarrow A^{qn}$ and we conclude that $T' \vdash A^m \Rightarrow \mathbf{c}_p^n$. Then, using modus ponens, we deduce that $T' \vdash_d \mathbf{c}_p^n$ for some $d \geq c'_m > \mathbf{0}$. However, $T' \vdash \mathbf{c}_p^n \Leftrightarrow \bot$ by the book-keeping axiom and therefore, by Corollary 4.2(b) T' is contradictory — a contradiction.

□

As a direct consequence of the previous lemmas, we obtain the following theorem.

THEOREM 4.12 (DEDUCTION THEOREM)
Let T be a fuzzy theory, A be a closed formula and $T' = T \cup \{\, ^{\mathbf{1}}/A\}$. Then to every formula $B \in F_{J(T)}$ there is an n such that

$$T \vdash_a A^n \Rightarrow B \quad \text{iff} \quad T' \vdash_a B$$

The theorem below demonstrates that it is possible to denote an element, which has a certain property represented by an existential formula, by a new constant not contained in the language of the theory in concern. Similarly as in classical logic, it turns out that such an extension of the theory is conservative.

THEOREM 4.13
Let T be a consistent fuzzy theory, $T \vdash (\exists x)A(x)$ and $\mathbf{c} \notin J(T)$ be a new constant. Then the theory

$$T' = T \cup \{\, ^{\mathbf{1}}/A_x[\mathbf{c}]\}$$

in the language $J(T) \cup \{\mathbf{c}\}$ is a conservative extension of the theory T.

PROOF: Let $\overline{T}$ be a conservative extension of T by Theorem 4.11. Let $B \in F_{J(T)}$. By Theorems 4.11 and 4.12,

$$T \vdash_b B \quad \text{iff} \quad \overline{T} \vdash_b A_x[\mathbf{c}]^n \Rightarrow B \quad \text{iff} \quad T \vdash A^n(x) \Rightarrow B$$

for some n. Furthermore, using (P13), we have $T \vdash ((\exists x)A)^n$. Then using (Q9) and modus ponens, we obtain $T \vdash_b B$. □

Reduction theorem. This theorem, which is an extension of the deduction theorem, makes possible to characterize contradictory extensions of fuzzy theories.

THEOREM 4.14 (REDUCTION FOR THE CONSISTENCY)
Let T be a fuzzy theory and $\Gamma \underset{\sim}{\subset} F_{J(T)}$ a fuzzy set of formulas. A fuzzy theory $T' = T \cup \Gamma$ is contradictory iff there are $m_1, \dots, m_n$ and $A_1, \dots, A_n \in \mathrm{Supp}(\Gamma)$ such that

$$T \vdash_c \neg A_1^{m_1} \nabla \cdots \nabla \neg A_n^{m_n}$$

and $a_1^{m_1} \otimes \cdots \otimes a_n^{m_n} > \mathbf{0}$ where $c >^* \neg(a_1^{m_1} \otimes \cdots \otimes a_n^{m_n})$, $a_1 = \Gamma(A_1), \dots, a_n = \Gamma(A_n)$.

PROOF: Let T' be contradictory. From it follows that $T' \vdash \bot$ and thus, given any value $e \in L$, we may always find a proof $w_\bot$ with the value $\mathrm{Val}_{T'}(w_\bot) \geq e$. Following the proof of Lemma 4.24(b) where $\bot$ takes the role of an axiom, we choose some suitable formulas $A_1, \dots, A_n \in \mathrm{Supp}(\Gamma)$ and $m_1, \dots, m_n$ such that $a_1^{m_1} \otimes \cdots \otimes a_n^{m_n} > \mathbf{0}$ and construct a proof w of the formula $A_1^{m_1} \Rightarrow (\cdots \Rightarrow (A_n^{m_n} \Rightarrow \mathbf{0}) \cdots)$ such that

$$\mathrm{Val}_T(w) \geq \mathrm{Val}_{T'}(w_\bot).$$

Using the tautology (P16) we get

$$T \vdash_c \neg A_1^{m_1} \nabla \cdots \nabla \neg A_n^{m_n}$$

where $c \geq \mathrm{Val}(w_\bot)$. Finally, take $w_\bot$ with $\mathrm{Val}_{T'}(w_\bot) \geq e = \neg(a_1^{m_1} \otimes \cdots \otimes a_n^{m_n})$.

Vice-versa, by the equivalence theorem we have $T \vdash_c \neg(A_1^{m_1} \,\&\, \cdots \,\&\, A_n^{m_n})$. But then there is a proof

$$\begin{aligned} w := \; & a_1/A_1 \; \{spec.\ axiom\}, \dots, \quad a_n/A_n \; \{spec.\ axiom\}, \dots, \\ & a_1^{m_1} \otimes \cdots \otimes a_n^{m_n} / A_1^{m_1} \,\&\, \cdots \,\&\, A_n^{m_n} \; \{r_{MP}\} \end{aligned}$$

in T' which follows that

$$T \vdash_d (A_1^{m_1} \,\&\, \cdots \,\&\, A_n^{m_n}) \,\&\, \neg(A_1^{m_1} \,\&\, \cdots \,\&\, A_n^{m_n})$$

for some $d \geq c \otimes (a_1^{m_1} \otimes \cdots \otimes a_n^{m_n}) > 0$, i.e. T' is contradictory. □

COROLLARY 4.4
A fuzzy theory $T' = T \cup \{\neg a/\neg A\}$ where $a > \mathbf{0}$ is contradictory iff $T \vdash_b mA$ for some m and $b >^* ma$.

PROOF: By Theorem 4.14, we have $T \vdash_b \neg(\neg A)^m$, $b >^* \neg(\neg a)^m$, but $\neg(\neg a)^m = \neg(\neg ma) = ma$ which gives also $\neg(\neg A)^m \Leftrightarrow mA$. □

We close this section by the analogue of the classical theorem on the introduction of a new predicate symbol. Its proof proceeds quite analogously as the classical one (see [121, 76]).

THEOREM 4.15
Let $A \in F_{J(T)}$ be a formula of the fuzzy theory T, P be a new predicate symbol not occurring in $J(T)$ and let T' be a new fuzzy theory obtained from T by adding a special axiom

$$T' = T \cup \{\, \mathbf{1}/P(x_1, \ldots, x_n) \Leftrightarrow A\}.$$

Then T' is a conservative extension of T and to every formula $B \in F_{J(T')}$ it is possible to construct a corresponding formula $B^* \in F_{J(T)}$ such that

$$T' \vdash B \Leftrightarrow B^*.$$

4.3.5 Henkin fuzzy theories

Analogously as in classical logic, we can introduce a set of special constants and special axioms for all the formulas of the form $(\forall x)A$. The obtained fuzzy theory will be called Henkin. We prove that every consistent fuzzy theory can be conservatively extended into the Henkin one.

DEFINITION 4.22
Given a formula $(\forall x)A(x)$ and let $\mathbf{r}$ be a special constant for it. A fuzzy theory T is called Henkin if for all formulas $A(x)$, the evaluated formulas

$$\mathbf{1}/A_x[\mathbf{r}] \Rightarrow (\forall x)A(x) \tag{4.33}$$

are among special axioms of T.

The evaluated formulas (4.33) are called Henkin axioms. Note that there is also the existential form

$$\mathbf{1}/(\exists x)A(x) \Rightarrow A_x[\mathbf{r}]. \tag{4.34}$$

Henkin theories are interesting because of the following property. Let $\mathcal{D}$ be a model of the Henkin fuzzy theory T and $\mathbf{r}$ a special constant for a formula $(\forall x)A$. Then there is an element $d_0 \in D$ such that $\mathcal{D}(\mathbf{r}) = d_0$ and

$$\mathcal{D}(A_x[\mathbf{r}])[d_0] = \bigwedge_{d \in D} \mathcal{D}(A)[\mathbf{d}] = \mathcal{D}((\forall x)A).$$

THEOREM 4.16
Let T be a consistent fuzzy theory, K a set of special constants for all the closed formulas $(\forall x)A$ and let Ax_H be a set of Henkin axioms (4.33) (a fuzzy set of Henkin axioms). Then the fuzzy theory $T_H = T \cup \mathrm{Ax}_H$ in the language $J(T_H) = J(T) \cup K$ is a conservative extension of the theory T.

PROOF: Similarly as in the classical proof, we construct sets of special constants $K_1, K_2, \ldots$ of the corresponding levels. Put $T_0 = T$ and

$$T_{i+1} = T_i \cup \{\, \mathbf{1}/A_x[\mathbf{r}] \Rightarrow (\forall x)A(x)\}$$

where $\mathbf{r} \in K_{i+1}$ is a special constant for $(\forall x)A$. Using the formal theorem (Q9) of Lemma 4.16, we obtain $T_i \vdash (\exists y)(A_x[y] \Rightarrow (\forall x)A)$. Then, using Theorem 4.13 and the formal theorem (Q10) we obtain that T_{i+1} is a conservative extension of T_i for every i. The theorem then follows from Lemma 4.23. □

The following lemma follows from Henkin and substitution axioms using the equivalence theorem.

LEMMA 4.28
Let T be a Henkin fuzzy theory and $\mathbf{r}$ a special constant for $(\forall x)A$. Then

$$T \vdash_a (\forall x)A \quad \textit{iff} \quad T \vdash_a A_x[\mathbf{r}].$$

4.3.6 Complete fuzzy theories

Recall that classical theory is complete if every formula or its negation is provable (a theorem) in it. A similar concept can be introduced also in fuzzy logic where complete fuzzy theories can be seen as fuzzy sets of formulas whose membership degrees (i.e. the degrees of provability) take the structure of truth values. Such theories are simpler than the general ones since we may easily compute the provability degree of more complex formulas from the provability degree of its subformulas. Similarly as in classical logic, every fuzzy theory can be extended into a complete one.

Characterization of complete theories.

DEFINITION 4.23
A fuzzy theory is complete if it is consistent and

$$T \vdash_a A \quad \textit{implies} \quad T \vdash A \Rightarrow \mathbf{a}$$

holds for every closed formula A and every $a \in L$.

The following is obvious.

LEMMA 4.29
Let a fuzzy theory T be complete. Then

$$T \vdash_a \quad \textit{iff} \quad T \vdash A \Leftrightarrow \mathbf{a}$$

holds for every formula A and every $a \in L$.

THEOREM 4.17
Let a fuzzy theory T be consistent. Then T is complete iff $T \vdash_a A$ and $T \vdash_{\neg a} \neg A$ holds for every formula A and some degree a.

PROOF: Let T be complete and $T \vdash_a A$. We write a proof

$$\begin{aligned} w := {} & a \to \mathbf{0}/\mathbf{a} \Rightarrow \bot \ \{\textit{log. axiom}\},\ b/A \Rightarrow \mathbf{a} \ \{\textit{some proof } w_{A\Rightarrow\mathbf{a}}\}, \\ & 1/(A \Rightarrow \mathbf{a}) \Rightarrow ((\mathbf{a} \Rightarrow \bot) \Rightarrow (A \Rightarrow \bot)) \ \{\textit{log. axiom}\}, \ldots \\ & \neg a \otimes b/A \Rightarrow \bot \ \{\textit{modus ponens}\} \end{aligned}$$

whence $T \vdash_c \neg A$, $\neg a \leq c$. Since T is consistent, it follows from Theorem 4.10(a) on page 131 that $c \leq \neg a$, i.e. $c = \neg a$.

Vice-versa, let w_A be a proof with the value $\mathrm{Val}(w_A) = b \leq a$ and $w_{\neg A}$ be a proof with $\mathrm{Val}(w_{\neg A}) = b' \leq \neg a$. Then

$$w := \quad {}^{b}/A \; \{w_A\}, \quad {}^{a \to b}/\mathbf{a} \Rightarrow A \; \{\mathit{Lemma\ 4.8}\},$$

and since $\bigvee_{w_A}(a \to b) = \mathbf{1}$, we conclude that $T \vdash \mathbf{a} \Rightarrow A$. By similar arguments, we obtain $T \vdash \neg\mathbf{a} \Rightarrow \neg A$. Then using logical axiom (R3) we obtain $T \vdash A \Rightarrow \mathbf{a}$. □

Hence, a complete theory may be defined also in the classical way using only negation.

The following theorem demonstrates that the degrees of theorem of complete theories exactly copy the structure of truth values. Hence, complete theories are in some sense the simplest fuzzy theories.

THEOREM 4.18
A fuzzy theory T is complete iff there is a model $\mathcal{D}$ fulfilling the condition: $T \vdash_a A$ iff $\mathcal{D}(A) = a$ for every A and $a \in L$.

PROOF: Let T be complete. We show that a canonical structure $\mathcal{D}_0$ for $J(T)$ defined on page 119 is a model of T such that

$$T \vdash_a A \quad \text{iff} \quad \mathcal{D}_0(A) = a \tag{4.35}$$

holds for every formula $A \in F_{J(T)}$.

Let T' be a conservative Henkin extension of T due to Theorem 4.16. We construct a canonical structure $\mathcal{D}_0$ for it, and by induction on the length of the formula we show (4.35) for T' and $\mathcal{D}_0$. As an evaluation, we use the provability degree in T, i.e. we put $\mathcal{D}_0(P(t_1, \ldots, t_n)) = a$ iff $T \vdash_a P(t_1, \ldots, t_n)$.

If A is atomic then (4.35) follows immediately from the definition of the canonical structure. If $A := B \Rightarrow C$ then we obtain (4.35) using the equivalence theorem and inductive assumption. If $A := (\forall x)B$ then we have

$$T' \vdash_a A \quad \text{iff} \quad T' \vdash_a B_x[\mathbf{r}] \quad \text{iff} \quad \mathcal{D}_0(B_x[\mathbf{r}]) \quad \text{iff} \quad \mathcal{D}_0((\forall x)B)$$

due to Lemma 4.28 and the inductive assumption. By closure theorem, (4.35) holds for every formula $A \in F_{J(T)}$. Obviously, $\mathcal{D}_0$ is a model of T' and so, it is a model of T by Lemma 4.22.

Conversely, let $\mathcal{D} \models T$ be such model and let $\mathcal{D}(A) = a$. Then $\mathcal{D}(A \Rightarrow \mathbf{a}) = \mathbf{1}$, i.e. $T \vdash A \Rightarrow \mathbf{a}$ which means that T is complete. □

The following example demonstrates that there exist incomplete fuzzy theories.

EXAMPLE 4.3
Let us consider a fuzzy theory

$$T = \{ {}^{a_1}/A_1, \; {}^{a_2}/A_2, \; {}^{c_1}/A_1 \Rightarrow B, \; {}^{c_2}/A_2 \Rightarrow B\}$$

where A_1, A_2, B are atomic formulas and $a_1 < a_2 < 1$, $c_1 < c_2 < 1$, $a_1 \otimes c_1 < a_2 \otimes c_2$. Then there is no model $\overline{\mathcal{D}}$ such that $\overline{\mathcal{D}}(A_1) = a_1$, $\overline{\mathcal{D}}(A_2) = a_2$, $\overline{\mathcal{D}}(A_1 \Rightarrow B) = c_1$, $\overline{\mathcal{D}}(A_2 \Rightarrow B) = c_2$. Indeed, the latter equalities follow that $\overline{\mathcal{D}}(B) = a_1 \otimes c_1 = a_2 \otimes c_2$ which is impossible. On the other hand if we put $\mathcal{D}(A_1) = a_1$, $\mathcal{D}(A_2) = a_2$ and $\mathcal{D}(B) = (a_1 \otimes c_1) \vee (a_2 \otimes c_2)$ then we obtain a model of T and thus, T is consistent. □

Completion of fuzzy theories. We will now demonstrate that every consistent fuzzy theory can be extended into a complete one.

THEOREM 4.19
Let T be a consistent fuzzy theory. Then there exists a complete theory $\bar{T}$ which is a simple extension of T.

PROOF: Let $\langle E_i \mid i < q \rangle$ be a chain of fuzzy sets of formulas such that $T_0 = T$ and $T_{i+1} = T_i \cup E_i$ is consistent. Using Zorn's lemma we construct a maximal fuzzy theory $\bar{T} = T \cup \bar{A}$ where $\bar{A}$ is a maximal fuzzy set, which is a simple consistent extension of T.

We show that $\bar{T}$ is complete. Let $\bar{T} \vdash_a A$, $b > a$ and put $\bar{T}' = \bar{T} \cup \{{}^{\mathbf{1}}/A \Rightarrow \mathbf{b}\}$. We know from Lemma 4.26 that $\bar{T}'$ is consistent and by maximality of $\bar{T}$ we conclude that $\{\,{}^{\mathbf{1}}/A \Rightarrow \mathbf{b}\} \subseteq \overline{\mathrm{SAx}}$. Hence, $\bar{T} \vdash A \Rightarrow \mathbf{b}$ for every $b > a$. Let $\bar{T} \vdash_c A \Rightarrow \mathbf{a}$ and write down the proof

$$w := {}^{\mathbf{1}}/A \Rightarrow \mathbf{b}\ \{spec.\ axiom\},\ {}^{b \to a}/\mathbf{b} \Rightarrow \mathbf{a})\ \{log.\ axiom\}, \ldots,$$
$$\qquad {}^{b \to a}/A \Rightarrow \mathbf{a}\ \{r_{MP}\},$$

From it follows that $c \geq b \to a$ for every $b > a$. But then

$$\bigvee\{b \to a \mid b \in L, b > a\} = \bigwedge\{b \mid b \in L, b > a\} \to a = a \to a = \mathbf{1} \leq c.$$

We obtain that $\bar{T} \vdash A \Rightarrow \mathbf{a}$, i.e. $\bar{T}$ is complete. □

Let us remark that, of course, the complete extension due to the previous theorem in general is not conservative, i.e. if $T \vdash_a A$ and $\bar{T} \vdash_b A$ then it may happen that $b > a$.

The following concept has been proposed by I. Perfilieva (cf. [100]). We say that fuzzy theories T_1, T_2 are *weakly equivalent* if they have the same complete extension.

LEMMA 4.30
If fuzzy theories T_1, T_2 are equivalent then they are weakly equivalent. If T_2 is a simple extension of T_1 then they are weakly equivalent. Every fuzzy theory is weakly equivalent with its complete extension.

4.3.7 Lindenbaum algebra of formulas and its properties

Let F_J be a set of formulas. In this section, we follow the classical Lindenbaum construction of an algebra of equivalence classes of formulas and show that it takes the basic structure of truth values, but does not copy it exactly. Our constructions will rely on the fact that we may conservatively extend a consistent theory into the Henkin one.

Construction of Lindenbaum algebra. Let us put $A \leq B$ iff $T \vdash A \Rightarrow B$. It follows from the axiom (R2) and the tautology (P2) that this is a preorder of $T_{J(T)}$. Furthermore, let us put $A \approx B$ iff $A \leq B$ and $B \leq A$. It is obvious that $\approx$ is an equivalence on $F_{J(T)}$ and thus, we can construct a factor set $F_{J(T)}| \approx$. Furthermore, we put

$$
\begin{aligned}
[A] \vee [B] &:= [A \vee B], & [A] \wedge [B] &:= [A \wedge B], \\
[A] \otimes [B] &:= [A \,\&\, B], & [A] \to [B] &:= [A \Rightarrow B], \\
\bigwedge_{t \in M_V} [A_x[t]] &:= [(\forall x)A], & \bigvee_{t \in M_V} [A_x[t]] &:= [(\exists x)A],
\end{aligned}
$$

where $[\cdot]$ denotes the equivalence class according to $\approx$. Obviously, the relation $\leq_T$ defined by

$$[A] \leq_T [B] \quad \text{iff} \quad A \leq B \tag{4.36}$$

is a partial ordering on $F_{J(T)}| \approx$.

THEOREM 4.20
Let T be a consistent Henkin fuzzy theory.

(a) *The relation $\approx$ is a congruence on $F_{J(T)}$.*

(b) *The Lindenbaum algebra*

$$\mathcal{L}(T) = \langle F_{J(T)}| \approx, \vee, \wedge, \otimes, \to, [\bot], [\top] \rangle \tag{4.37}$$

on $F_{J(T)}$ is a complete residuated lattice.

(c) *The function $h : a \mapsto [\mathbf{a}]$ is a homomorphism $h : \mathcal{L} \longrightarrow \mathcal{L}(T)$.*

PROOF: (a) follows from the equivalence theorem since $A \approx B$ iff $T \vdash A \Leftrightarrow B$.

(b) The tautology (P2) and Lemma 4.8 give that $[\bot]$ and $[\top]$ are the smallest as well as the greatest elements, respectively.

Furthermore, the adjointness of $\otimes$ and $\to$ follows from the tautology ($P12$). From Lemma 4.6 on page 123 it follows that $\otimes$ is isotonic in both arguments, and $\to$ is isotonic in the second and antitonic in the first arguments.

The instance $T \vdash \top \Rightarrow (A \Rightarrow (A \,\&\, \top))$ of (P13) and $T \vdash (\top \Rightarrow (A \Rightarrow A)) \Rightarrow ((\top \,\&\, A) \Rightarrow A)$ of (P12) imply that $[\top]$ is a unit with respect to $\otimes$.

Starting from $T \vdash ((A \,\&\, B) \,\&\, C) \Rightarrow ((A \,\&\, B) \,\&\, C)$ and using (P12) and (R2) we obtain associativity of $\otimes$. Using (P4), (P9) and (P10) we prove that $\wedge$ is infimum and using (R4), (P6) and (P11) we prove that $\vee$ is supremum. This proves that (4.37) is a residuated lattice. To prove that it is complete, we use the substitution axiom (T1) and Henkin axioms (4.33) and (4.34).

(c) The fact that h is a homomorphism follows from the book-keeping axiom (B1). Indeed, then e.g.

$$h(a \to b) = [\overline{\mathbf{a} \to \mathbf{b}}] = [\mathbf{a} \Rightarrow \mathbf{b}] = [\mathbf{a}] \to [\mathbf{b}] = h(a) \to h(b).$$

□

THEOREM 4.21

(a) A fuzzy theory T is contradictory iff $\mathfrak{L}(T)$ is a degenerated algebra.

(b) If a fuzzy theory T is consistent then the mapping $h : a \mapsto [\mathbf{a}]$ is and injection.

PROOF: (a) If T is contradictory then $T \vdash A \Leftrightarrow B$ for any two formulas $A, B \in F_{J(T)}$ and thus, $\mathfrak{L}(T)$ is degenerated.

Conversely, let $\mathfrak{L}(T)$ be degenerated and take $a < b$. Then $[\overline{\mathbf{b} \to \mathbf{a}}] = [\top]$ whence $T \vdash \overline{\mathbf{b} \to \mathbf{a}}$. Since $b \to a < \mathbf{1}$, T is contradictory by Corollary 4.2(b) on page 130.

(b) Let h be not injection. Then there are $a \neq b$ such that $h(a) = h(b)$, i.e. $T \vdash \mathbf{a} \Leftrightarrow \mathbf{b}$. Since either $a < b$ or $b < a$, we obtain that T is contradictory by Corollary 4.2 (b). □

Filters. We will now introduce the concept of filter also in Lindenbaum algebra of formulas. In comparison with Definition 2.13 on page 35, however, the definition of filter here must be extended since we work in predicate fuzzy logic.

DEFINITION 4.24

A nonempty set $H \subseteq F_{J(T)} | \approx$ is a filter if it fulfils the following conditions:

(i) If $[A] \in H$ and $[A] \leq_T [B]$ then $[B] \in H$.

(ii) If $[A], [B] \in H$ then $[A] \otimes [B] \in H$.

(iii) If $[A] \in H$ then $\bigwedge_{t \in M_V} [A_x[t]] \in H$.

Recall that the maximal filter is called the *ultrafilter*.

LEMMA 4.31

Let T be a consistent Henkin fuzzy theory.

(a) Every filter $H \subseteq F_{J(T)} | \approx$ with the property

$$H \cap h(L) = h(\mathbf{1}) \tag{4.38}$$

can be extended into an ultrafilter with the same property.

(b) Let $G \subseteq F_{J(T)} | \approx$ be an ultrafilter with the property (4.38) and $A \in F_{J(T)}$ a closed formula. Then $A \notin G$ iff there are a closed formula B, $[B] \in G$, $a < \mathbf{1}$ and $n \in \mathbb{N}^+$ such that

$$[A]^n \otimes [B] \leq_T h(a).$$

(c) If G is an ultrafilter of $\mathcal{L}(T)$ having the property (4.38) then

$$[A] \vee [B] \in G \quad \text{iff} \quad ([A] \in G \quad \text{or} \quad [B] \in G)$$

holds for all closed formulas $A, B \in F_{J(T)}$.

PROOF: (a) is a consequence of Zorn's lemma.

(b) If $[A] \in G$ then it follows from the assumption that the only a such that $h(a) \in G$ is $a = \mathbf{1}$.

Conversely, let there be no such B, $a < \mathbf{1}$ and n. Then

$$G' = \left\{[C] \in F_{J(T)}| \approx \,\middle|\, [A]^n \otimes [B] \leq_T [C] \text{ for some } n \text{ and } [B] \in G\right\}$$

is a filter fulfilling (4.38). Indeed, the properties (i) and (ii) follow from the algebraic properties of $\mathcal{L}(T)$. To prove (iii), let $[C] \in G'$. Then $T \vdash A^n \mathbin{\&} B \Rightarrow C$ and by Lemma 4.14 we obtain $T \vdash A^n \mathbin{\&} B \Rightarrow C_x[t]$ for all $t \in M_V$. But this means that $[C_x[t]] \in G'$ for all $t \in M_V$ which yields

$$\bigwedge_{t \in M_V} [C_x[t]] \in G'.$$

The property (4.38) follows from the assumption. Finally, $[A], [B] \in G'$, i.e. $G \cup \{[A]\} \subseteq G'$ and since G is maximal, we have $[A] \in G$.

(c) Let $[A], [B] \notin G$. Due to (b), there are C, C', a, b, m, n such that

$$[A]^m \otimes [C] \leq_T h(a) \quad \text{and} \quad [A]^n \otimes [C'] \leq_T h(b).$$

Put $k = \max\{m, n\}$. Using the tautology (P18) we obtain

$$([A] \vee [B])^k \otimes ([C] \otimes [C']) = ([A]^k \vee [B]^k) \otimes ([C] \otimes [C']) \leq_T \\ ([A]^m \otimes [C]) \vee ([B])^n \otimes [C']) \leq_T h(a) \vee h(b) = h(a \vee b).$$

Since $[C], [C'] \in G$, we have $[C] \otimes [C'] \in G$. But $a \vee b < \mathbf{1}$ and so $[A] \vee [B] \notin G$ by (b). The converse is obvious. □

LEMMA 4.32

Let T be a consistent Henkin fuzzy theory and $T \vdash_a A$. Let $b >^* a$. Then

$$H = \{[B] \mid ([A] \rightarrow [\mathbf{b}])^n \leq_T [B] \text{ for some } n > 0\} \tag{4.39}$$

is a filter with the property (4.38).

PROOF: The properties (i) and (ii) of the filter are obvious. Let $[C] \in H$. Then

$$T \vdash (A \Rightarrow \mathbf{b})^n \Rightarrow (\forall x) C$$

i.e. $[(\forall x)C] \in H$ which means that $\bigwedge_{t \in M_V} [C_x[t]] \in H$ and thus, H is a filter.

It remains to prove that H has the property (4.38). Let $b > a$ and suppose that there are $c < \mathbf{1}$ and $n > 0$ such that $T \vdash (A \Rightarrow \mathbf{b})^n \Rightarrow \mathbf{c}$. The tautologies (P17), (P18) and (P13) yield

$$T \vdash (A \Rightarrow \mathbf{b})^n \vee (\mathbf{b} \Rightarrow A)^n.$$

From the assumptions, (R2) and (P4) we obtain

$$T \vdash (A \Rightarrow \mathbf{b})^n \vee (\mathbf{c} \vee (\mathbf{b} \Rightarrow A)^n.$$

Then using (P11) we obtain $T \vdash \mathbf{c} \vee (\mathbf{b} \Rightarrow A)^n$ and finally, using (R2), (R4), (P6), (P11) and (P5) we obtain $T \vdash \mathbf{c} \vee (\mathbf{b} \Rightarrow A)$. Since c is nilpotent, find m such that $c^m = \mathbf{0}$. Finally, using (P13), (P18), (P11), (P5) and the fact that $T \vdash \mathbf{c}^m \Leftrightarrow \bot$ we conclude that

$$T \vdash \mathbf{b} \Rightarrow A.$$

From this and the logical axiom $^{b}/\mathbf{b}$ we finally obtain $T \vdash_{b'} A$ where $a < b \leq b'$ — a contradiction. From it follows that $c = \mathbf{1}$ and for any n

$$T \vdash (A \Rightarrow \mathbf{b})^n \Rightarrow \top,$$

i.e. only $[\top] \in H$. □

In the proof of the following two lemmas, we directly suppose that the truth values form the Łukasiewicz algebra $\mathcal{L}_{\text{Ł}}$.

LEMMA 4.33

(a) Let $a, a', b, b' \in L$, $a' < a$, $b < b' < \mathbf{1}$ and $0 < a^n \oplus b \leq a'^n \oplus b'$, $n \in \mathbb{N}^+$. Then

$$\models (\mathbf{a} \Rightarrow B)^n \Rightarrow \mathbf{b}) \Rightarrow ((\mathbf{a}' \Rightarrow B)^n \Rightarrow \mathbf{b}') \tag{4.40}$$

is a tautology.

(b) Let $a, a', b, b' \in L$, $a < a'$, $b < b' < \mathbf{1}$ and $0 < a'^n \otimes \neg b' \leq a^n \otimes \neg b$, $n \in \mathbb{N}^+$. Then

$$\models ((B \Rightarrow \mathbf{a})^n \Rightarrow \mathbf{b}) \Rightarrow ((B \Rightarrow \mathbf{a}')^n \Rightarrow \mathbf{b}') \tag{4.41}$$

is a tautology.

The formulas (4.40) and (4.41) are below supposed to be *additional logical axioms* with the degree $\mathbf{1}$.

LEMMA 4.34

Let T be a consistent Henkin fuzzy theory, $T \vdash_a A$ and let G be an ultrafilter with the property (4.38) obtained from the filter H (4.39). Let (4.40) and (4.41) be additional logical axioms in the degree $\mathbf{1}$. Then to every closed formula A there is $a \in L$ such that

$$[A] \leftrightarrow [\mathbf{a}] \in G. \tag{4.42}$$

PROOF: Put

$$D_A = \{a \mid [\mathbf{a}] \to [A] \in G\} \quad \text{and} \quad U_A = \{a \mid [A] \to [\mathbf{a}] \in G\}.$$

Let $a' \le a$. Then $[\mathbf{a}'] \le_T [\mathbf{a}]$ which implies

$$[\mathbf{a}] \to [A] \le_T [\mathbf{a}'] \to [A]$$

i.e. $[\mathbf{a}'] \to [A] \in G$. Analogously we proceed for U_A and thus, we conclude that D_A is the initial and U_A the terminal segment of L. Moreover, on the basis of Lemma 4.31(c) and the tautology (P17) we know that $D_A \cup U_A = L$.

We show that both D_A and U_A are closed. Then, since L is connected we will have $D_A \cap U_A \neq \emptyset$. Therefore, there is an element a such that both $[\mathbf{a}] \to [A] \in G\}$ and $[A] \to [\mathbf{c}] \in G$, i.e. (4.42) holds.

We will now demonstrate that $L \backslash D_A$ is open. To do it, we will suppose that $a \notin D_A$ and show that $a' \notin D_A$ holds for every $a' < a$.

The assumption $a \notin D_A$ and Lemma 4.31(b) imply that there are a closed formula B, $b < 1$ and n such that $[B] \in G$ and

$$([\mathbf{a}] \to [A])^n \otimes [B] \le_T [\mathbf{b}].$$

Then $([\mathbf{a}] \to [A])^n \to [\mathbf{b}] \in G$ by the adjunction. Now we choose b' such that $b < b' < 1$ and a' such that

$$a > a' > a - \frac{b' - b}{n}.$$

We can verify that $a^n \oplus b < a'^n \oplus b'$, i.e. the conditions under which the axiom (4.40) can be used are fulfilled and thus, using modus ponens we conclude that $b' \notin D_A$.

Using (4.41) we similarly show that $L \setminus U_A$ is also open, which means that D_A and U_A are closed. □

Using the ultrafilter G from Lemma 4.34, we can construct a congruence $\cong$ on $\mathcal{L}(T)$ using

$$[A] \cong [B] \quad \text{iff} \quad [A] \leftrightarrow [B] \in G. \tag{4.43}$$

The corresponding factoralgebra will be denoted by $F_{J(T)}|G$.

4.3.8 Completeness theorems

In this section, we will prove generalizations of the classical Gödel completeness theorems of first-order fuzzy logic. By the analogy with classical logic, they have two forms. We will keep this and start with the second form from which the first one follows. It is surprizing that the formulation of the second form of the completeness theorem is exactly the same as in classical logic. However, its proof is not a trivial analogy, and we present two kinds of it, namely syntactical and algebraic ones. Both are interesting since each of them throws a different light on the structure of fuzzy logic. Moreover, as will be discussed later, an important consequence of the algebraic proof is the possibility to extend the language by the wide class of the additional connectives.

Syntactical proof of the completeness theorem II. The syntactical proof of the completeness proceeds more or less in the classical way and has been given independently by V. Novák in [93] and P. Hájek in [38].

THEOREM 4.22 (COMPLETENESS THEOREM II)
A fuzzy theory T is consistent iff it has a model.

PROOF: If T has a model then it is consistent by Lemma 4.18.

Conversely, using Theorems 4.16 and 4.19 we construct a conservative Henkin extension T_H of the fuzzy theory T and complete it to a theory $\bar{T}_H$. Since $\bar{T}_H$ is complete then by Theorem 4.18, there is model $\bar{\mathcal{D}}_H \models \bar{T}_H$ which is a model of T_H. However, T_H is an extension of T and, therefore, $\bar{\mathcal{D}}_H$ is a model of T. □

Algebraic proof of the completeness theorem II. The algebraic proof of the completeness presented below has been first published by J. Pavelka in [103] for the propositional fuzzy logic. For the predicate fuzzy logic, it has been adapted by V. Novák in [88]. The algebraic proof is based on the methods developed by H. Rasiowa and R. Sikorski in [112].

LEMMA 4.35
Let T be a consistent Henkin fuzzy theory, A a closed formula, $T \vdash_a A$ and $b >^* a$. Then there is a homomorphism $R : F_{J(T)} \longrightarrow L$ preserving all suprema and infima such that

$$\mathrm{SAx}(B) \leq R(B) \tag{4.44}$$

for every $B \in F_{J(T)}$.

PROOF: Put $I = \{[B] \in F_{J(T)} \mid \approx \mid B \text{ is a closed formula}\}$ and let

$$g : F_{J(T)} \longrightarrow F_{J(T)} | \approx \quad \text{and} \quad F_{J(T)} | \approx \longrightarrow F_{J(T)} | G$$

be a canonical epimorphisms.

Since T is consistent, the function h from Theorem 4.20(c) is a homomorphism. By Lemma 4.32, $h(L) \cap G = \mathbf{1}$ and so,

$$[\mathbf{a}] \leftrightarrow [\mathbf{b}] \notin G$$

whenever $[\mathbf{a}] \neq [\mathbf{b}]$, i.e. $f([\mathbf{a}]) \neq ([\mathbf{b}])]$. From it follows that $fh : L \longrightarrow F_{J(T)}|G$ is a monomorphism. Lemma 4.34 implies $(fh)(L) = f(I)$. Indeed, $h(L) \subseteq I$ and, therefore, $(fh)(L) \subseteq f(I)$.

Conversely, if $f([A]) \in f(I)$ then there is $a \in L$ such that $[A] \in f([A]) \in (fh)(L)$. Then $f([A]) = f([\mathbf{a}]$, i.e. $f([A]) \in (fh)(L)$. This yields $f(I) \subseteq (fh)(L)$.

The function $fh : L \longrightarrow f(I)$ is an isomorphism since $f(I)$ is a subalgebra of $\mathcal{L}(T)$. Set

$$R(A) = \begin{cases} (fh)^{-1} fg(A) & \text{if } A \text{ is a closed formula,} \\ (fh)^{-1} fg(A') & \text{where } A' \text{ is a closure of } A \text{ otherwise.} \end{cases}$$

We will demonstrate that R is a homomorphism which preserves all suprema and infima. Indeed,

$$R(\mathbf{a}) = (fh)^{-1}fg(\mathbf{a}) = (fh)^{-1}f([\mathbf{a}]) = (fh)^{-1}fh(a) = a.$$

If A, B are closed formulas then

$$R(A \Rightarrow B) = (fh)^{-1}fg(A \Rightarrow B) = (fh)^{-1}f([A] \to [B]) = \\ = R(A) \to R(B).$$

Let $A(x)$ has the only free variable x. Then $T \vdash (\forall x)A \Rightarrow A_x[t]$, $t \in M_V$ and we obtain

$$R((\forall x)A) = (fh)^{-1}f([(\forall x)A]) \le (fh)^{-1}f([A_x[t]]) = R([A_x[t]]).$$

Finally, using the Henkin axiom, we obtain $R([A_x[\mathbf{r}]]) \le R((\forall x)A)$, which yields $R((\forall x)A) = \bigwedge_{t \in M_V} R([A_x[t]])$. If A has more free variables, we take its closure and proceed as above.

To prove the inequality (4.44), let $T \vdash_a B$. Then $\mathrm{SAx}(B) \le a$ and $h(a) = [\mathbf{a}] \le_T [B]$ since $T \vdash \mathbf{a} \Rightarrow B$. Then

$$R(B) = (fh)^{-1}f([\bar{B}]) \ge (fh)^{-1}f([\mathbf{a}]) = a \ge \mathrm{SAx}(B)$$

where $\bar{B} := B$ if it is a closed formula and $\bar{B}$ is the closure of B otherwise. □

THEOREM 4.23 (COMPLETENESS THEOREM II)
A fuzzy theory T is consistent iff it has a model.

PROOF: If T has a model then it is consistent by Lemma 4.18.

Conversely, let T be consistent and $\bar{T}$ its conservative Henkin extension. By Lemma 4.35 there is a homomorphism $R : F_{J(T)} \longrightarrow L$ using which we can construct a canonical structure $\mathcal{D}_0$ for which $\overline{\mathrm{SAx}}(A) \le \mathcal{D}_0(A)$ for all $A \in F_{J(T)}$. This means that $\mathcal{D}_0$ is a model of $\bar{T}$ and, therefore, of T as well. □

Completeness theorem I. In this section, the analogy of the first form of the completeness theorem is presented. It is a nice generalization of the classical completeness theorem since it gives the provability and truth degrees of a formula into an exact balance, thus concluding the proclaim that these concepts nontrivially generalize the classical concepts of provability and validity of a formula.

THEOREM 4.24 (COMPLETENESS THEOREM I)

$$T \vdash_a A \quad \textit{iff} \quad T \models_a A$$

holds for every formula $A \in F_J$ and every consistent fuzzy theory T.

PROOF: The implication left to right is a validity theorem.

Vice versa, let $T \vdash_a A$, $a > 0$. We have to show that there is a model $\mathcal{D} \models T$ such that $\mathcal{D}(A) \leq b$ holds for every b, $a <^* b$. By Lemma 4.26, $T' = T \cup \{ ^{\mathbf{1}}/A \Rightarrow \mathfrak{b}\}$ is a consistent theory. By Theorem 4.22 (or 4.23) it has a model $\mathcal{D} \models T'$. But then $\mathcal{D} \models T$ and $\mathcal{D}(A \Rightarrow \mathfrak{b}) = \mathbf{1}$ which gives $\mathcal{D}(A) \leq b$. □

REMARK 4.3
All the reasoning concerning completeness theorems can be done also in the case we deal with a many-sorted predicate language with finite number of sorts and prove a completeness theorem for the many-sorted predicate fuzzy logic. We will not give an explicit proof of this fact, but use it in the sequel, especially in Chapter 6.

4.3.9 Sorites fuzzy theories

The completeness theorem is a deep result with many consequences. In this subsection, we will demonstrate its application to the sorites paradox which has been discussed in Chapter 1. We will consider in this subsection that $L = [0, 1]$.

THEOREM 4.25
Let T be a fuzzy theory in which all Peano axioms are accepted in the degree 1. Furthermore, let $1 \geq \varepsilon > 0$ and $\mathbb{HR} \notin J(T)$ be a new predicate. Then the fuzzy theory

$$T^+ = T \cup \{ ^1/\mathbb{HR}(0),\ ^{1-\varepsilon}/(\forall x)(\mathbb{HR}(x) \Rightarrow \mathbb{HR}(x+1)),\ ^1/(\exists x)\neg\mathbb{HR}(x)\}$$

is a conservative extension of T.

PROOF: Let $\mathcal{D}$ be a classical model of Peano arithmetics with $D = \mathrm{N}$. Furthermore, we define a fuzzy set $\mathbb{HR}_D \underset{\sim}{\subset} \mathrm{N}$ assigned to $\mathbb{HR}$ as follows: $\mathbb{HR}_D(0) = 1$ and

$$\mathbb{HR}_D(n+1) = \mathbb{HR}_D(n) \otimes (1 - \varepsilon)$$

for all $n \in \mathrm{N}$. It is not difficult to show that

$$\bigwedge_{n \in \mathrm{N}} (\mathbb{HR}_D(n) \rightarrow \mathbb{HR}_D(n+1)) = 1 - \varepsilon$$

and $\neg\mathbb{HR}(m) = 1$ for all $m > 1/\varepsilon$. Consequently, $\mathbb{HR}_D \models T^+$. The conservativeness is obvious. □

Few things may be deduced from this simple theorem. Recall that R. Parikh in [102] has introduced a similar theorem for classical arithmetics where he proved that it is possible to construct a number θ such that if we restrict only to proofs of the length smaller than θ then the theory T^+ is consistent relatively to this. Our result demonstrates analogous property with respect to the degree of provability, i.e. the formula $(\forall x)\mathbb{HR}(x)$ is provable in the degree 0. In other way, each sorites chain of inferences decreases the provability value when increasing its length.

There is also other, more pragmatic aspect of the problem, apparent when changing the magnitude of ε. If we put $\varepsilon = 1$ then we immediately obtain that the only small number is 0. At the same time, we may find any number $n \in \mathbb{N}\backslash\{0\}$ to be small, at least in some nonzero degree when setting ε appropriately, or when defining appropriately the fuzzy set $\mathbb{HR}_D$ (recall that only the inequality $1-\varepsilon \leq \mathbb{HR}_D(n) \rightarrow \mathbb{HR}_D(n+1)$ must be fulfilled). Though seemingly disqualifying the above theorem, this only verifies the context-dependency of the predicate "to be small". Namely, the same number n may be small in one context, but not small in another one.

4.3.10 Partial inconsistency and the consistency threshold

In Section 4.3.3 we have introduced the concepts of consistent and contradictory fuzzy theory and demonstrated that contradictory fuzzy and classical theories coincide. The definition was based on the use of the Łukasiewicz conjunction. A natural question arises whether an ordinary conjunction $\wedge$, which is weaker, may lead to some degrees of the consistency of fuzzy theories. We will show in this section that in most cases, we again obtain a contradictory theory. The results are due to S. Gottwald and V. Novák from [36].

Partial inconsistency.

DEFINITION 4.25
The degree of inconsistency of the fuzzy theory T is the value

$$\text{Incons}(T) := \sup\{a \mid T \vdash_a A \wedge \neg A \textit{ and } A \in \mathcal{F}_{\mathcal{L}(T)}\}. \tag{4.45}$$

LEMMA 4.36
Let T be a fuzzy theory. If there is a formula A and a proof $w_{A\wedge\neg A}$ such that $\text{Val}_T(w_{A\wedge\neg A}) > 1/2$ then T is contradictory.

PROOF: Using (P4) and the symmetry of $\wedge$ we obtain from the proof $w_{A\wedge\neg A}$ proofs w_A and $w_{\neg A}$ such that $\text{Val}_T(w_A) > 1/2$ and $\text{Val}_T(w_{\neg A}) > 1/2$ which gives $\text{Val}_T(w_A) \otimes \text{Val}_T(w_{\neg A}) > \mathbf{0}$. Therefore T is contradictory. □

The previous lemma leads to the following theorem, which demonstrates that the degrees of consistency are fairly limited.

THEOREM 4.26
A fuzzy theory T is consistent iff $\text{Incons}(T) \leq 1/2$.

PROOF: Let $\text{Incons}(T) > 1/2$. Then there is a formula A such that $T \vdash_c A \wedge \neg A$ and $c > 1/2$. Using Lemma 4.12(b), we obtain $T \vdash_a A$ and $T \vdash_b \neg A$ where $a, b > 1/2$. Then T is contradictory by Lemma 4.36.

Vice versa, let $\text{Val}_T(w_{A\wedge\neg A}) \leq 1/2$ hold true for all formulas A and all proofs $w_{A\wedge\neg A}$. Using Lemma 4.12(b) we obtain that $T \vdash_a A$ and $T \vdash_b \neg A$

and $a \wedge b \leq 1/2$. From it follows that for all proofs w_A and $w_{\neg A}$ we have $\mathrm{Val}_T(w_A) \wedge \mathrm{Val}_T(w_{\neg A}) \leq 1/2$ and hence, $\mathrm{Val}_T(w_A) \otimes \mathrm{Val}_T(w_{\neg A}) = \mathbf{0}$. □

We see that the inconsistency degree greater than 1/2 makes all the formulas provable in the degree one. What about inconsistency degrees smaller than 1/2, but different from zero? The answer follows.

LEMMA 4.37
There is no nonzero lower bound for the provability degrees of formulas of the form $A \wedge \neg A$.

PROOF: Let us assume that $c > \mathbf{0}$ would be such a lower bound. Furthermore, assume that B is a tautology. Then the formula $B \wedge \neg B$ has the truth degree $\mathbf{0}$ in any model. By completeness, it follows that $T \vdash_{\mathbf{0}} B \wedge \neg B$ which contradicts to the assumption $c > 0$. □

It follows from this theorem that a nonzero inconsistency degree (but smaller than 1/2) for a fuzzy theory T does not imply that there are nonzero provability degrees for all the formulas of the language of T.

Consistency threshold. The consistency threshold of a fuzzy theory is an interesting possibility specific for the fuzzy logic, which, on the other hand, has no sense in classical logic. In some proofs below, we have used the completeness theorem.

LEMMA 4.38
Let T be a consistent fuzzy theory and $T \vdash_a A$ where A is a closed formula. Let $T' = T \cup \{ {}^{b}/\neg A\}$. Then T' is consistent iff $b \leq \neg a$.

PROOF: If $b > \neg a$ then $b \otimes a > 0$. At the same time $T' \vdash_{b'} \neg A$ where $b' \geq b$. Thence, $T' \vdash_d A \,\&\, \neg A$ for $d > 0$. Consequently, T' is contradictory.

Conversely, let T' be contradictory. We have to show that $b > \neg a$. It is sufficient to suppose that $b = \neg a$ since then, for $b < \neg a$, the lemma will follow from the properties of extension of fuzzy theories.

We will use the completeness theorem and suppose two cases. Let there be a model $\mathcal{D} \models T$ such that $\mathcal{D}(A) = a$. Then $\mathcal{D}(\neg A) = \neg a$, i.e. $\mathcal{D} \models T'$ and thus, T' is consistent.

Let $\mathbf{1} > \mathcal{D}(A) > a$ hold for each model $\mathcal{D} \models T$. Since by the assumption, T' is contradictory then by Lemma 4.25, given $B := \bot$ there is m_0 such that $T' \vdash (\neg A)^{m_0} \Rightarrow \bot$. The latter is equivalent to $T' \vdash m_0 A$. Choose $\mathcal{D}$ and m_0 such that $\mathcal{D}(m_0 A) = m_0 \mathcal{D}(A) = \mathbf{1}$. Then for each $\mathcal{D}'$ such that $a < \mathcal{D}'(A) < \mathcal{D}(A)$ we have $m_0 \mathcal{D}'(A) < \mathbf{1}$ which, by completeness, implies that $T' \vdash_c m_0 A$ for some $c < \mathbf{1}$ and consequently, T' is not contradictory. We conclude that $b > \neg a$. □

The following is immediate.

COROLLARY 4.5
If $T \vdash_a A$ then $T' = T \cup \{ {}^{b}/\neg A\}$ is contradictory iff $b > \neg a$.

Lemma 4.38 is the main motivation for the concept of the consistency threshold.

DEFINITION 4.26
Let T be a consistent theory and A a formula. Then an element $\neg a$ is a consistency threshold for A in T if $T' = T \cup \{ {}^{b}/\neg A\}$ is contradictory for all $b > \neg a$ and consistent otherwise.

An immediate consequence of Lemma 4.38 is the following.

LEMMA 4.39
Let T be a consistent fuzzy theory. For every formula A, $T \vdash_a A$ iff $\neg a$ is a consistency threshold for A in T.

4.3.11 Additional connectives

As discussed in Section 4.2, it is possible to enrich the structure of truth values by additional operations, which may then serve us as interpretations of new logical connectives. Recall that the reasons for that follow from the local character of fuzzy logic and the necessity to find a more realistic model of some propositions of natural language. Besides extending the language and the definition of the formula, we have to add new logical axioms characterizing properties of the new connectives. The most important seems to be preservation of the logical equivalence.

DEFINITION 4.27
A connective $\square$ is logically fitting if it is assigned a logically fitting operation $\square$ (i.e. fulfilling condition (4.24)).

On the level of syntax, we now make the following steps.

1. The language J is extended into a language J_e by a finite set $\Theta = \{\square_j \mid j = 1, \dots, m\}$ of additional n_j-ary connectives.

2. Definition of a formula is extended by the following item: If $A_1, \dots, A_n$ are formulas and $\square \in \Theta$ then $\square(A_1, \dots, A_n)$ is a formula.

3. The fuzzy set LAx of logical axioms is extended by the following scheme (FC) of logical axioms.

   (FC) If $A_1, \dots, A_n, B_1, \dots, B_n$ are formulas then there are natural numbers $k_1 > 0, \dots, k_n > 0$ such that the evaluated formula

$$\mathbf{1}/(A_1 \Leftrightarrow B_1)^{k_1} \,\&\cdots\&(A_n \Leftrightarrow B_n)^{k_n} \Rightarrow (\square(A_1, \dots, A_n) \Leftrightarrow \square(B_1, \dots, B_n))$$

   is a logical axiom for every additional connective $\square \in \Theta$.

LEMMA 4.40
Let T_e be a fuzzy theory in the language J_e and $\mathcal{L}(T_e)$ be a corresponding Lindenbaum algebra.

(a) $\mathcal{L}(T_e)$ *is an enriched residuated lattice, i.e. the induced operation*

$$\square([A_1], \ldots, [A_n]) := [\square(A_1, \ldots, A_n)] \tag{4.46}$$

is logically fitting.

(b) *Let* $\cong$ *be a congruence (4.43) on* $\mathcal{L}(T)$. *Then it is also a congruence on* $\mathcal{L}(T_e)$.

PROOF: (a) The axiom (FC) and (P5) follow that (4.46) is a well defined operation. Furthermore, since $[A] \leftrightarrow [B] = ([A] \rightarrow [B]) \wedge ([A] \rightarrow [B]) = [A \Leftrightarrow B]$, we obtain using again (FC) that there are $k_1, \ldots, k_n$ such that

$$([A_1] \leftrightarrow [B_1])^{k_1} \otimes \cdots \otimes ([A_n] \leftrightarrow [B_n])^{k_n} \leq_T$$
$$\leq_T \square([A_1], \ldots, [A_n]) \leftrightarrow \square([B_1], \ldots, [B_n]),$$

i.e., $\square$ is a logically fitting operation on $\mathcal{L}(T_e)$.

(b) is a consequence of (a) and Theorem 4.4. □

COROLLARY 4.6 (COMPLETENESS IN THE ENRICHED LANGUAGE)
Let T *be a consistent fuzzy theory in the enriched language* J_e. *Then*

$$T \vdash_a A \quad \textit{iff} \quad T \models_a A$$

holds for every formula $A \in F_{J_e}$.

PROOF: It follows from Lemma 4.40 that the congruence $\cong$ in (4.43) is respected also by the operation $\square$ in (4.46). Consequently, Lemma 4.35 on page 148 holds also in the case that A contains some additional logically fitting connective. We may construct a model $\mathcal{D} \models T$ and then proceed in the same way as in the proof of Theorem 4.24. □

This corollary assures us that the completeness theorem holds also for formulas containing additional logically fitting connectives. However, to be able to work with the logically fitting connectives, we must characterize their properties. This can be done by accepting special axioms for them. A typical additional connective is $\odot$ being interpreted by the ordinary product of reals. One may easily find axioms which would characterize this connective. Note, however, that due to Theorem 4.2 on page 105 and Lemma 4.3 on page 112, we may add neither Gödel nor Goguen implications since both of them are discontinuous.

4.3.12 Disposing irrational logical constants

When introducing the language of fuzzy logic, we have added logical constants for all $a \in L$, which in the case of $L = [0, 1]$ gives an uncountable language and uncountable set of formulas F_J. Consequently, FLn is, in general, algorithmically non-tractable. We will show in this section that this is unnecessary since all the logical constants for irrational truth values can be disposed. Hence, we can take all the logical constants only as a technical trick simplifying the proofs[12].

Theories without irrational logical constants. The language of fuzzy logic is supposed to contain logical constants for all $a \in L$ which means that it may be uncountable. In this subsection, we will demonstrate that it is possible to confine only to logical constants forming a countable dense subset of L.

Let J be a language of fuzzy logic containing logical constants for all the truth values $a \in L$ and let $Z \subseteq L$ be a countable dense subset. Furthermore, let $J' \subset J$ be a language obtained from J by omitting logical constants for all $a \in L \setminus Z$.

Now, we will consider two fuzzy theories, namely T in the language J and a restricted fuzzy theory T' in the language J', which have the same non-zero special axioms, $\mathrm{Supp}(\mathrm{SAx}') = \mathrm{Supp}(\mathrm{SAx})$, i.e. formulas from $\mathrm{Supp}(\mathrm{SAx})$ do not contain logical constants for $a \in L \setminus Z$.

LEMMA 4.41
To every proof w in a fuzzy theory T of some formula A there is a set of proofs W in a fuzzy theory T' of formulas A' obtained from A by replacing logical constants $\mathbf{a}$ for $a \in L \setminus Z$ by certain logical constants $\mathbf{a}$ for $a \in Z$ such that

$$\mathrm{Val}_T(w) = \bigvee \{\mathrm{Val}_{T'}(w') \mid w' \in W\}.$$

PROOF: Let length of the proof w be 1. Then $w := \; a/A$. If A is a special axiom then $W = \{w\}$ and similarly for A if it is a tautology, or $A := \mathbf{a}$, $\mathbf{a} \in Z$. Otherwise

$$W = \{w' := \; a'/\mathbf{a}' \; \{SA\} \mid a' < a, a' \in Z\}.$$

Obviously, $a = \bigvee\{\mathrm{Val}(w') \mid w' \in W\}$ and each $w' \in W$ is a proof of the formula $\mathbf{a}'$, $a' < a$.

Let the proposition hold for every proof w of the length smaller than n. Let

$$\begin{aligned} w := \dots, \; & a_1/A_1 \; \{w_{A_1}\}, \dots, \; a_m/A_m \; \{w_{A_m}\}, \\ \dots, \; & a_n = r^{evl}(a_1, \dots, a_m)/A_n = r^{syn}(A_1, \dots, A_m) \; \{r\}. \end{aligned}$$

Due to the inductive assumption, there are sets $W_1, \dots, W_m$ of proofs of the formulas $A'_1, \dots, A'_m$ in T' which have been obtained form the corresponding $A_1, \dots, A_m$ as follows.

[12]P. Hájek in [41] demonstrates the completeness proof directly in the language containing only rational logical constants. Moreover, from the metamathematical point of view, he calls fuzzy logic with the evaluated syntax Rational Pavelka logic (RPL). Our approach is different and takes the former as the initial concept, which uses graded approach to deal with the vagueness phenomenon. As demonstrated, it nontrivially generalizes all the classical concepts and is open to extension by the additional connectives. Therefore, we do not support the name RPL.

For every $i = 1, \dots, m$ let $\mathbf{b}_1, \dots, \mathbf{b}_s$, $b_j \in L \setminus Z$, $j = 1, \dots, s$ be all the logical constants occurring in the formula A_i. Put $Q_j = \{b'_j \in Z \mid b'_j < b_j\}$, $j = 1, \dots, s$. Then the formulas A'_i are results of replacing $\mathbf{b}_j$ by $\mathbf{b}'_j$, $b'_j \in Q_j$ for $j = 1, \dots, s$. Similarly we derive the formulas A'_n, which are results of the following proofs w' in T':

$$\begin{aligned} w' := \dots, \quad & a'_1/A'_1 \ \{w'_1\}, \dots, \quad a'_m/A'_m \ \{w'_m\}, \\ \dots, \quad & a'_n = r^{evl}(a'_1, \dots, a'_m)/A'_n = r^{syn}(A'_1, \dots, A'_m) \ \{r\} \end{aligned} \tag{4.47}$$

where $w'_i \in W_i$, $i = 1, \dots, m$. Let $W = \{w' \mid w'_1 \in W_1, \dots, w'_m \in W_m\}$ be the set of all the proofs (4.47). Then

$$\mathrm{Val}_T(w) = a_n = \bigvee \{\mathrm{Val}_{T'}(w') \mid w' \in W\}$$

since the operation r^{evl} is lower semicontinuous. □

COROLLARY 4.7

Let $A \in F_{J'}$ be a formula, which contains no logical constants $\mathbf{a}$ for $a \in L \setminus Z$. Then to every proof w_A in a fuzzy theory T there is a set of proofs W_A in T' such that

$$\mathrm{Val}_T(w_A) = \bigvee \{\mathrm{Val}_{T'}(w') \mid w' \in W_A\}. \tag{4.48}$$

PROOF: Since A contains only logical constants for $a \in Z$, every proof $w \in W_A$ is obtained from w_A by replacing all the occurrences of $a \in L \setminus Z$ by the procedure given in the proof of Lemma 4.41. However, this does not concern the formula A which remains unchanged. □

This lemma is used in the proof of the following theorem.

THEOREM 4.27

Let J be a language of fuzzy logic, $Z \subset L$ be a countable dense subset and $J' = J \setminus \{\mathbf{a} \mid a \in L \setminus Z\}$. Let T be a fuzzy theory in J and T' be a fuzzy theory in J' such that $\mathrm{Supp}(\mathrm{SAx}') = \mathrm{Supp}(\mathrm{SAx})$. *Then*

$$T \vdash_a A \quad \textit{iff} \quad T' \vdash_a A$$

holds for every formula $A \in F_{J'}$.

PROOF: Let w_A be a proof of A in T. By Corollary 4.7, there is a set of proofs W_A of A in T' such that (4.48) holds. If $T \vdash_a A$ then

$$\begin{aligned} a &= \bigvee \{\mathrm{Val}_T(w_A) \mid \text{all proofs } w_A\} = \\ &= \bigvee \Big\{ \bigvee \{\mathrm{Val}_{T'}(w') \mid w' \in W_A\} \mid W_A \text{ is a set from (4.48)} \Big\}, \end{aligned}$$

i.e., $T' \vdash_b A$, $a \leq b$.

Conversely, every proof of A in T' is a proof of A in T. Hence, if $T' \vdash_a A$ then $T \vdash_b A$, $a \leq b$ which gives $a = b$. □

Because the set of all rationals $\mathbb{Q} \subset [0, 1]$ is dense in $[0, 1]$, by this theorem we may confine only to rational logical constants.

Square roots. Using square roots is an exciting possibility for disposing of most of the logical constants. We will show that it is sufficient to introduce one logical constant only, and a new connective of *square root* $\sqrt{\ }$ interpreted by the operation of square root introduced in Section 2.1. The safety of introducing square roots follows from the discussion about introducing additional connectives due to the following lemma, whose proof follows from Lemma 2.8 on page 28.

LEMMA 4.42
Let $\mathcal{L}$ be a residuated lattice. Then the operation $\sqrt{\ }$ is logically fitting.

Furthermore, we introduce a new special rule of *rough extension*

$$r_{RE}: \frac{a/A}{\sqrt{a}/\sqrt{A}}.$$

The following is easy to prove.

LEMMA 4.43
The rule r_{RE} is sound.

It is also necessary to introduce a special book-keeping logical axiom for $\sqrt{\ }$ (cf. Definition 4.14 on page 116)

(B2) $\sqrt{\mathbf{a}} \Leftrightarrow \overline{\sqrt{a}}$
where $\overline{\sqrt{a}}$ denotes the logical constant for the truth value $\sqrt{a}$ when a is given.

Similarly as (B1), this axiom belongs to LAx in the degree $\mathbf{1}$.

Let us now introduce special formulas $\mathbf{b}_{n,k}$ of the form

$$\mathbf{b}_{n,k} := \neg\left(\sqrt[2^{k-m_1}]{\bot} \,\&\cdots\&\, \sqrt[2^{k-m_s}]{\bot}\right) \tag{4.49}$$

where $\sum_{i=1}^{s} 2^{m_i} = n$. As special cases, we put

$$\mathbf{b}_{0,k} := \bot, \qquad k \in \mathbb{N},$$
$$\mathbf{b}_{1,1} := \bot \Rightarrow \bot = \top.$$

In the sequel, by J we denote a language containing the connective $\sqrt{\ }$. The following is a consequence of the book-keeping axiom (B2) and the equivalence theorem.

LEMMA 4.44
Let T be a fuzzy theory in J. Then

$$T \vdash \mathbf{b}_{n,k} \Leftrightarrow \overline{\mathbf{b}}_{n,k}$$

holds true for every $b_{n,k} \in L$ where $n, k \in \mathbb{N}$.

Put $Z = \{b_{n,k} \mid n, k \in \mathbb{N}\}$. It follows from Example 2.14 on page 32 that Z is a set of the dyadic numbers and thus, it is dense in $L = [0,1]$. Let

$$J' = J \setminus \{\mathbf{a} \mid a \in [0,1] \setminus Z\},$$
$$J'' = J \setminus \{\mathbf{a} \mid a \in (0,1]\}.$$

Furthermore, let T be a fuzzy theory in J whose fuzzy set of special axioms contains only formulas of the language J', i.e. $\mathrm{Supp}(\mathrm{SAx}) \subset F_{J'}$. Now, given a formula $\bar{A} \in F_{J''}$, we construct from it a formula $A \in F_J$ by replacing all the occurrences of the formulas $\mathbf{b}_{n,k}$ in $\bar{A}$ by the logical constants $\mathbf{b}_{n,k} \in J$ for the corresponding $b_{n,k} \in Z$.

THEOREM 4.28
Let T'' be a fuzzy theory in the language J'' and T be a fuzzy theory in J such that $\mathrm{SAx}''(\bar{B}) = \mathrm{SAx}(B)$ for every formula $\bar{B} \in F_{J''}$ and the formula $B \in F_J$ constructed from it. Then

$$T'' \vdash_a \bar{A} \quad \textit{iff} \quad T \vdash_a A$$

holds for every formula $\bar{A} \in J''$.

PROOF: Let $T'' \vdash_a \bar{A}$ and $w_{\bar{A}}$ be its proof. Further, let T' be a fuzzy theory in the language J' such that $\mathrm{SAx}''(\bar{B}) = \mathrm{SAx}'(B)$ for every formula $\bar{B} \in F_{J''}$ (this is assumption is correct since, by the construction of B from $\bar{B}$, $B \in F_{J'}$).

If we replace each formula $\mathbf{b}_{n,k}$ in $w_{\bar{A}}$ by the logical constant $\mathbf{b}_{n,k}$ then due to Lemma 4.44, we obtain a proof w_A of the corresponding formula A constructed in the above way with the same value, i.e. $\mathrm{Val}(w_A) = \mathrm{Val}(w_{\bar{A}})$. But the same can be done also vice versa, i.e. from the proof w_A we can construct the proof $w_{\bar{A}}$ with the same value. From it follows that $T'' \vdash \bar{A}$ iff $T' \vdash_a A$. But since $Z \subset L$ is countable dense subset, T' and T fulfil conditions of Theorem 4.27 which gives the proposition of our theorem. □

By this theorem, we may restrict the language J of fuzzy logic to the language with the only logical constant $\bot$, i.e., J is at most countable. Since the logical constants have an auxiliary role, this restriction is not principal. Theorem 4.28 states that the expression power of fuzzy theories in this restricted language is preserved (of course, with respect only to formulas not containing logical constants for the truth values $L \setminus Z$).

REMARK 4.4
Introducing square roots is an example of the possibility to extend FLn by new connectives. Another such possibility has been demonstrated in Subsection 4.3.11. Note that besides adding new connectives, we may also introduce additional (sound) inference rules, such as the rule r_{RE}.

4.4 Fuzzy Theories with Equality and Open Fuzzy Theories

Most of the results presented thus far concern closed formulas. However, characterization of theories given by open axioms may cast a different light on fuzzy theories. We will follow the classical way because we want fuzzy logic to be developed, besides other, in parallel with classical one to demonstrate that many classical results are special cases of ours. On the other hand, methods used in their proofs are often different from those of the classical ones since we have to use weaker properties. An important role will be played by Henkin fuzzy theories as they make possible to dispose quantifiers from formulas. Recall that they can easily be obtained by conservative extension of the non-Henkin fuzzy theories.

4.4.1 *Fuzzy theories with equality*

To be able to study open fuzzy theories, we must have possibility to compare elements. In classical logic, there is only one possibility to do that — to introduce the equality predicate. In fuzzy logic we should be able to compare elements being equal only to some degree. Quite naturally we arrive at the fuzzy equality predicate. Its properties are generalization of the classical equality axioms.

DEFINITION 4.28
The following schemes of evaluated formulas are logical fuzzy equality axioms:

(E1) $\mathbf{1}/x \doteq x$.

(E2) If f is a functional symbol then

$$\mathbf{1}/(x_1 \doteq y_1) \Rightarrow (\dots \Rightarrow (x_n \doteq y_n) \Rightarrow (f(x_1, \dots, x_n) \doteq f(y_1, \dots, y_n))\dots).$$

(E3) If P is a predicate symbol then

$$\mathbf{1}/(x_1 \doteq y_1) \Rightarrow (\dots \Rightarrow (x_n \doteq y_n) \Rightarrow (P(x_1, \dots, x_n) \Rightarrow P(y_1, \cdots, y_n))\dots).$$

A special kind of fuzzy equality is the sharp (classical) one $=$ which, in addition, fulfils also the crispness axiom $\mathbf{1}/(\forall x)(\forall y)((x = y) \vee \neg(x = y))$.

LEMMA 4.45
Let T be a fuzzy predicate calculus with fuzzy equality. Then the following properties of the fuzzy equality are provable in T:

(a) Symmetry

$$T \vdash (x \doteq y) \Rightarrow (y \doteq x),$$

(b) transitivity

$$T \vdash ((x \doteq y) \mathbin{\&} (y \doteq z)) \Rightarrow (x \doteq z).$$

PROOF: (a) The evaluated formula

$$\mathbf{1}/(x \doteq y) \Rightarrow ((x \doteq x) \Rightarrow ((x \doteq x) \Rightarrow (y \doteq x)))$$

is an instance of (E3). The symmetry follows from this, (E1) and (P20) using modus ponens.

(b) Similarly, the evaluated formula

$$\mathbf{1}/(x \doteq x) \Rightarrow ((y \doteq z) \Rightarrow ((x \doteq y) \Rightarrow (x \doteq z)))$$

is an instance of (E3). The transitivity then follows from this, (E1), (P20) and (P12) using modus ponens. □

The following theorem holds for every fuzzy equality.

THEOREM 4.29 (EQUALITY)
Let T be a fuzzy theory with fuzzy equality and $A \in F_{J(T)}$ a formula. If $T \vdash_{a_i} t_i \doteq s_i$, $i = 1, \ldots, n$ then there are $m_1, \ldots, m_n$ such that

$$T \vdash_b A \Leftrightarrow A', \qquad b \geq a_1^{m_1} \otimes \cdots \otimes a_n^{m_n}$$

where A' is a formula which is a result of replacing of the terms t_i by the term s_i in A, respectively.

PROOF: By induction on the length of the formula.

If $A := P(t_1, \ldots, t_n)$ where P is an n-ary predicate symbol then the proposition follows from the equality axiom (E3) and modus ponens. The rest is analogous to the classical proof of the equality theorem and the proof of the equivalence Theorem 4.6 on page 125. □

REMARK 4.5
This theorem is not as strong as in classical logic. For small a_i it is practically trivial and becomes interesting only for a_i close to $\mathbf{1}$. The numbers m_i depend on the complexity of the given formula. The magnitude of b depends also on the number of replacements of t_i by s_i.

4.4.2 Consistency theorem in fuzzy logic

In this subsection, we prove the fuzzy analogy of the classical Hilbert-Ackermann consistency theorem.

Formulas related to special constants. Let T be a fuzzy theory and T_H its conservative Henkin extension. Let $\mathbf{r}$ be the special constant for $(\forall x)A$. We say that a formula A is in relation with a special constant $\mathbf{r}$ if it is either of the formulas

$$A_x[\mathbf{r}] \Rightarrow (\forall x)A \quad \text{or} \quad (\forall x)A \Rightarrow A_x[t]$$

where t is a closed term.

DEFINITION 4.29
Let T be a Henkin fuzzy theory. We define a fuzzy set of formulas $\Delta(T)$ as follows:

$$\Delta(T)(A) = \begin{cases} \mathbf{1} & \text{if } A := Q, \\ (\mathrm{SAx} \cup \mathrm{LAx})(B) & \text{if } A := B_{x_1,\ldots,x_n}[t_1, \ldots, t_n] \end{cases}$$

where Q is any of the formulas

$$A_x[\mathbf{r}] \Rightarrow (\forall x)A, \qquad (\forall x)A \Rightarrow A_x[t], \qquad t \doteq t,$$
$$(t_1 \doteq s_1)^{m_1} \Rightarrow (\cdots \Rightarrow ((t_n \doteq s_n)^{m_n} \Rightarrow (f(t_1, \ldots, t_n) \doteq f(s_1, \ldots, s_n)) \ldots),$$
$$(t_1 \doteq s_1)^{m_1} \Rightarrow (\cdots \Rightarrow ((t_n \doteq s_n)^{m_n} \Rightarrow (P(t_1, \ldots, t_n) \Rightarrow P(s_1, \ldots, s_n)) \ldots),$$

t_i, s_i are closed terms, $m_i \geq 1$, $i = \ldots, n$, $\mathbf{r}$ is the Henkin constant for $(\forall x)A$ and B is a formula not being any of the previous cases.

LEMMA 4.46
Let T be a fuzzy theory. Then

$$\mathfrak{D} \models T_H \quad \text{iff} \quad \Delta(T_H) \leq \mathfrak{D}$$

holds for every structure $\mathfrak{D}$ for the language $L(T_H)$.

PROOF: Let $\mathfrak{D} \models T_H$. Then $\mathrm{SAx}, \mathrm{LAx} \leq \mathfrak{D}$ and $\mathfrak{D}(A_H) = \mathbf{1}$ for every Henkin axiom A_H. If B is a special axiom then

$$\mathrm{SAx}(B) = \Delta(T_H)(B_{x_1,\ldots,x_n}[t_1, \ldots, t_n]) \leq \mathfrak{D}(B).$$

Using the substitution axiom and the definition of the closure we obtain

$$\mathfrak{D}(B) \leq \mathfrak{D}(B_{x_1,\ldots,x_n}[t_1, \ldots, t_n]).$$

The opposite implication follows from the definition of $\Delta(T_H)$. □

LEMMA 4.47
Let w be a proof of the formula A in a fuzzy theory T, $\mathrm{Val}_T(w) = a$ and $A' \in L(T_H)$ be a closed instance of A where T_H is a Henkin extension of T. Then there are formulas $C_1, \ldots, C_m \in \Delta(T)$ such that A' is their tautological consequence and

$$\Delta(T)(C_1) \otimes \cdots \Delta(T)(C_m) \leq a.$$

PROOF: Consider the proof

$$w := \quad a_1/A_1 \; \{\textit{some proof } w_{A_1}\}, \ldots \quad a_n/A_n := A \; \{\textit{some proof } w_{A_n}\}.$$

By induction on the length of the proof we prove that there always is the desired tautology. In what follows, we will denote $\Delta(T)(C), \ldots$ by the corresponding small letters $c, \ldots,$ possibly with the subscript and the closed instances of formulas by adding prime to the corresponding letters.

(a) Let $A := \mathbf{a}$. Then $A' := \mathbf{a}$, $\mathbf{a} \in \Delta(T)$ and $\models \mathbf{a} \Rightarrow \mathbf{a}$ where $a \leq a$.

(b) Let A' be an instance of the special axiom. Then $A' \in \Delta(T)$ and $\models A' \Rightarrow A'$. By the assumption, $\Delta(T)(A') = \mathrm{SAx}(A) \leq a$.

(c) If A' is an instance of the axioms (R1)–(R4) then A' is a tautological consequence of the empty set of formulas.

(d) Let $A := (\forall x)B \Rightarrow B$. Then $A' := (\forall x)B' \Rightarrow B'_x[t'] \in \Delta(T)$ and we have $\models A' \Rightarrow A'$. Similarly for the equality axioms.

(e) Let $A := (\forall x)(C \Rightarrow D) \Rightarrow (C \Rightarrow (\forall x)D)$. Then $A' := (\forall x)(C' \Rightarrow D') \Rightarrow (C' \Rightarrow (\forall x)D')$ where C', D' are closed instances of the formulas C, D. Furthermore, let $\mathbf{r}$ be a special constant for $(\forall x)D'$. Then $(\forall x)(C' \Rightarrow D') \Rightarrow (C' \Rightarrow D'_x[\mathbf{r}]) \in \Delta(T)$ and $D'_x[\mathbf{r}] \Rightarrow (\forall x)D' \in \Delta(T)$. The desired tautology is then

$$\models ((\forall x)(C' \Rightarrow D') \Rightarrow (C' \Rightarrow D'_x[\mathbf{r}])) \Rightarrow ((D'_x[\mathbf{r}] \Rightarrow (\forall x)D') \Rightarrow ((\forall x)(C' \Rightarrow D') \Rightarrow (C' \Rightarrow (\forall x)D'))).$$

In cases (c)–(e), $a = \mathbf{1}$.

Let the inductive assumption holds.

(f) Consider the proof

$$^{b}/B\ \{\textit{some proof } w_B\},\quad ^{e}/B \Rightarrow A\ \{\textit{some proof } w_{B\Rightarrow A}\},\quad ^{b\otimes e}/A\ \{r_{MP}\}.$$

By the inductive assumption

$$\models (C_1 \,\&\cdots\&\, C_m) \Rightarrow B',$$
$$\models (D_1 \,\&\cdots\&\, D_m) \Rightarrow (B' \Rightarrow A')$$

where $c_1 \otimes \cdots c_m \leq b$ and $d_1 \otimes \cdots d_n \leq e$ (recall that $c_1 = \Delta(T)(C_i), d_1 = \Delta(T)(D_i)$). Then

$$\mathcal{D}(C_1) \otimes \cdots \otimes \mathcal{D}(C_m) \leq \mathcal{D}(B'),$$
$$\mathcal{D}(D_1) \otimes \cdots \otimes \mathcal{D}(D_n) \otimes \mathcal{D}(B') \leq \mathcal{D}(A')$$

for every $\mathcal{D}$, which gives

$$\models ((C_1 \,\&\cdots\&\, C_m) \,\&\, (D_1 \,\&\cdots\&\, D_m)) \Rightarrow A'$$

and $c_1 \otimes \cdots c_m \otimes d_1 \otimes \cdots d_n \leq b \otimes e$.

(g) Let $A := (\forall x)B$ and consider the proof

$$^{b}/B\ \{w_1\},\quad ^{b}/(\forall x)B\ \{r_G\}.$$

By the inductive assumption $\models (C_1 \,\&\cdots\&\, C_m) \Rightarrow B'_x[\mathbf{r}]$ where $\mathbf{r}$ is a special constant for $(\forall x)B'$, i.e.

$$\mathcal{D}(C_1) \otimes \cdots \otimes \mathcal{D}(C_m) \leq \mathcal{D}(B'_x[\mathbf{r}])$$

for every structure $\mathcal{D}$. However, the formula $(B'_x[\mathbf{r}] \Rightarrow (\forall x)B') \in \Delta(T)$ and we have

$$\mathcal{D}(B'_x[\mathbf{r}]) \otimes \mathcal{D}(B'_x[\mathbf{r}] \Rightarrow (\forall x)B') \leq \mathcal{D}((\forall x)B')$$

for every structure $\mathcal{D}$, from which it follows that

$$\models ((C_1 \,\&\cdots\&\, C_m) \,\&\, (B'_x[\mathbf{r}] \Rightarrow (\forall x)B')) \Rightarrow (\forall x)B'.$$

If $A := \mathfrak{b} \Rightarrow B$ is obtained using r_{LC} then we proceed analogously using the inductive assumption and the adjointness property. □

Special sequences of formulas.

DEFINITION 4.30
A sequence of formulas $A_1^{m_1}, \ldots, A_n^{m_n}$, $m_i \geq 1$, $i = 1, \ldots, n$, *is called special if*

$$\models \neg A_1^{m_1} \nabla \cdots \nabla \neg A_n^{m_n},$$

i.e. $\models \neg(A_1^{m_1} \mathbin{\&} \cdots \mathbin{\&} A_n^{m_n})$.

The following is obvious.

LEMMA 4.48
Let $A_1^{m_1}, \ldots, A_n^{m_n}$ *be a special sequence. Then*

$$\mathcal{D}(A_1^{m'_1}) \otimes \cdots \otimes \mathcal{D}(A_n^{m'_n}) = \mathbf{0}$$

holds true in every structure $\mathcal{D}$ *and for all* $m'_i \geq m_i$, $i = 1, \ldots, n$.

COROLLARY 4.8
If $A_1^{m_1}, \ldots, A_n^{m_n}$ *is a special sequence then* $A_1^{m_1}, \ldots, A_n^{m_n}$, $B_1^{p_1}, \ldots,$ $B_q^{p_q}$ *is a special sequence.*

The *order* of the special constant $\mathbf{r}$ for $(\forall x)A$ is the number of occurrences of $\forall$ in the latter. Analogously as in classical logic, we define the fuzzy set $\Delta_m(T) \subseteq \Delta(T)$, which is obtained from $\Delta(T)$ by omitting all the formulas A from its support which are in relation to a special constants with the order greater than m.

LEMMA 4.49
Let T *be a Henkin fuzzy theory and* $A_1^{m_1}, \ldots, A_n^{m_n} \in \Delta_m(T)$, $m > 0$, *be a special sequence. Then there is a special sequence* $B_1^{s_1}, \ldots, A_p^{s_p} \in \Delta_{m-1}(T)$.

PROOF: Analogously as in the classical proof (cf. [121], Lemma 2 in Section 4.3) we will consider a special sequence consisting either of the formulas $A_1^{m_1}, \ldots, A_p^{m_p} \in \Delta_{m-1}(T)$, or those being in relation to special constants $\mathbf{r_1}, \ldots, \mathbf{r_s}$ with the same order as a special constant $\mathbf{r}$ with the highest order. Let the remaining formulas in the considered special sequence be

$$B_x[\mathbf{r}] \Rightarrow (\forall x)B, \quad (\forall x)B \Rightarrow B_x[t_i], \qquad i = 1, \ldots, q.$$

Let $A_i := C_y[\mathbf{s}] \Rightarrow (\forall x)C$ or $A_i := (\forall x)C \Rightarrow C_y[t]$. Since the order of $\mathbf{s}$ is smaller than or equal to that of $\mathbf{r}$, $(\forall x)C$ can occur neither in $C_y[t]$ nor in $C_y[\mathbf{s}]$. Furthermore, $(\forall x)B \neq (\forall x)C$ since $\mathbf{s} \neq \mathbf{r}$. Hence, no A_i contains $(\forall x)B$ and thus, we have a special sequence

$$A_1^{m_1}, \ldots, A_p^{m_p}, (B_x[\mathbf{r}] \Rightarrow (\forall x)B)^m, ((\forall x)B \Rightarrow B_x[t_1])^{m_{p_1}}, \\ \ldots, ((\forall x)B \Rightarrow B_x[t_q])^{m_{p_q}},$$

which means that

$$\mathcal{D}(A_1)^{m_1} \otimes \cdots \otimes \mathcal{D}(A_p)^{m_p} \otimes \mathcal{D}(B_x[\mathbf{r}] \Rightarrow (\forall x)B)^m \otimes \\ \otimes \mathcal{D}((\forall x)B \Rightarrow B_x[t_1])^{m_{p1}} \otimes \cdots \otimes \mathcal{D}((\forall x)B \Rightarrow B_x[t_q])^{m_{pq}} = 0$$

holds for every $\mathcal{D}$. Furthermore, in every $\mathcal{D}$,

$$\mathcal{D}(A_1^{m_1}) \otimes \cdots \otimes \mathcal{D}(A_p^{m_p}) \otimes \mathcal{D}((B_x[\mathbf{r}] \Rightarrow (\forall x)B)^m) \otimes \\ \otimes \mathcal{D}((\forall x)B \Rightarrow (B_x[t_1]^{m_{p1}} \,\&\cdots\&\, B_x[t_q]^{m_{pq}})) = \mathbf{0}$$

and thus, this is a special sequence as well.

Let now, as the first step, $(\forall x)B$ be replaced by $B_x[\mathbf{r}]$ and, as the second step, the latter by $B_x[t_i]$, $i = 1, \ldots, q$. Furthermore, we will replace all the occurrences of $\mathbf{r}$ in all the formulas by the term t_i. Then we obtain two special sequences

$$A_1^{m_1}, \ldots, A_p^{m_p}, (B_x[\mathbf{r}] \Rightarrow B_x[t_1]^{m_{p1}} \,\&\cdots\&\, B_x[t_q])^{m_{pq}}, \tag{4.50}$$

$$(A_1^{(i)})^{m_1}, \ldots, (A_p^{(i)})^{m_p}, (B_x[t_i] \Rightarrow B_x^{(i)}[t_1]^{m_{p1}} \,\&\cdots\&\, B_x^{(i)}[t_q])^{m_{pq}}, \tag{4.51}$$

$i = 1, \ldots, q$. We want to demonstrate that

$$A_1^{m'_1}, \ldots, A_p^{m'_p}, (A_1^{(1)})^{m'_1}, \ldots, (A_p^{(1)})^{m'_p}, (A_1^{(q)})^{m'_1}, \ldots, (A_p^{(q)})^{m'_p}, \tag{4.52}$$

is a special sequence for some m'_j, $j = 1, \ldots, p$.

Assume the opposite, i.e. for every m'_j, m'_{p_j} there exists $\mathcal{D}$ such that

$$\bigotimes_{j=1}^{p} \mathcal{D}(A_j)^{m'_j} \otimes \bigotimes_{\substack{j=1,\ldots,p \\ i=1,\ldots,q}} \mathcal{D}(A_j^{(i)})^{m'_j} > \mathbf{0}. \tag{4.53}$$

Let us find m'_{p_j} such that

$$\mathcal{D}(B_x[t_i]) \to (\mathcal{D}(B_x^{(i)}[t_1])^{m'_{p1}} \otimes \cdots \otimes \mathcal{D}(B_x^{(i)}[t_q])^{m'_{pq}}) = \mathbf{0}$$

(this is possible due to the nilpotency of $\otimes$ and the inequality (4.53)). This may hold true only if $\mathcal{D}(B_x[t_i]) = 1$ and $\mathcal{D}(B_x^{(i)}[t_1])^{m'_{p1}} \otimes \cdots \otimes \mathcal{D}(B_x^{(i)}[t_q]))^{m'_{pq}} = \mathbf{0}$. But then $\mathcal{D}(B_x[\mathbf{r}] \Rightarrow (B_x[t_1]^{m_{p1}} \,\&\cdots\&\, B_x[t_q]^{m_{pq}}) = 1$, from which follows that (4.50) is not special — a contradiction.

Finally, analogously as in the classical proof, we may demonstrate that each $A_j^{(i)}$ belongs to $\Delta_{m-1}(T)$ or is in relation with some of the constants $\mathbf{r}_1, \mathbf{r}_s$. Then the previous procedure can be repeated to get rid of all the special constants with the order greater than $m - 1$. □

Hilbert-Ackermann consistency theorem. A serious problem in the proof of the fuzzy analogy of the Hilbert-Ackermann theorem is the fact that we have no direct (i.e. syntactical) proof of the tautology theorem (saying that every tautology is a theorem). Therefore, we are forced to use the completeness theorem, which means that some parts of the proof are non-constructive.

DEFINITION 4.31
We say that a formula A is a fuzzy quasitautology in the degree a if it is a tautological consequence of some closed instances $B_1, \dots, B_k$ of the equality axioms in the degree a, i.e. if

$$\models_a B_1 \,\&\, \cdots \,\&\, B_k \Rightarrow A \tag{4.54}$$

is an a-tautology. We will write $\models^Q_a A$ if (4.54) holds.

THEOREM 4.30 (CONSISTENCY)
An open fuzzy theory T is contradictory iff there are $p_1, \dots, p_n$ and special axioms $A_1, \dots, A_n$ of the theory T such that

$$\models^Q_b \neg \bar{A}_1^{p_1} \nabla \cdots \nabla \neg \bar{A}_n^{p_n}$$

where $\bar{A}_i$ are instances of the special axioms and $b > \neg(a_1^{p_1} \otimes \cdots \otimes a_n^{p_n})$ where $a_i = \mathrm{SAx}(A_i)$, $i = 1, \dots, n$.

PROOF: Let T be contradictory. Then $T \vdash x \neq x$ and $\mathbf{r} \neq \mathbf{r}$ is an instance of this formula. Hence, by Lemma 4.47 and the fact that there is a proof of $\mathbf{r} \neq \mathbf{r}$ with the value $\mathbf{1}$, there are formulas $A_1, \dots, A_{n-1} \in \mathrm{Supp}(\Delta(T))$ such that

$$\models_b A_1 \,\&\, \cdots \,\&\, A_{n-1} \Rightarrow \mathbf{r} \neq \mathbf{r}.$$

for b fulfilling the above condition. As $\mathbf{r} \neq \mathbf{r} \in \Delta(T)$, we conclude that

$$\models_b \neg A_1 \nabla \cdots \nabla \neg A_n$$

where $A_1, \dots, A_n \in \Delta(T)$. By Lemma 4.49, there are $p_1, \dots, p_n$ and a special sequence $A_1^{p_1}, \dots, A_m^{p_m}$ of formulas from $\Delta_0(T)$. Then

$$\models_b B_1 \,\&\, \cdots \,\&\, B_k \Rightarrow (\neg A_1^{p_1} \nabla \cdots \nabla \neg A_n^{p_n}) \tag{4.55}$$

where $B_1, \dots, B_k$ are instances of the equality axioms occurring in $A_1^{p_1}, \dots, A_m^{p_m}$ (the exponents at B_i are equal to 1 as all these instances are theorems in the degree $\mathbf{1}$), i.e. $\neg A_1^{p_1} \nabla \cdots \nabla \neg A_n^{p_n}$ is the required quasitautology.

Vice-versa, because if $a_i = \mathrm{SAx}(A_i)$ then it follows from Lemma 4.14 on page 127 that $T \vdash_{\bar{a}_i} \bar{A}_i$ where $\bar{a}_i \geq a_i$, $i = 1, \dots, n$. Then we obtain

$$T \vdash_a \bar{A}_1^{p_1} \,\&\, \cdots \,\&\, \bar{A}_n^{p_2}, \quad a \geq a_1^{p_1} \otimes \cdots \otimes a_n^{p_n},$$

using the tautology (P13).

Let $\mathcal{D} \models T$. Then $\mathcal{D}(\bar{A}_1^{p_1} \,\&\, \cdots \,\&\, \bar{A}_n^{p_2}) \geq a$. But $\mathcal{D} \in \mathcal{C}^{sem}(\Delta(T))$ which follows

$$\mathcal{D}(\neg \bar{A}_1^{p_1} \nabla \cdots \,\&\, \neg \bar{A}_n^{p_2}) = \mathcal{D}(\neg(\bar{A}_1^{p_1} \,\&\, \cdots \,\&\, \bar{A}_n^{p_2})) \geq b$$

and $a \otimes b > 0$. But no such structure $\mathcal{D}$ may exist and thus, T is contradictory by the completeness theorem. □

In Chapter 5 on page 208 we will demonstrate that in general, we cannot put the exponents $p_1, \dots, p_n$ equal to 1.

4.4.3 Herbrand theorem in fuzzy logic

One of the crucial theorems in classical logic is the Herbrand one. On the basis of it, the resolution principle in classical predicate logic has been developed, which leads to the theory of automatic proving, and also to logic programming. Thus, it may be interesting to have an analogy of the Herbrand theorem also in fuzzy logic. However, due to multiple truth degrees, its formulation is more complicated.

The following theorem is fundamental. We use here the concept of the consistency threshold introduced in Section 4.3.

THEOREM 4.31
Let T be fuzzy predicate calculus with the fuzzy equality, A be a closed existential formula $A := (\exists x_1)\cdots(\exists x_n)B$. Then $T \vdash_a A$ iff to every $b < a$ there are positive integers $p_1, \ldots, p_n$ such that

$$\models^Q_d p_1 B_1 \nabla \cdots \nabla p_n B_n$$

is a fuzzy quasitautology where $B_1, \ldots, B_n$ are instances of the formula B and $d >^* (p_1 + \cdots + p_n)b$.

PROOF: First, note that

$$T \vdash \neg(\exists x)B \Leftrightarrow (\forall x)\neg B.$$

By Lemma 4.39 on page 153, $T \vdash_a A$ iff $\neg a$ is a consistency threshold for A in T, i.e. iff

$$T' = T \cup \{ {}^{\neg b}/(\forall x_1)\cdots(\forall x_n)\neg B\}$$

is contradictory for all $\neg b > \neg a$ and thus, for all $b < a$. However, by Lemma 4.38, this is equivalent to $T' = T \cup \{ {}^{\neg b}/\neg B\}$ being contradictory.

Since T is consistent, T' is consistent for all $b \geq a$ by Lemma 4.13 on page 127. Choose $b < a$. By Theorem 4.30 there is a quasitautology of instances of $\neg B$

$$\models^Q_d \neg(\neg B_1)^{p_1} \nabla \cdots \nabla \neg(\neg B_n)^{p_n}$$

where $d >^* \neg((\neg b)^{p_1} \otimes \cdots \otimes (\neg b)^{p_n})$. But this is equivalent with

$$\models^Q_d p_1 B_1 \nabla \cdots \nabla p_n B_n$$

where $d >^* \neg\neg(p_1 a \oplus \cdots \oplus p_n a) = (p_1 + \cdots + p_n)a$. □

Now we are ready to prove the Herbrand theorem in FLn.

THEOREM 4.32
Let T be a fuzzy predicate calculus and $A \in F_{J(T)}$ be a closed formula in prenex form. Then

$$T \vdash_a A$$

iff to every $b < a$ there are $p_1, \dots, p_n$ such that $\models_d^Q p_1 A_H^{(1)} \nabla \cdots \nabla p_n A_H^{(n)}$ is a fuzzy quasitautology, where $d >^* (p_1 + \cdots + p_n)b$ and $A_H^{(i)}$ are instances of the matrix of the Herbrand existential formula A_H constructed from A.

PROOF: First, we extend the language $J(T)$ by the new functional symbols from A_H and construct a fuzzy theory T_H being a Henkin extension of T. This is a conservative extension of T. Then we prove that

$$T \vdash_a A \quad \text{iff} \quad T_H \vdash_a A_H \tag{4.56}$$

where A_H is the Herbrand existential formula.

If $T \vdash_a A$ then $T_H \vdash_{a'} A$ where $a \leq a'$. Using the substitution axiom of the form $\vdash (\forall a)B(y) \Rightarrow B_y[f(x_1, \dots, x_n)]$ and Lemma 4.13 we successively obtain $T_H \vdash_b A_H$, $b \geq a$.

Conversely: to simplify the description of the proof, we assume, in the same way as in [121], that

$$A := (\exists x)(\forall y)(\exists z)(\forall w)B(x, y, z, w)$$

and $A_H := (\exists x)(\exists z)B(x, f(x), z, g(x, z))$. Furthermore, if u, v are closed terms then we denote by $\mathbf{r}_\forall(u)$ a special constant for $(\forall y)(\exists z)(\forall w)B(u, y, z, w)$ and similarly, $\mathbf{r}_\forall(u, v)$ that for $(\forall w)B(u, \mathbf{r}_\forall(u), v, w)$, and omit the notation of the replaced variables.

First, we show that

$$T_H \vdash B(u, \mathbf{r}_\forall(u), v, \mathbf{r}_\forall(u, v)) \Rightarrow A \tag{4.57}$$

for all terms u, v. Indeed, using the Henkin axioms, as well as the substitution axiom, we have

$$\begin{aligned}
&T_H \vdash B(u, \mathbf{r}_\forall(u), v, \mathbf{r}_\forall(u, v)) \Rightarrow (\forall w)B(u, \mathbf{r}_\forall(u), v, w),\\
&T_H \vdash (\forall w)B(u, \mathbf{r}_\forall(u), v, w) \Rightarrow (\exists z)(\forall w)B(u, \mathbf{r}_\forall(u), z, w),\\
&T_H \vdash (\exists z)(\forall w)B(u, \mathbf{r}_\forall(u), z, w) \Rightarrow (\forall y)(\exists z)(\forall w)B(u, y, z, w),\\
&T_H \vdash (\forall y)(\exists z)(\forall w)B(u, y, z, w) \Rightarrow A.
\end{aligned}$$

Furthermore, we must show that

$$T_H \vdash (\exists x)(\exists z)B(x, f(x), z, g(x, z)) \Rightarrow B(\mathbf{r}_A, \mathbf{r}_\forall(\mathbf{r}_A), \mathbf{r}_\exists, \mathbf{r}_\forall(\mathbf{r}_A, \mathbf{r}_\exists)) \tag{4.58}$$

where $\mathbf{r}_A$ is a special constant for A and $\mathbf{r}_\exists$ is a special constant for

$$(\exists z)(\forall w)B(\mathbf{r}_A, \mathbf{r}_\forall(\mathbf{r}_A), z, w).$$

This follows from the completeness theorem. Indeed, let $\mathcal{D} \models T_H$. Then (4.58) is equivalent with

$$\bigvee_{d \in D} \bigvee_{d' \in D} \mathcal{D}(B[\mathbf{d}, f(\mathbf{d}), \mathbf{d}', g(\mathbf{d}, \mathbf{d}')]) \leq \mathcal{D}(B[\mathbf{d}_A, \mathbf{d}_\forall(\mathbf{d}_A), \mathbf{d}_\exists, \mathbf{d}_\forall(\mathbf{d}_A, \mathbf{d}_\exists)]) \tag{4.59}$$

where $\mathbf{d}_A$ is a constant for $d_A = \mathcal{D}(\mathbf{r}_A)$, and similarly the other constants represent $d_\forall(d_A) = \mathcal{D}(\mathbf{r}_\forall(\mathbf{r}_A))$, $d_\exists = \mathcal{D}(\mathbf{r}_\exists)$ and $d_\forall(d_A, d_\exists) = \mathcal{D}(\mathbf{r}_\forall(\mathbf{r}_A, \mathbf{r}_\exists))$. Since T_H is a Henkin theory, we may suppose that the interpretation of the functional symbol f in $\mathcal{D}$ is such that $\mathcal{D}(f(\mathbf{d})) = d_\forall(d_A)$ and $\mathcal{D}(g(\mathbf{d}, \mathbf{d}')) = d_\forall(d_A, d_\exists)$. For the same reason the left-hand side of (4.59) is equal to

$$\mathcal{D}(B[\mathbf{d}_A, f(\mathbf{d}_A), \mathbf{d}_\exists, g(\mathbf{d}_A, \mathbf{d}_\exists)])$$

and, consequently, we have even equality in (4.59). Combining (4.58) and (4.57) we obtain

$$T_H \vdash A_H \Rightarrow A. \tag{4.60}$$

Let $T_H \vdash_c A_H$. Using (4.60) we conclude that $T_H \vdash_d A$ where $c \leq d$ and since T_H is a conservative extension of T, we obtain $T \vdash_d A$. The equivalence (4.56) then follows from Lemma 4.7 on page 123.

Finally, let $T_H \vdash_a A_H$. Using Theorem 4.31, to every $b < a$ there are positive integers $p_1, \ldots, p_n$ and a fuzzy quasitautology

$$\models^Q_d \; p_1 A_H^{(1)} \nabla \cdots \nabla p_n A_H^{(n)}$$

where $d >^* (p_1 + \cdots + p_n)b$ and $A_H^{(1)}, \ldots, A_H^{(n)}$ are instances of the matrix of the formula A_H. □

Another, a little weaker form of the Herbrand theorem in FLn has been proved in [98].

4.5 Model Theory in Fuzzy Logic

Model theory in fuzzy logic, or briefly a theory of fuzzy models, is the theory of structures, in which we interpret evaluated formulas. The situation is analogous to classical model theory and thus, various concepts can be naturally generalized. On the other hand, introduction of degrees opens promising possibilities for solving problems having no counterpart in classical model theory. Let us mention, for example, the degrees of elementary equivalence, which can be interpreted as degrees of similarity between models and thus, we can develop a more intricate semantics for new kinds of connectives (e.g. modal ones).

Some attempts to elaborate a theory of concrete models of fuzzy theories occurred already in the early works on fuzzy set theory. We can rank to them, e.g. the theory of fuzzy orderings introduced by L. A. Zadeh in [137], or fuzzy groups by A. Rosenfeld in [116].

The first formulation of the model theory in predicate fuzzy logic (but without knowing the completeness theorem), has been given by G. Gerla and A. di Nola in [21]. A very general presentation of some concepts of the model theory, including the ultraproduct theorem, has been provided by M. Ying in [134] and independently by V. Novák in [94]. Further contribution to the model theory has been elaborated by I. Perfilieva and V. Novák in [100].

4.5.1 Basic concepts of fuzzy model theory

In this subsection, we present some direct generalizations of the classical notions.

Relations between models.

DEFINITION 4.32
Two models $\mathcal{D}$ and $\mathcal{D}'$ are isomorphic in the degree c, $\mathcal{D} \cong_c \mathcal{D}'$, if there is a bijection $g : D \longrightarrow D'$ such that the following holds for all $d_1, \dots, d_n \in D$:

(i) For each couple of functions f_D in $\mathcal{D}$ and $f_{D'}$ in $\mathcal{D}'$ assigned to a functional symbol $f \in J$,

$$g(f_D(d_1, \dots, d_n)) = f_{D'}(g(d_1), \dots, g(d_n)).$$

(ii) For each predicate symbol $P \in J$,

$$c = \bigwedge_{P \in J} \bigwedge_{d_1, \dots, d_n \in D} (P_D(d_1, \dots, d_n) \leftrightarrow P_{D'}(g(d_1), \dots, g(d_n))). \tag{4.61}$$

(iii) For each couple of constants u in $\mathcal{D}$ and u' in $\mathcal{D}'$ assigned to a constant symbol $\mathbf{u} \in J$,

$$g(u) = u'.$$

Due to the properties of biresiduation, $\cong_1$ leads to equality of truth values for all the fuzzy relations assigned to the corresponding predicate symbols $P \in J$.

Let us consider some model $\mathcal{D}$. In the sequel, we will suppose that we work in the extended language $J(\mathcal{D}) = J \cup \{\mathbf{d} \mid d \in D\}$. Furthermore, let a formula $A(x_1, \dots, x_n) \in F_J$ and elements $d_1, \dots, d_n \in D$ be given. Then we will denote by $A_{x_1,\dots,x_n}[\mathbf{d}_1, \dots, \mathbf{d}_n]$ the instance of the formula $A(x_1, \dots, x_n)$, in which all free occurrence of the variables $x_1, \dots, x_n$ have been replaced by the corresponding constants $\mathbf{d}_1, \dots, \mathbf{d}_n \in J(\mathcal{D})$ being assigned to the elements $d_1, \dots, d_n \in D$, respectively.

DEFINITION 4.33

(i) Two models $\mathcal{D}$ and $\mathcal{D}'$ are elementary equivalent in the degree c, $\mathcal{D} \equiv_c \mathcal{D}'$, if

$$c = \bigwedge \{\mathcal{D}(A) \leftrightarrow \mathcal{D}'(A) \mid A \in F_J\}. \tag{4.62}$$

(ii) A model $\mathcal{D}$ is a (fuzzy) submodel of $\mathcal{D}'$, in symbols $\mathcal{D} \subset \mathcal{D}'$, if $D \subseteq D'$, $f_D = f_{D'}|D^n$ holds for each couple of functions $f_D, f_{D'}$ assigned to the functional symbol $f \in J$ and $P_D = P_{D'}|D^n$ holds for each couple of fuzzy relations $P_D \underset{\sim}{\subseteq} D^n$, $P_{D'} \underset{\sim}{\subseteq} (D')^n$ assigned to the predicate symbol $P \in J$. Furthermore, for each pair of the constants u in $\mathcal{D}$ and u' in $\mathcal{D}'$ assigned to a constant symbol $\mathbf{u} \in J$ is $u = u'$. The model $\mathcal{D}'$ is the extension of the model $\mathcal{D}$.

(iii) *The $\mathcal{D}$ is an elementary fuzzy submodel of $\mathcal{D}'$ in the degree c, $\mathcal{D} \prec_c \mathcal{D}'$, if $\mathcal{D} \subset \mathcal{D}'$ and*

$$c = \bigwedge\{\mathcal{D}(A_{x_1,\dots,x_n}[\mathbf{d}_1,\dots,\mathbf{d}_n]) \leftrightarrow \mathcal{D}'(A_{x_1,\dots,x_n}[\mathbf{d}_1,\dots,\mathbf{d}_n]) \mid A(x_1,\dots,x_n) \in F_J \quad \text{and} \quad d_1,\dots,d_n \in D\}. \quad (4.63)$$

The model $\mathcal{D}'$ is an elementary extension of $\mathcal{D}$.

(iv) *A model $\mathcal{D}$ is isomorphically embedded in $\mathcal{D}'$ in the degree c if there is a submodel $\mathcal{E}$ of $\mathcal{D}'$ such that $\mathcal{E} \cong_c \mathcal{D}$.*

We will write $\mathcal{D} \cong \mathcal{D}'$, $\mathcal{D} \equiv \mathcal{D}'$, $\mathcal{D} \prec \mathcal{D}'$, instead of $\mathcal{D} \prec_{\mathbf{1}} \mathcal{D}'$, $\mathcal{D} \equiv_{\mathbf{1}} \mathcal{D}'$ and $\mathcal{D} \prec_{\mathbf{1}} \mathcal{D}'$, respectively.

The following is immediate.

LEMMA 4.50

(a) *If $\mathcal{D} \cong \mathcal{D}'$ then $\mathcal{D} \equiv \mathcal{D}'$.*

(b) *If $\mathcal{D} \prec_c \mathcal{D}'$ then $\mathcal{D} \equiv_d \mathcal{D}'$ and $c \leq d$.*

(c) *If $\mathcal{D}_1 \prec_a \mathcal{D}_2$ and $\mathcal{D}_2 \prec_b \mathcal{D}_3$ then $\mathcal{D}_1 \prec_c \mathcal{D}_3$ and $a \otimes b \leq c$.*

Diagrams. The model

$$\mathcal{D}_D = \langle \mathcal{D}, \{\mathbf{d} \mid d \in D\}\rangle.$$

is the expanded model of the language $J(\mathcal{D})$ in which the constants $\mathbf{d}$ are interpreted by the corresponding elements $d \in D$ (cf. (3.10) on page 75).

DEFINITION 4.34

Let a model $\mathcal{D}_D$ be given. A diagram $\Delta_{\mathcal{D}}$ is a set of evaluated formulas

$$\Delta_{\mathcal{D}} = \{\, {}^{a}/P(t_1,\dots,t_n),\ {}^{\neg a}/\neg P(t_1,\dots,t_n) \mid \mathcal{D}_D(P(t_1,\dots,t_n)) = a, \\ P \in J, t_1,\dots,t_n \in M_V\},$$

i.e. it is a set of evaluated closed atomic formulas and their negations, where the evaluation is equal to the truth degree of the respective formula in the expanded model $\mathcal{D}_D$.

The following theorem is a generalization of the similar classical one. We suppose that the language J contains the fuzzy equality predicate $\doteq$ interpreted by a fuzzy equality $\doteq \mathrel{\underset{\sim}{\subset}} D^2$. We will write $\doteq$ instead of $\doteq_1$ (i.e. the equality degree is unit).

THEOREM 4.33

A model $\mathcal{D}$ is isomorphic embedding in $\mathcal{E}$ in the degree $\mathbf{1}$ iff there is an expanded model of $\mathcal{E}$ being a model of the diagram $\Delta_{\mathcal{D}}$.

PROOF: Let f be isomorphic embedding of $\mathcal{D}$ in $\mathcal{E}$ in the degree $\mathbf{1}$. Furthermore, let ${}^{a}/A \in \Delta_{\mathcal{D}}$. Put

$$\mathcal{E}_D = \langle \mathcal{E}, \{\mathbf{f}(\mathbf{d}) \mid d \in D\}\rangle \quad (4.64)$$

Because the formula A contains constants from $\mathcal{D}_D$ we obtain $\mathcal{E}_D(A) = a = \mathcal{D}_D(A)$ from $\mathcal{E}_D(A) \leftrightarrow \mathcal{D}_D(A) = \mathbf{1}$ due to the fact that f is isomorphic embedding.

Vice-versa, let $f : D \longrightarrow E$ be a function and $\mathcal{E}_D \models \Delta_{\mathcal{D}}$ be the expanded model (4.64). Because $^{a}/A \in \Delta_{\mathcal{D}}$ as well as $^{\neg a}/\neg A \in \Delta_{\mathcal{D}}$ for some atomic formula A, we have

$$\mathcal{E}_D(A) = b \geq a$$
$$\mathcal{E}_D(\neg A) = \neg b \geq \neg a$$

which gives $a = b$ due to the fact that $a \leftrightarrow b = \neg a \leftrightarrow \neg b$. Let $f(d_1) \doteq f(d_2)$. Then $\mathcal{E}_D(\mathbf{f(d_1)} \doteq \mathbf{f(d_2)}) = \mathbf{1}$ and because $\mathcal{E}_D \models \Delta_{\mathcal{D}}$, we have $\mathcal{D}_D(\mathbf{d_1} \doteq \mathbf{d_2}) = \mathbf{1}$ which means that $d_1 \doteq d_2$. □

Compactness and cardinality. Fuzzy logic has also a compactness property. However, unlike classical logic, this concerns only its model part. We also present the upward and downward Lö wenheim-Skolem theorems which are in FLn identical with the classical ones.

THEOREM 4.34 (COMPACTNESS)
A fuzzy theory T has model iff each its finite subtheory $T' \subseteq T$ has a model.

PROOF: The implication left to right is immediate. Conversely, let each T' be consistent and have a model $\mathcal{D}' \models T'$. We prove that T is also consistent. Let T be contradictory. Then there is and $a \in L \backslash \{\mathbf{1}\}$ and a proof w of $\mathbf{a}$ such that $\mathrm{Val}(w) > a$. Let $B_S = \{B_1, \ldots, B_n\}$ be all the special axioms occurring in w. Put $\mathrm{SAx}' = \mathrm{SAx}|B_S$. Obviously, $\mathrm{Supp}(\mathrm{SAx}') = B_S$ is finite, i.e., the corresponding fuzzy theory T' is contradictory and has no model — a contradiction. Hence, we conclude that T has a model by the completeness theorem. □

The classical Löwenhein-Skolem theorems characterize cardinality of models of the given theory. It turns out that they hold also in FLn. Their proofs are verbatim repetition of the classical ones and are omitted.

THEOREM 4.35 (DOWNWARD LÖWENHEIN-SKOLEM)
Let T be a fuzzy theory in the language $J(T)$ of cardinality κ and let $\kappa < \lambda$. If T has a model of cardinality λ then it has a model of cardinality κ' where $\kappa \leq \kappa' < \lambda$.

THEOREM 4.36 (UPWARD LÖWENHEIN-SKOLEM)
Let T be a fuzzy theory in the language $J(T)$ of cardinality κ and let $\mathcal{D}$ be a model of T with the cardinality λ, $\kappa \leq \lambda$. Then T has a model of cardinality μ for each $\lambda < \mu$.

4.5.2 Chains of models

DEFINITION 4.35

Let $\mathcal{D}_1 \subset \mathcal{D}_2 \subset \ldots \subset \mathcal{D}_\alpha \subset \ldots$ be a chain of models, $\alpha < \xi$ for some ordinal number ξ. Put $D = \bigcup_{\alpha<\xi} D_\alpha$ and each $f_D = \bigcup_{\alpha<\xi} f_{D_\alpha}$. Furthermore, we extend P_{D_α} into D by putting

$$P_D(d_1, \ldots, d_n) = \bigvee_{\beta<\xi} P_{D_\alpha}(d_1, \ldots, d_n) \tag{4.65}$$

for each predicate symbol P and each sequence $d_1, \ldots, d_n \in D$, where β is the first ordinal such that $d_1, \ldots, d_n \in D_\beta$. Then the union of the above chain of models is the model

$$\mathcal{D} = \bigcup_{\alpha<\xi} \mathcal{D}_\alpha = \langle D, P_D, \ldots, f_D, \ldots, u, \ldots \rangle$$

where the constants $u, \ldots,$ of $\mathcal{D}$ are taken from all the $\mathcal{D}_\alpha$, $\alpha < \xi$.

Using the maximality principle, we obtain that $\mathcal{D}_\alpha(A) \leq \mathcal{D}(A)$ holds for every closed formula A.

THEOREM 4.37

Let $\mathcal{D}_1 \prec \mathcal{D}_2 \prec \ldots \prec \mathcal{D}_\alpha \prec \ldots$ be an elementary chain of models, $\alpha < \xi$ for some ordinal number ξ. Then

$$\mathcal{D} = \bigcup_{\alpha<\xi} \mathcal{D}_\alpha$$

is an elementary extension of each $\mathcal{D}_\alpha$, i.e., $\mathcal{D}_\alpha \prec \mathcal{D}$ for every α.

PROOF: Since we consider the degree of elementary submodel to be equal to $\mathbf{1}$, we must prove that given α, if $d_1, \ldots, d_n \in D_\alpha$ then $\mathcal{D}(A_{x_1,\ldots,x_n}[\mathbf{d}_1, \ldots, \mathbf{d}_n]) = \mathcal{D}_\alpha(A_{x_1,\ldots,x_n}[\mathbf{d}_1, \ldots, \mathbf{d}_n])$.

For atomic formulas, the statement follows from the assumption and (4.65). For $A := B \Rightarrow C$ we obtain the statement by the inductive assumption.

Finally, let $A := (\forall x)B$ and $\alpha < \xi$. Then

$$\mathcal{D}(A) = \bigwedge_{d\in\bigcup_{\alpha<\xi} D_\alpha} \mathcal{D}(B_x[\mathbf{d}]) = \bigwedge_{d\in\bigcup_{\alpha<\xi} D_\alpha} \mathcal{D}_\alpha(B_x[\mathbf{d}]) \leq \bigwedge_{d\in D_\alpha} (\mathcal{D}_\alpha(B_x[\mathbf{d}])$$

where the second equality follows form the inductive assumption. On the other hand, $\mathcal{D}_\alpha(A) \leq \mathcal{D}(A)$ for every $\alpha < \xi$ since $\mathcal{D}_\alpha \subset \mathcal{D}$ and we conclude that $\mathcal{D}(A) = \mathcal{D}_\alpha$. □

4.5.3 Ultraproduct theorem

In this section, we will deal with sets of models and prove an analogy of the famous Los' ultraproduct theorem for fuzzy logic.

DEFINITION 4.36
Let $\mathcal{L}$ be a residuated lattice and I an index set. By $\mathcal{L}^I$ we denote its direct product

$$\mathcal{L}^I = \langle L^I, \vee, \wedge, \otimes, \rightarrow, \mathbf{0}^I, \mathbf{1}^I \rangle. \tag{4.66}$$

Note that $\mathcal{L}^I$ is a complete residuated lattice. By $a^I \in L^I$ we denote an element such that $a(i) = a$ for all $i \in I$.

Remember the concepts of filter (Definition 2.13 on page 35) and ultrafilter introduced in Section 2.2. Sometimes, we will require that the filter is closed also with respect to arbitrary infima, i.e. if G is a filter then it, moreover, fulfils the condition

(F4) If $P \subseteq G$ then $\bigwedge P \in G$.

Another property we have to consider is the following one:

$$\mathbf{1}^I \in G \quad \text{and} \quad a^I \notin G \quad \text{for any} \quad a < \mathbf{1}. \tag{4.67}$$

LEMMA 4.51
Let G be a filter with the property (4.67). Then it is possible to extend it into an ultrafilter with the same property.

The proof of the following lemma is analogous to the proof of Lemma 4.31(b) on page 144.

LEMMA 4.52
Let $G \subseteq \mathcal{L}^I$ be an ultrafilter with the property (4.67). Then $a \notin G$ iff there are $b \in G$, $c < \mathbf{1}$ and $n > 0$ such that $a^n \otimes b \leq c^I$.

LEMMA 4.53
Let $G \subseteq L^I$ be an ultrafilter with the property (4.67). Then to every $a \in G$ there is $i \in I$ such that $a(i) = \mathbf{1}$.

Let $G \subseteq L^I$ be an ultrafilter with the property (4.67). Put

$$a \sim b \quad \text{iff} \quad a \leftrightarrow b \in G. \tag{4.68}$$

LEMMA 4.54
The relation $\sim$ given in (4.68) is a congruence on $\mathcal{L}^I$. If G has, moreover, a property (F4) then $\sim$ is also a congruence with respect to $\bigwedge$.

PROOF: The first part of this lemma follows from the properties of G and biresiduation $\leftrightarrow$ (cf. [87, 88, 103]).

Let $a_j \leftrightarrow b_j \in G$, $j \in J$ hold for some index set J. Then $\bigwedge_{j\in J}(a_j \leftrightarrow b_j) \in G$ and since $\bigwedge_{j\in J}(a_j \leftrightarrow b_j) \leq \bigwedge_{j\in J} a_j \leftrightarrow \bigwedge_{j\in J} b_j$ we obtain that $\bigwedge_{j\in J} a_j \sim \bigwedge_{j\in J} b_j$. □

We denote by $L^I|G$ the factor set on L^I with respect to $\sim$. The equivalence class of $a \in L^I$ is denoted by $[a]$.

Note that G must be an ultrafilter with the property (4.67) since otherwise Lemma 4.52 would not hold and moreover, since $b^I \leftrightarrow a^I = c^I$ for some $c \leq 1$, $[a^I] = [b^I]$ might hold for some $a \neq b$.

The operations $\vee, \wedge, \otimes, \rightarrow, \bigwedge$ can be defined on $L^I|G$ as usual. We obtain a residuated lattice which will be denoted by $\mathcal{L}^I|G$.

THEOREM 4.38
Let G be an ultrafilter of $\mathcal{L}^I$ with the properties (F4) and (4.67). Then $\mathcal{L}^I|G \cong \mathcal{L}$.

PROOF: First, we prove that to every $[b] \in L^I|G$ there is $a \in L$ such that $[b] = [a^I]$.

Let $b \leftrightarrow a^I \notin G$ for every $a \in L$. Then there are $e_a \in G$, $n_a > 0$ and $c_a < 1$ such that

$$(b \leftrightarrow a^I)^{n_a} \otimes e_a \leq c_a^I.$$

Due to the property (F4), $e = \bigwedge\{e_a \mid a \in L\} \in G$ and $(b \leftrightarrow a^I)^{n_a} \otimes e \leq (b \leftrightarrow a^I)^{n_a} \otimes e_a \leq c_a^I$. for every a. By Lemma 4.53, there is $i \in I$ such that $e(i) = \mathbf{1}$. Choose $a \in L$ such that $b(i) = a$. Then $(b(i) \leftrightarrow a)^{n_a} \otimes e(i) = \mathbf{1} > c_a$ — a contradiction. Hence, there is a such that $b \leftrightarrow a^I \in G$, i.e., $b \in [a^I]$.

Now, let us define two functions $f : L \longrightarrow L^I|G$ and $g : L^I|G \longrightarrow L$ by $f(a) = [a^I]$, $a \in L$ and $g([a^I]) = a$. Then $g(f(a)) = g([a^I]) = a$ and $f(g([a^I])) = f(a) = [a^I]$. We have obtained that $L^I|G \cong L$. Furthermore, for every $a, b \in L$ we have

$$g([a^I] \vee [b^I]) = g([a^I \vee b^I]) = g([(a \vee b)^I]) = a \vee b$$

and analogously for the operations $\wedge, \otimes$ and $\rightarrow$ because $\sim$ is a congruence with respect to these operations.

Let $P \subseteq L^I|G$ and denote $P' = \{a \mid [a^I] \in P\}$. Then

$$g(\bigwedge P) = g([\bigwedge\{a^I \mid [a^I] \in P\}]) = g([(\bigwedge\{a \mid a \in P'\})^I]) = \bigwedge\{a \mid a \in P'\}$$

because $\sim$ is a congruence also with respect to $\bigwedge$. □

Let us now demonstrate the connection between filters of $\mathcal{L}^I$ and I. Let $a \in \mathcal{L}^I$ and $B_a = \{i \in I \mid a(i) = \mathbf{1}\}$ Obviously, $B_0 = \emptyset$. If $G \subseteq L^I$ then we denote

$$H_G = \{B_a \mid a \in G\}. \tag{4.69}$$

LEMMA 4.55
If $G \subseteq L^I$ is a filter then H_G is a set filter. If G is a maximal filter then H_G is also a maximal filter.

PROOF: First, let us realize that given $B \subseteq I$, there is $a \in L^I$ such that $B_a = B$. Furthermore, if $B \subseteq B' \subseteq I$ then there are $d \leq e$ such that $B_d = B$ and $B_e = B'$. Indeed, if $j \notin B'$ then $d(j) < \mathbf{1}$ and thus, it will do to put $e(j) = a \in L$ such that $d(j) \leq a < \mathbf{1}$.

Now, $\emptyset \notin H_G$ since $\mathbf{0}^I \notin G$. Let $d \in G$, $B_d \in H_G$ and $B_d \subseteq B$. Then there is $e \geq d$ such that $B_e = B$. Since $e \in G$, we obtain $B \in H_G$.

Let $B_d, B_e \in H_G$. Then

$$B_d \cap B_e = \{i \mid d(i) = \mathbf{1} \text{ and } e(i) = \mathbf{1}\} = \{i \mid (d \wedge e)(i) = \mathbf{1}\} = B_{d \cap e}.$$

Since $d \wedge e \in G$, we have $B_d \cap B_e \in H_G$.

Let G be maximal and H be a filter on I such that $H_G \subseteq H$. Choose a set $B \in H$. Then $B \subseteq I$ and there is e such that $B = B_e$. By Lemma 4.52, $e \notin G$ if there are $z \in G$, $n > 0$ and $a < 1$ such that

$$e^n \otimes z \leq a^I. \tag{4.70}$$

But $B_z \in H$ and since H is a filter, $B_z \cap B_e \neq \emptyset$, i.e., there is $i \in I$ such that $z(i) = \mathbf{1}$ and $e(i) = \mathbf{1}$ which contradicts to (4.70). □

The rest of this section will be devoted to the formulation and proof of the ultraproduct theorem. By T, we denote a fixed consistent first-order fuzzy theory.

Let us be given a set of models

$$\{\mathcal{D}_i \mid \mathcal{D}_i \models T, i \in I\} \tag{4.71}$$

where I is some index set. Analogously as in classical logic, we construct a new structure $\mathcal{D}_G$.

Let $D = \prod_{i \in I} D_i$ where D_i, $i \in I$ are supports of the corresponding models $\mathcal{D}_i$. Given a term t, we denote by $\mathcal{D}(t)$ an I-tuple

$$\mathcal{D}(t) = \langle \mathcal{D}_i(t) \mid i \in I \rangle$$

and

$$\mathcal{D}(P(t_1, \dots, t_n)) = \langle P_{D_i}(\mathcal{D}_i(t_1), \dots, \mathcal{D}_i(t_n)) \mid i \in I \rangle \in L^I$$

where P_{D_i} are fuzzy relations assigned to the predicate symbol P in the respective models $\mathcal{D}_i$. Similarly, $\mathcal{D}(A)$ denotes an element $\langle \mathcal{D}_i(A) \mid i \in I \rangle \in L^I$.

Let $G \subseteq L^I$ be an ultrafilter and H_G the adjoint set ultrafilter from (4.69). We define

$$d \approx d' \quad \text{iff} \quad \{i \mid d(i) = d'(i)\} \in H_G$$

for every d, $d' \in D$. Obviously, $\approx$ is an equivalence on D and we put $D_G = D| \approx$. This factor set will serve us as a support of the structure $\mathcal{D}_G$. Its elements, equivalence classes, will be denoted by $|d|$, $d \in D$ (to distinguish them from $[a] \in L^I|G$ introduced on page 173).

Let a function symbol f be given. We put

$$f_{D_G}(|d_1|, \dots, |d_n|) = |\langle f_{D_i}(d_1(i), \dots, d_n(i)) \mid i \in I \rangle| = |d| \quad \text{iff}$$
$$\{i \mid f_{D_i}(d_1(i), \dots, d_n(i)) = d(i)\} \in H_G. \tag{4.72}$$

Let P be a predicate symbol. Then we interpret P in $\mathcal{D}_G$ by a fuzzy relation

$$P_{D_G}(|d_1|, \ldots, |d_n|) = a \quad \text{iff} \quad P_D(d_1, \ldots, d_n) \leftrightarrow a^I \in G \tag{4.73}$$

where P_{D_G}, P_D are interpretations of the predicate symbol P (i.e., fuzzy relations) in D_G and D, respectively and $|d_j| = \mathcal{D}_G(t_j) = |\mathcal{D}(t_j)|$, $j = 1, \ldots, n$. It can be verified that the definitions (4.72) and (4.73) do not depend on the choice of the representatives from $|d|$.

The model $\mathcal{D}_G$ corresponding to the ultrafilter G is defined by

$$\mathcal{D}_G = \langle D_G, P_{D_G}, \ldots, f_{D_G}, \ldots, u_G, \ldots \rangle \tag{4.74}$$

where P_{D_G} are fuzzy relations defined in (4.73), f_{D_G} functions defined by (4.72), and $u_G = |\mathcal{D}(\mathbf{u})|$ for a constant $\mathbf{u}$. The model $\mathcal{D}_G$ will be called the *ultraproduct* of $\{\mathcal{D}_i \mid i \in I\}$.

The following is a fuzzy analogy of the Los' ultraproduct theorem.

THEOREM 4.39
Let $G \subseteq L^I$ be an ultrafilter of $\mathcal{L}^I$ with the properties (F4) and (4.67) and $\mathcal{D}_G$ the structure (4.74). Then

(a) $\mathcal{D}_G(t) = |d|$ *iff* $\{i \mid \mathcal{D}_i(t) = d(i)\} \in H_G$ *holds for every term* t.

(b) $\mathcal{D}_G(A) = a$ *iff* $\mathcal{D}(A) \leftrightarrow a^I \in G$ *holds for every formula* A.

PROOF: Due to Lemma 4.55, H_G is a filter and so the proof of (a) is identical with the proof of the analogous statement in classical logic.

(b) Let a formula A be given. Then $\mathcal{D}(A) = \langle \mathcal{D}_i(A) \mid i \in I \rangle \in L^I$. Since by Lemma 4.54, $\sim$ is a congruence with respect to all the operations of $\mathcal{L}^I$ (including arbitrary infima), we obtain from the definition of $\mathcal{D}$ (using the induction on the complexity of the formula A) that $\mathcal{D}(A) \in [a^I]$, i.e. that $\mathcal{D}(A) \leftrightarrow a^I \in G$. Then we conclude that $\mathcal{D}_G(A) = a$ since it follows from Theorem 4.38 that the assignment $[a^I] \mapsto a^I$ is one-to-one. □

REMARK 4.6
Theorem 4.39 holds, provided that the ultrafilter G has the property (F4). This assumption may be avoided when dealing with Henkin extension of the fuzzy theory as well as with that of the given models. Unfortunately, such a result is not fully general.

4.6 Recursive Properties of Fuzzy Theories

Recursive properties of mathematical theories is an important problem which gives us information about their tractability. In 1962, B. Scarpellini published the paper [118] in which he proved that the set of theorems of the infinite-valued predicate Łukasiewicz logic with a recursively enumerable set of axioms is not recursively enumerable, i.e. there does not exist an effective method for finding their proofs. Fuzzy logic presented in this book has much in common with Łukasiewicz logic and, therefore, it takes many of its properties including the recursive ones. Namely, FLn is not axiomatizable in the above sense. A more precise result has been obtained by P. Hájek in [38, 40] who proved that the set of all theorems in the degree **1** is Π_2-complete and thus, provability in fuzzy logic is highly ineffective. Because we replace vagueness by introducing degrees, i.e. we add complexity, such a result is not surprizing and we cannot expect more.

At first sight, this contradicts to the proclaim that fuzzy logic offers simple solutions to complex problems. However, be aware that the mentioned result is general and so, there are various situations in which fuzzy logic is indeed much simpler than the classical techniques, especially if natural language is involved. Second, since we start with imprecise description, we cannot expect nothing more than an imprecise (though objective) solution. It is a challenge to search concrete fuzzy logic solutions, which do not meet the general ineffectiveness discussed above. An example of these are famous fuzzy controllers which are part of fuzzy logic in broader sense described in Chapter 6.

In this short section, we present only the main results about recursive properties of fuzzy logic. A more detailed elaboration can be found, e.g. in [41].

Recall that a relation $R(n_1, \ldots, n_k)$, $R \subseteq \mathbb{N}^k$, is recursive if there is an algorithm deciding in finite time about each sequence $\langle n_1, \ldots, n_k \rangle$ whether it belongs to R or not. A relation $S(n_1, \ldots, n_k)$ is Σ_1, i.e. recursively enumerable if there is a recursive relation $R(n, n_1, \ldots, n_k)$ such that

$$S(n_1, \ldots, n_k) \quad \text{iff} \quad (\exists n)R(n, n_1, \ldots, n_k),$$

and Π_1 if

$$S(n_1, \ldots, n_k) \quad \text{iff} \quad (\forall n)R(n, n_1, \ldots, n_k).$$

The relation S is Δ_1 if it is both Σ_1 and Π_1. Furthermore, S is Π_2 if there is a recursive relation $R(m, n, n_1, \ldots, n_k)$ such that

$$S(n_1, \ldots, n_k) \quad \text{iff} \quad (\forall m)(\exists n)R(m, n, n_1, \ldots, n_k).$$

A set $M \subset \mathbb{N}$ is Π_2 complete if it is Π_2 and for each Π_2 set P there is a recursive function $f : P \longrightarrow M$ such that $P = \{n \mid f(n) \in M\}$.

To be able to characterize recursive properties of fuzzy theories, we suppose that the language contains only a countable number of (rational) logical constants $\mathbf{a}$, $a \in I = \mathbb{Q} \cap [0, 1]$ (cf. Theorem 4.27 on page 156).

DEFINITION 4.37
A fuzzy theory T is recursive if the set $\{\, {}^{a}/A \mid \mathrm{SAx}(A) = a, a \in I, A \in F_{J(T)}\}$ is recursive.

The following has been proved in [40].

LEMMA 4.56
For each axiomatizable fuzzy theory T the set $\{\, {}^{a}/A \mid T \vdash_a A, a \in I\}$ is Π_2.

THEOREM 4.40
There is a recursive fuzzy theory T such that the set of its theorems $\{A \mid T \vdash A\}$ is Π_2-complete.

From the point of view of the recursion theory, it is possible to introduce a weaker concept of the recursive enumerability as done in the works of L. Biancino and G. Gerla [6, 7]. Unlike the classical concept where we strive for finding one effective method leading to a concrete element, the weak recursive enumerability means finding an effective method leading to some element being *near* to that we strive for. Thus, we may have only a sequence of still better effective methods at disposal, but the sole effective method may not exist.

DEFINITION 4.38
A fuzzy set $B \underset{\sim}{\subset} \mathbb{N}$ is weakly recursively enumerable if there is a recursive function $f : \mathbb{N} \times \mathbb{N} \longrightarrow [0, 1] \cap \mathbb{Q}$ such that

$$B(m) = \bigvee \{f(m, n) \mid n \in \mathbb{N}\}.$$

On the basis of this definition, it is possible to prove the following theorem (cf. [94]).

THEOREM 4.41
Let T be a fuzzy theory. If SAx is weakly recursively enumerable then the fuzzy set of syntactic consequences $\mathcal{C}^{syn}(\mathrm{LAx} \cup \mathrm{SAx}) \underset{\sim}{\subset} F_{J(T)}$ is weakly recursively enumerable.

By this theorem, fuzzy logic has analogous recursive properties as classical logic, but in a weaker sense. Actually, recursive properties of fuzzy theories are characterized here only in the discussed "limit sense".

5 FUNCTIONAL SYSTEMS IN FUZZY LOGIC THEORIES

In the preceding chapters, we have been guided by the traditional way of presentation of formal logical systems. This chapter relates to what is known as many-valued functional systems, i.e. systems of functions, whose domain and range are given by the same set of truth values. Functional systems relate to logical theories in such a way that formulas of the latter correspond to elements of the former. The inverse correspondence is possible (but not necessary) when the set of truth functions is finite.

In some cases, investigation of the functional systems can suggest additional insight into "pure" logical problems. We will mention first, the problem of deciding whether a formula is a theorem based on the completeness theorem and second, the problem of automation of proving formulas recognized as theorems. In FLn, both problems exist as well, but the interest in the investigation of the functional systems related to special theories has been caused by other reasons namely, by the applications of the so called fuzzy logic reasoning.

Recall that FLn took its origin in description of vague events, which cannot be done precisely (see, e.g. the approximate reasoning). From the other side, certain level of precision should be kept and it should correspond either to our objective knowledge, or our subjective wish. In any case, choosing FLn as a syntactical tool we would like to control and evaluate the level of precision. Thus, we necessarily fall to the semantics and in general, to the functional systems.

In this section, we make an attempt to give a systematic investigation of the latter from logical, algebraic and functional points of view. The following are special problems to be considered:

(i) characterization of the functional systems related to propositional and predicate calculi,

(ii) construction of special formulas, which generalize Boolean disjunctive and conjunctive normal forms,

(iii) investigation of what can be approximately described by fuzzy logic normal forms in both calculi,

(iv) investigation of the quality of an approximation, which can be achieved by the special choice of a defuzzification.

Let us briefly outline content of the subsequent sections. In Section 5.1, the functional system related to propositional fuzzy logic will be presented. The result is close to the known McNaughton theorem [75]. In our case, however, the result is constructive and leads to the special representation of every formula.

In Section 5.2, the disjunctive and conjunctive normal forms are introduced for FL-functions [1] and, as a special case, for piecewise linear FL-functions defined on $[0, 1]$. It is proved that the subclass of FL- functions represented by the normal forms serves us as an approximation set for continuous functions.

Section 5.3 deals with functional system related to predicate fuzzy logic. Similarly as in the propositional case, we introduce the class of FL-relations, which, as a subclass, includes the functional system mentioned above. The disjunctive and conjunctive normal forms are introduced for FL-relations and their approximation capability is established. At the end of this section, it is proved that the direct analogy of the classical consistency theorem is not valid in fuzzy logic. The proof is essentially based on the form of representation of FL-relations elaborated in this section.

Section 5.4 is concerned with approximation of any continuous function on a compact set by some FL-relation represented by a normal form. Two solutions are suggested. The first one is a direct consequence of the result obtained in the preceding section. The second one provides an algorithmic construction of an FL-relation represented by a normal form, which approximates the given function. Finally, the approximation is performed by applying a defuzzification operation to the approximating FL-relation. In this case, the property of the best approximation has happened to be proved.

In Section 5.5, we are interested in the precise representation of continuous functions by FL-relations. In addition, the FL- relation is supposed to be represented by one of the normal forms. It turned out that the disjunctive normal form is not suitable for this. The theorem concerning the representation of a continuous and piecewise monotonous real valued function has been proved for an FL-relation expressed in the conjunctive normal form.

[1] FL stands for fuzzy logic so that an FL-function means the one defined and valued on the set of truth values.

The theory of functional systems in fuzzy logic presented here has been elaborated by I. Perfilieva. All the results are original. The exception is Section 5.1, which is based on the joint paper [104].

5.1 Fuzzy Logic Functions and Their Representation by Formulas

This section presents the notion of an FL-function as a generalization of a Boolean function and considers the question how it can be characterized. In parallel, we introduce the notion of an FL- function represented by a formula of propositional fuzzy logic, which is essentially a truth function with respect to some interpretation.

We will find a characterization of all the FL-functions admitting representation by formulas of propositional fuzzy logic and will see that not every FL-function can be represented by a formula. Throughout this section, we assume that a set of truth values $L = [0, 1]$ and the interpretation of logical connectives is given by the operations of the Łukasiewicz algebra $\mathcal{L}_{\text{L}}$.

The problem of a characterization of all the functions admitting representation by formulas of Łukasiewicz algebra with two constants 0 and 1 has been solved by McNaughton in [75]. More precisely, he proved that a function f_A is represented by the formula A iff it is continuous, piecewise linear, and each its piece is defined by integer coefficients. Such a function is now usually called the *McNaughton function*. The proof given in [75] was not constructive and it gave rise to further investigations. The known constructive proof has been given by D. Mundici in [83]. It is specified by the extension of the set of two constants to that of all rationals in $[0, 1]$. Another one has been suggested by I. Perfilieva and A. Tonis in [104] for the case when the set of constants is further extended to $[0, 1]$. Below, we reproduce the essential parts of the latter proof. We will use the term formula exclusively in the meaning "formula of propositional fuzzy logic".

5.1.1 *Formulas and their relation to fuzzy logic functions*

Let us begin with the definition of an L-valued fuzzy logic function, which generalizes the notion of a 2-valued Boolean function.

DEFINITION 5.1
Let L be a support of an MV-algebra. A function $f(x_1, \ldots, x_n)$, $n \geq 0$, defined on the set L^n and taking values from L is called an L-valued fuzzy logic function (or, shortly, an L-valued FL-function).

The set of all the L-valued FL-functions will be denoted by P_L. In the case that $L = \{\mathbf{0}, \mathbf{1}\}$, $P_{\{\mathbf{0},\mathbf{1}\}}$ is a set of Boolean functions. It is known that each Boolean function can be represented by a certain logical formula of the classical propositional calculus. We will show that this is not true for any L-valued FL-function when $L = [0, 1]$ even if the set of constants is extended to $L = [0, 1]$.

Throughout this section, we will consider the language of fuzzy propositional calculus with countable set of propositional variables $X = \{x_1, \ldots, x_n, \ldots\}$, the set of logical connectives $\{\Rightarrow, \neg, \&, \nabla, \Leftrightarrow, \wedge, \vee\}$ and the set of constants $L = [0,1]$. Moreover, suppose that the propositional variables are evaluated in $L = [0,1]$ and the interpretation of the logical connectives is given by the operations of the Łukasiewicz algebra $\mathcal{L}_{\text{L}}$. In this section, we restrict ourselves to $[0,1]$- valued FL-functions and therefore, drop the attribute "L – *valued*" when dealing with them.

DEFINITION 5.2
We say that an FL-function f_A is a function represented by a formula A or equivalently, a formula A represents the function f_A if their connection is based on the following inductive procedure:

(i) If A is an atomic formula, i.e. $A := x_i$ or $A := \mathbf{a}$, $a \in [0,1]$, then either $f_A(x_i) \equiv x_i$ or $f_A \equiv a$.

(ii) If A, B are formulas and f_A, f_B are functions represented by them then, in accordance with the definition of formula, the following couples explicate the correspondence between formula and the function represented by it:

$$\neg A \quad \textit{represents} \quad \neg f_A,$$

$$\begin{array}{llllll} A \,\&\, B & \textit{represents} & f_A \otimes f_B, & A \nabla B & \textit{represents} & f_A \oplus f_B, \\ A \wedge B & \textit{represents} & f_A \wedge f_B, & A \vee B & \textit{represents} & f_A \vee f_B, \\ A \Rightarrow B & \textit{represents} & f_A \rightarrow f_B, & A \Leftrightarrow B & \textit{represents} & f_A \leftrightarrow f_B. \end{array}$$

Let us denote by RF the class of all FL-functions represented by formulas and by RF_n the class of all FL-functions represented by formulas and depending on n variables. As mentioned, the elements of RF are essentially truth functions with respect to the interpretation determined by the Łukasiewicz algebra.

We aim at characterizing FL-functions represented by formulas as piecewise linear functions with integer coefficients. Two definitions should be given to introduce the terms.

DEFINITION 5.3
A non-empty set $D \subseteq [0,1]^n$ is a convex polyhedron if it can be given by the system of linear inequalities

$$D = \{\mathbf{x} \in [0,1]^n \mid \mathbf{A}\mathbf{x} \leq \mathbf{b}\}$$

where $\mathbf{A}$ is an integer $(m \times n)$-matrix, $\mathbf{x}$ is a real $(n \times 1)$-vector, $\mathbf{b}$ is a real $(m \times 1)$-vector and $\leq$ is defined componentwise. The convex polyhedron will be denoted by the pair $\langle \mathbf{A}, \mathbf{b} \rangle$.

EXAMPLE 5.1
The n-cube $[0,1]^n$ is a convex polyhedron because it can be represented in accordance with the previous definition using the matrix and the vector

$$\mathbf{A} = \begin{pmatrix} \mathbf{E}_n \\ -\mathbf{E}_n \end{pmatrix}, \qquad \mathbf{b} = \begin{pmatrix} \mathbf{1}_n \\ \mathbf{0}_n \end{pmatrix},$$

respectively where $\mathbf{E}_n$ is the identity matrix, $\mathbf{1}_n$ and $\mathbf{0}_n$ are $(n \times 1)$-vectors of ones and zeroes, respectively. □

REMARK 5.1
The given definition of a convex polyhedron is equivalent with the common one (where the convex polyhedron is defined as a convex hull of some finite set of points) by the known Weyl-Minkowski theorem. We prefer the former because of its algebraic nature.

DEFINITION 5.4
A continuous function $f(x_1, \dots, x_n) : [0,1]^n \longrightarrow [0,1]$ *is piecewise linear (with integer coefficients) if there exists a finite number of convex polyhedra* $D_1, \dots, D_r$ *such that* $\bigcup_{i=1}^r D_i = [0,1]^n$ *and* $f|D_1, \dots, f|D_r$ *are linear functions with integer coefficients, i.e.*

$$f(\mathbf{x}) = \mathbf{c}_i \mathbf{x} + d_i \quad \textit{if} \quad \mathbf{x} \in D_i,$$

where $\mathbf{x} = (x_1, \dots, x_n)$ *is a real vector,* $\mathbf{c}_i$ *is an* n*-dimensional vector of integers and* d_i *is a real number,* $1 \leq i \leq r$. *The convex polyhedra* $D_1, \dots, D_r$ *will be called the linearity domains of* f.

EXAMPLE 5.2
As an example of a piecewise linear function let us consider the function $f_{\Rightarrow}(x, y)$ represented by the formula $(x \Rightarrow y)$. Due to the definition,

$$f_{\Rightarrow}(x, y) = \min(1, 1 - x + y) = \begin{cases} 1 & \text{if } x \leq y, \\ 1 - x + y & \text{otherwise.} \end{cases}$$

Choosing convex polyhedra $D_1 = \{(x, y) \mid x - y \leq 0\}$, $D_2 = \{(x, y) \mid -x + y \leq 0\}$, and two linear functions with integer coefficients $f_1(x, y) = 1, f_2(x, y) = 1 - x + y$, we easily see that $f_{\Rightarrow}|D_1 = f_1$ and $f_{\Rightarrow}|D_2 = f_2$. Analogous considerations lead us to the following lemma. □

LEMMA 5.1
Each formula of the form $x \Rightarrow y$, $\neg x$, $x \mathbin{\&} y$, $x \nabla y$, $x \Leftrightarrow y$,$x \wedge y$, $x \vee y$ *represents a piecewise linear function, which is equal to* $x \to y$, $1 - x$, $x \otimes y$, $x \oplus y$, $x \leftrightarrow y$, $x \wedge y$, $x \vee y$, *respectively.*

Let us denote the class of all piecewise linear functions by PL and the class of all the n-ary piecewise linear functions by PL_n. Note that PL_0 consists of constants. By $\mathbf{x}$, we denote the vector $(x_1, \dots, x_n)$ where $n \geq 1$.

The proof of the subsequent lemma is not difficult and therefore, it is omitted.

LEMMA 5.2
Let $f_1, \dots, f_p \in PL_n$. *Then there exist convex polyhedra* $D_1, \dots, D_r$ *such that* $\bigcup_{i=1}^r D_i = [0,1]^n$ *and each* D_i, $1 \leq i \leq r$, *is a linearity domain for all* $f_1, \dots, f_p$.

The following theorem is fundamental for this subsection as it gives the required characterization of FL-functions represented by formulas.

THEOREM 5.1
Let f_A be an FL-function represented by a formula A then f_A is a piecewise linear, i.e. $f_A \in PL$.

PROOF: The proof proceeds by induction on the number of connectives in the formula A.

1. If A is an atomic formula then $f_A(x) \equiv x$ or $f_A \equiv a$, $a \in [0,1]$, and thus, $f_A \in PL_0 \cup PL_1$.

2. Suppose that $A := \neg B$ or $A := B \circ C$ where $\circ \in \{\Rightarrow, \&, \nabla, \Leftrightarrow, \wedge, \vee\}$, and $f_B, f_C \in PL$. By virtue of Lemma 5.1, we need to prove that $1 - f_B, f_B \bullet f_C$ where $\bullet \in \{\to, \otimes, \oplus, \leftrightarrow, \wedge, \vee\}$, respectively, are piecewise linear functions.

If $f_A = 1 - f_B$ the assertion is evidently true. Suppose that $f_A = f_B \bullet f_C$. We will prove the following more general proposition. Namely, if $f_1, \dots, f_p \in PL_n$ and $g \in PL_p$ for $n, p \geq 1$, then $g(f_1(x_1, \dots, x_n), \dots, f_p(x_1, \dots, x_n)) \in PL_n$. To simplify the notation, let us use a vector-function $\mathbf{f}$ of the form

$$\mathbf{f}(\mathbf{x}) := (f_1(\mathbf{x}), \dots, f_p(\mathbf{x})).$$

In accordance with Lemma 5.2, choose a convex polyhedron D_1 being a linearity domain for all $f_1, \dots, f_p$ and a convex polyhedron D_2 being a linearity domain for g. Then there exist integer matrices $\mathbf{C}_1$ $(p \times n)$ and $\mathbf{C}_2$ $(1 \times p)$, vector of reals $\mathbf{d}_1(p \times 1)$ and a real number d_2 such that

$$\begin{aligned} \mathbf{f}(\mathbf{x}) &= \mathbf{C}_1\mathbf{x} + \mathbf{d}_1 \quad \text{if} \quad \mathbf{x} = (x_1, \dots, x_n) \in D_1, \\ g(\mathbf{y}) &= \mathbf{C}_2\mathbf{y} + d_2 \quad \text{if} \quad \mathbf{y} = (y_1, \dots, y_p) \in D_2. \end{aligned}$$

Put $D = D_1 \cap \mathbf{f}^{-1}(D_2)$ and verify that D is a convex polyhedron. Indeed, suppose that the pairs $\langle \mathbf{A}_1, \mathbf{b}_1 \rangle$ and $\langle \mathbf{A}_2, \mathbf{b}_2 \rangle$ specify D_1 and D_2 respectively. Then it is easy to see that D is specified by the pair $\langle \mathbf{A}, \mathbf{b} \rangle$ where

$$\mathbf{A} = \begin{pmatrix} \mathbf{A}_1 \\ \mathbf{A}_2 \cdot \mathbf{C}_1 \end{pmatrix}, \qquad \mathbf{b} = \begin{pmatrix} \mathbf{b}_1 \\ \mathbf{b}_2 - \mathbf{A}_2 \cdot \mathbf{d}_1 \end{pmatrix}.$$

Assume that $D \neq \emptyset$ and $\mathbf{x} \in D$. Then

$$g(\mathbf{f}(\mathbf{x})) = g(\mathbf{C}_1\mathbf{x} + \mathbf{d}_1) = \mathbf{C}_2(\mathbf{C}_1\mathbf{x} + \mathbf{d}_1) + d_2 := \mathbf{C}\mathbf{x} + d$$

where $\mathbf{C} = \mathbf{C}_2\mathbf{C}_1$, $d = \mathbf{C}_2\mathbf{d}_1 + d_2$.

Finally,

$$\bigcup_{(D_1, D_2)} (D_1 \cap f^{-1}(D_2)) = [0,1]^n \cap f^{-1}([0,1]^p) = [0,1]^n$$

where the union is taken over all couples (D_1, D_2) described above. Thus, $g(\mathbf{f}(\mathbf{x}))$ is a piecewise linear function. □

COROLLARY 5.1
The class of FL-functions represented by formulas is a subset of the class of piecewise linear functions, i.e. $RF \subseteq PL$.

5.1.2 *Piecewise linear functions and their representation by formulas*

In this subsection, we are going to prove the result converse to that of Theorem 5.1. Namely, $PL \subseteq RF$, or any piecewise linear function can be represented by a formula. As discussed in the beginning of this section, it is our special interest to give the constructive proof of this result. Strictly speaking, we will show how any piecewise linear function can be represented by certain algebraic formula. Its transformation into a formula of propositional fuzzy logic can be performed when following the proofs of the subsequent propositions.

In a lot of cases below, in order to obtain FL-functions represented by formulas over the ordinary algebra of reals we will use the following truncation operation '*' on the set of real numbers:

$$x^* := \begin{cases} 0 & \text{if } x < 0, \\ x & \text{if } 0 \leq x \leq 1, \\ 1 & \text{if } x > 1. \end{cases} \tag{5.1}$$

The idea of the proof is the following. First, we will show how all the "truncated linear" functions of the form $(\mathbf{cx} + d)^*$ ("linear pieces" of piecewise linear function) can be represented by formulas and second, how to "glue" thus obtained formulas. For technical reasons, it is convenient to begin with the second part.

Let $n \geq 1$ be a natural number. Define a class $UL(n)$ whose elements are unit level sets of n-ary FL-functions represented by formulas. Formally, $E \in UL(n)$ iff there exists a function $g \in RF_n$ such that

$$E = \{\mathbf{x} \in [0,1]^n \mid g(\mathbf{x}) = 1\}.$$

For each $E \in UL(n)$, let RF_E be a class of n-ary FL-functions coinciding on E with some function represented by a formula. Formally, $f \in RF_E$ iff there exists a function $h \in RF_n$ such that $g(\mathbf{x}) = h(\mathbf{x})$ whenever $\mathbf{x} \in E$. Obviously, $RF_{[0,1]^n} = RF_n$.

LEMMA 5.3

(a) *Let $E_1, E_2 \in UL(n)$ and $f \in RF_{E_1} \cap RF_{E_2}$. Then $f \in RF_{E_1 \cup E_2}$ and therefore, $RF_{E_1} \cap RF_{E_2} = RF_{E_1 \cup E_2}$.*

(b) *Let $E_1, \dots, E_r \in UL(n)$ and $f \in RF_{E_1} \cap \dots \cap RF_{E_r}$. Moreover, let $\bigcup_{i=1}^{r} E_i = [0,1]^n$. Then $f \in RF_n$.*

PROOF: (a) Since $f \in RF_{E_1} \cap RF_{E_2}$ then by the definition, there exist $h_1, h_2 \in RF_n$ such that $f|E_1 = h_1|E_1$, $f|E_2 = h_2|E_2$. On the other hand, since $E_1, E_2 \in UL(n)$ then there exist $g_1, g_2 \in RF_n$ such that $g_1|E_1 = 1$, $g_2|E_2 = 1$.

Let us define the function $h^{p,q} : [0,1]^n \longrightarrow [0,1]$ by

$$h^{p,q}(\mathbf{x}) := (h_1(\mathbf{x}) \otimes (g_1(\mathbf{x}))^p) \vee (h_2(\mathbf{x}) \otimes (g_2(\mathbf{x}))^q) \tag{5.2}$$

where p and q are natural numbers.

Obviously, $h^{p,q} \in RF_n$. Now, we aim at proving that there exist natural numbers $\bar{p}, \bar{q}$, such that $f|(E_1 \cup E_2) = h^{\bar{p},\bar{q}}|(E_1 \cup E_2)$. Thus, the formula (5.2) "glues" h_1 and h_2 into one function $h^{\bar{p},\bar{q}}$ so that $h^{\bar{p},\bar{q}}|E_1 = h_1|E_1$, $h^{\bar{p},\bar{q}}|E_2 = h_2|E_2$.

It follows from Theorem 5.1 that $g_1, g_2, h_1, h_2 \in PL_n$. Due to Lemma 5.2, there exist convex polyhedra $D_1, \dots, D_r$ ($\bigcup_{i=1}^r D_i = [0,1]^n$) specified by the pairs $\langle \mathbf{A}_1, \mathbf{b}_1 \rangle, \dots, \langle \mathbf{A}_r, \mathbf{b}_r \rangle$, which are linearity domains for all the functions g_1, g_2, h_1, h_2. It follows that, for example, $g_1(\mathbf{x}) = \mathbf{c}_i\mathbf{x} + d_i$ if $\mathbf{x} \in D_i$, $1 \le i \le r$. Consider new convex polyhedra $D'_1, \dots, D'_r$ specified by

$$\mathbf{A}'_i = \begin{pmatrix} \mathbf{A}_i \\ -\mathbf{c}_i \end{pmatrix}, \qquad \mathbf{b}'_i = \begin{pmatrix} \mathbf{b}_i \\ d_i - 1 \end{pmatrix}, \qquad 1 \le i \le r.$$

It is not difficult to see that $D'_1, \dots, D'_r$ are also linearity domains for all the functions g_1, g_2, h_1, h_2 and that $D'_i = D_i \cap E_1$, $1 \le i \le r$, whence $\bigcup_{i=1}^r D'_i = E_1$.

Let $D'_i \neq \emptyset$. At first, we will show that there exists a natural number q_i such that

$$h_2(\mathbf{x}) \otimes (g_2(\mathbf{x}))^{q_i} \le h_1(\mathbf{x}), \qquad \mathbf{x} \in D'_i. \tag{5.3}$$

For this, we will estimate the difference between the right- hand and the left-hand sides of (5.3)

$$h_1(\mathbf{x}) - h_2(\mathbf{x}) \otimes (g_2(\mathbf{x}))^{q_i} = h_1(\mathbf{x}) - (h_2(\mathbf{x}) - q_i(1 - g_2(\mathbf{x})))^*. \tag{5.4}$$

Denote the right-hand side of (5.4) by $\varphi_i(\mathbf{x})$, $\mathbf{x} \in D'_i$. If $\mathbf{x} \in D'_i \cap E_2$ then $h_1(\mathbf{x}) = f(\mathbf{x}) = h_2(\mathbf{x})$ and $g_2(\mathbf{x}) = 1$, whence $\varphi_i(\mathbf{x}) = 0$. On the other hand, if $\mathbf{x} \in D'_i \setminus E_2$ then $1 - g_2(\mathbf{x}) > 0$, and q_i could be chosen so large that the inequality $\varphi_i(\mathbf{x}) \ge 0$ holds true.

Now, eliminate the dependence of the choice of q_i on $\mathbf{x}$. It is known that any linear function on a convex polyhedron attains its minimum value at one of its vertices. Let $\mathbf{x}_{i1}, \dots, \mathbf{x}_{is}$ be the vertices of D'_i. Choose q_i so large that $\varphi_i(\mathbf{x}_{ij}) \ge 0$ holds true for any $j = 1, \dots, s$. Below, we will consider $\varphi_i(\mathbf{x})$ with that chosen value of q_i.

By the definition, $\varphi_i(\mathbf{x})$ is a linear function on D'_i and therefore, it attains its minimum on D'_i at some of $\mathbf{x}_{ij}$ ($1 \le j \le s$). Since $\varphi_i(\mathbf{x}_{ij}) \ge 0$, we have $\varphi_i(\mathbf{x}) \ge 0$ for any $\mathbf{x} \in D'_i$, which gives (5.3).

Let $\bar{q} = \max_{1 \le i \le r} q_i$ where q_i, $1 \le i \le r$, is the value validating (5.3). Since $\bigcup_{i=1}^r D'_i = E_1$, then for all $\mathbf{x} \in E_1$ and any natural number p

$$\begin{aligned} h^{p,\bar{q}}(\mathbf{x}) =& (h_1(\mathbf{x}) \otimes (g_1(\mathbf{x}))^p) \vee (h_2(\mathbf{x}) \otimes (g_2(\mathbf{x}))^{\bar{q}}) = \\ =& (h_1(\mathbf{x}) \otimes 1) \vee (h_2(\mathbf{x}) \otimes (g_2(x))^{\bar{q}}) = h_1(\mathbf{x}) \vee (h_2(\mathbf{x}) \otimes (g_2(\mathbf{x}))^{\bar{q}}) = \\ =& h_1(\mathbf{x}) = f(\mathbf{x}). \end{aligned}$$

Analogously as above, it can be proved that there exists a natural number $\bar{p}$, such that for any $\mathbf{x} \in E_2$ $h^{\bar{p},\bar{q}}(\mathbf{x}) = f(\mathbf{x})$. Thus, $f|(E_1 \cup E_2) = h^{\bar{p},\bar{q}}|(E_1 \cup E_2)$. Therefore, $f \in RF_{E_1 \cup E_2}$.

(b) Given (a), (b) can be obtained by induction. □

In the following two lemmas we will realize the first part of our proof, i.e. we will show, that the "truncated linear" function of the form $(\mathbf{cx} + d)^*$ can be represented by a formula. We will consider first, the case of non-negative integer coefficients and second, the general case.

Let n, k be non-negative integers and $n \neq 0$. Define the function $s_n^k : [0,1]^n \longrightarrow [0,1]$ by putting

$$s_n^k(\mathbf{x}) = s_n^k(x_1, \dots, x_n) := (\sum_{i=1}^{n} x_i - k)^*.$$

It is easy to see that $s_2^0(x_1, x_2) = x_1 \oplus x_2$ and $s_2^1(x_1, x_2) = x_1 \otimes x_2$.

LEMMA 5.4
Each function s_n^k, $n \geq 1$, $k \geq 0$, can be represented by a formula, i.e. $s_n^k \in RF_n$.

PROOF: The lemma will be proved by induction on n and k. Given n, the proof will proceed by induction on k.

- $s_1^0 = x_1$, $s_1^k \equiv 0$, $k \geq 1$. Thus, $s_1^k \in RF_1$ for all $k \geq 0$.
- Suppose that $s_{n-1}^k \in RF_{n-1}$ holds for all $k \geq 0$. We will show that $s_n^k \in RF_n$, $k \geq 0$. By the definition, $s_n^0(\mathbf{x}) = x_1 \oplus \cdots \oplus x_n$, which follows $s_n^0 \in RF_n$. Now, let $k \geq 1$, $\mathbf{x}^{(n)} = (x_1, \dots, x_n)$, $\mathbf{x}^{(n-1)} = (x_1, \dots, x_{n-1})$. Then

$$s_n^k(\mathbf{x}^{(n)}) = \begin{cases} (x_n + s_{n-1}^{k-1}(\mathbf{x}^{(n-1)}) - 1)^* = x_n \otimes s_{n-1}^{k-1}(\mathbf{x}^{(n-1)}) \\ \quad \text{if } \neg s_{n-1}^k(\mathbf{x}^{(n-1)}) = 1, \\ (x_n + s_{n-1}^k(\mathbf{x}^{(n-1)}))^* = x_n \oplus s_{n-1}^k(\mathbf{x}^{(n-1)}) \\ \quad \text{if } \quad s_{n-1}^{k-1}(\mathbf{x}^{(n-1)}) = 1. \end{cases} \tag{5.5}$$

  Using the induction assumption and Lemma 5.3(a), we obtain that $s_n^k \in RF_n$.

□

REMARK 5.2
The following recurrent formula for s_n^k can be obtained by applying "gluing" formula (5.2) to (5.5):

$$\begin{aligned} s_n^k(\mathbf{x}^{(n)}) = &(x_n \otimes s_{n-1}^{k-1}(\mathbf{x}^{(n-1)}) \otimes (\neg s_{n-1}^k(\mathbf{x}^{(n-1)}))^p) \vee ((x_n \oplus s_{n-1}^k(\mathbf{x}^{(n-1)})) \otimes \\ &\otimes (s_{n-1}^{k-1}(\mathbf{x}^{(n-1)}))^q). \end{aligned} \tag{5.6}$$

LEMMA 5.5
Let $f : [0,1]^n \longrightarrow [0,1]$ be a truncated linear function, i.e. $f(\mathbf{x}) = (\mathbf{c}\mathbf{x} + d)^*, \mathbf{c} = (c_1, \dots, c_n) \in \mathbb{Z}^n, d \in \mathbb{R}$. Then $f \in RF_n$.

PROOF: Put

$$x_i^{c_i} := \begin{cases} x_i & \text{if } c_i \geq 0, \\ \neg x_i & \text{if } c_i < 0, \end{cases}$$

$$k := -\left[d + \sum_{c_i<0} c_i\right], \qquad d' := \left\{d + \sum_{c_i<0} c_i\right\}$$

where square and curly brackets are used to denote integer and fraction parts, respectively. Then $f(\mathbf{x})$ can be rewritten in such a way that

$$f(\mathbf{x}) = \left(\sum_{i=1}^{n} |c_i| x_i^{c_i} + d' - k\right)^* . \tag{5.7}$$

If $k < 0$ then $f(\mathbf{x}) \equiv 1$ and thus, $f \in RF$. Otherwise, let $l = \sum_{i=1}^{n} |c_i| + 1$. Rewrite the expression (5.7) for $f(\mathbf{x})$ with the external function s_l^k:

$$f(\mathbf{x}) = s_l^k(\underbrace{x_1^{c_1}, \dots, x_1^{c_1}}_{|c_1|\text{- times}}, \dots, \underbrace{x_n^{c_n}, \dots, x_n^{c_n}}_{|c_n|\text{-times}}, d'). \tag{5.8}$$

Now, apply the previous Lemma 5.4 to the right-hand side of 5.8 and obtain that $f \in RF_n$. □

The following theorem summarizes all things being proved earlier and establishes the final result.

THEOREM 5.2
Each piecewise linear function can be represented by a formula.

PROOF: Let $f(x_1, \dots, x_n)$ be a piecewise linear function. By the definition, there exists a convex polyhedron $D \subseteq [0,1]^n$ specified by the pair $\langle \mathbf{A}, \mathbf{b} \rangle$ such that $f|D$ is a linear function. We will show that $f \in RF_D$, i.e. $D \in UL(n)$ and there exists a function $h \in RF_n$ such that $f|D = h|D$. Taking into account Lemma 5.5 it is sufficient to prove only that $D \in UL(n)$.

Let $\mathbf{a}_1, \dots, \mathbf{a}_m$ be rows of $\mathbf{A}$ and $\mathbf{b} = (b_1, \dots, b_m)$. Define the function $g : [0,1]^n \longrightarrow [0,1]$ by

$$g(\mathbf{x}) := \neg(\mathbf{a}_1\mathbf{x} - b_1)^* \wedge \cdots \wedge \neg(\mathbf{a}_m\mathbf{x} - b_m)^* . \tag{5.9}$$

By virtue of Lemma 5.5, $g \in RF_n$. Since $D = \{\mathbf{x} \in 0,1]^n \mid g(\mathbf{x}) = \mathbf{1}\}$, then $D \in UL(n)$ and thus, $f \in RF_D$. Recall that D is an arbitrary linearity domain of f. It remains only to apply Lemma 5.3(b) and obtain that $f \in RF_n$. □

COROLLARY 5.2 (MCNAUGHTON THEOREM)
The class of $[0,1]$-valued FL-functions represented by formulas of propositional fuzzy logic coincides with the class of piecewise linear functions with integer coefficients defined on $[0,1]$, i.e. $RF = PL$.

PROOF: It follows from Theorems 5.1 and 5.2. □

As an interesting consequence of Theorem 5.2 and the preceding lemmas we can deduce a uniform representation of any piecewise linear function by a formula. This formula is in some sense a *canonical form of a piecewise linear function.* We will refer to it when speaking about the normal forms in the next section.

COROLLARY 5.3
Let $f \in PL_n$ be a piecewise linear function with linearity domains $D_1, \dots, D_r$, so that $\bigcup_{i=1}^{r} D_i = [0,1]^n$ and $f(\mathbf{x}) = \mathbf{c}_i\mathbf{x} + d_i$ if $\mathbf{x} \in D_i$, $1 \leq i \leq r$. Let $D_1, \dots, D_r$ be specified by pairs $\langle \mathbf{A}_1, \mathbf{b}_1 \rangle, \dots, \langle \mathbf{A}_r, \mathbf{b}_r \rangle$ respectively, so that $\mathbf{a}_{i1}, \dots, \mathbf{a}_{im_i}$ are rows of $\mathbf{A}_i$, and $b_{i1}, \dots, b_{im_i}$ are elements of $\mathbf{b}_i$. Then $f(\mathbf{x})$ can be represented in a canonical form :

$$f(\mathbf{x}) = \bigvee_{i=1}^{r} ((\mathbf{c}_i\mathbf{x} + d_i)^* \otimes (\bigwedge_{j=1}^{m_i} \neg(\mathbf{a}_{ij}\mathbf{x} - b_{ij})^*)^{p_i}) \tag{5.10}$$

where p_i, $1 \leq i \leq r$, are some natural numbers.

PROOF: Follow the proof of Theorem 5.2. First, apply the "gluing" formula (5.2) $n-1$ times using (5.9) to represent linearity domains of f as unit level sets of functions from *RF*. The result is (5.10). □

Observe, that (5.10) is not a formula of propositional fuzzy logic but it can be transformed into the latter when following the proofs of Theorem 5.2 and lemmas 5.3–5.5.

5.2 Normal Forms for FL-Functions and Formulas of Propositional Fuzzy Logic

In comparison with classical logic where the class of functions represented by formulas of propositional logic coincides with the class of Boolean functions, the situation in FLn is different. The results of the previous section demonstrate that not every FL-function can be represented by a formula of propositional fuzzy logic. Therefore, if we suppose that normal forms are given by formulas of propositional fuzzy logic, then the obvious conclusion follows. Having this on mind, we will distinguish two algebras: MV-algebra of L-valued FL-functions and the algebra of formulas of propositional fuzzy logic. Moreover, the latter will be replaced by the algebra of piecewise linear functions defined on $[0,1]$. The normal forms will be introduced as formulas over certain algebras. In both cases they coincide with Boolean ones in the special case of two truth values.

Let us now recall the disjunctive and conjunctive normal forms for Boolean functions and explain our way of generalization. Let $f(x_1, \dots, x_n)$, $n \geq 1$, be a Boolean function, i.e. its domain and range is the set $\{\mathbf{0}, \mathbf{1}\}$. To represent the propositional variable x and its negation, we will write

$$x^\sigma = \begin{cases} x' & \text{if } \sigma = \mathbf{0}, \\ x & \text{if } \sigma = \mathbf{1}. \end{cases}$$

It is easy to see that $x^\sigma = \mathbf{1}$ iff $x = \sigma$. Then $f(x_1, \ldots, x_n)$ can be represented in the disjunctive normal form

$$\begin{aligned} f(x_1, \ldots, x_n) &= \bigvee_{f(\sigma_1, \ldots, \sigma_n) = \mathbf{1}} (x_1^{\sigma_1} \wedge \cdots \wedge x_n^{\sigma_n}) = \\ &= \bigvee_{(\sigma_1, \ldots, \sigma_n)} (x_1^{\sigma_1} \wedge \cdots \wedge x_n^{\sigma_n} \wedge f(\sigma_1, \ldots, \sigma_n)) \end{aligned} \tag{5.11}$$

provided that $f \not\equiv \mathbf{0}$, and in the conjunctive normal form

$$\begin{aligned} f(x_1, \ldots, x_n) &= \bigwedge_{f(\tau_1, \ldots, \tau_n) = \mathbf{0}} (x_1^{\bar{\tau}_1} \vee \cdots \vee x_n^{\bar{\tau}_n}) = \\ &= \bigwedge_{(\tau_1, \ldots, \tau_n)} (x_1^{\bar{\tau}_1} \vee \cdots \vee x_n^{\bar{\tau}_n} \vee f(\tau_1, \ldots, \tau_n)) \end{aligned} \tag{5.12}$$

provided that $f \not\equiv \mathbf{1}$.

A special interest will be payed to the right-hand sides of both expressions since they are prototypes of the suggested generalization. Analyzing each elementary conjunction $x_1^{\sigma_1} \wedge \cdots \wedge x_n^{\sigma_n} \wedge f(\sigma_1, \ldots, \sigma_n)$ in the disjunctive normal form we see that it contains two parts joined by conjunction, namely the characterization $x_1^{\sigma_1} \wedge \cdots \wedge x_n^{\sigma_n}$ of the set $\{(\sigma_1, \ldots \sigma_n)\} \subset \{\mathbf{0}, \mathbf{1}\}^n$ consisting of one point, and the value of the represented function defined on it. Analogously, analyzing each elementary disjunction $x_1^{\bar{\tau}_1} \vee \cdots \vee x_n^{\bar{\tau}_n} \vee f(\tau_1, \ldots, \tau_n)$ in the conjunctive normal form we see that it contains a negative characterization $x_1^{\bar{\tau}_1} \vee \cdots \vee x_n^{\bar{\tau}_n}$ of the set $\{(\tau_1, \ldots \tau_n)\} \subset \{\mathbf{0}, \mathbf{1}\}^n$ also consisting of one point, which is joined by disjunction with the value of the represented function.

Thus, in the generalization of the disjunctive and conjunctive normal forms given below we will preserve two parts in each elementary construction (conjunction or disjunction), which consists of the characterization (positive or negative respectively) of a certain set joined with the average value of a certain function on it.

In this section, we will construct normal forms for L-valued FL-functions and in particular, for FL-functions represented by formulas of propositional fuzzy logic. In both cases, we aim at establishing the relation between each uniformly continuous FL-function and its normal form.

5.2.1 *Normal forms for L-valued FL-functions*

We start with the construction of normal forms for L-valued FL-functions where L denotes a support of some MV-algebra $\mathcal{L}$. Throughout this subsection we will deal solely with L- valued FL-functions and, therefore, we will omit the attribute "L-valued".

Let us take into the consideration the algebra of FL-functions

$$\mathcal{P}_L = \langle P_L, \oplus, \otimes, \neg, \mathbf{1}_L, \mathbf{0}_L \rangle,$$

which is an induced MV-algebra where each operation is defined pointwise and constants are the respective constant FL-functions (cf. also Example 2.9). The subscript L is used here to distinguish constant functions from constants in the support set. In some cases we will also employ the residuation (implication) operation, which is a derived operation in MV-algebra.

Let F be some fixed ultrafilter of $\mathcal{L}$ and $\cong_F$ be the corresponding congruence relation on $\mathcal{L}$. Moreover, let $\mathcal{L}|F$ be the corresponding factor algebra with elements denoted by $[v]$, so that $[v] = \{u \in L \mid u \cong_F v\}$. On account of Section 2.2, $\mathcal{L}|F$ is locally finite and linearly ordered. This means that for each element $v \in L$, $v \notin [\mathbf{0}]$, there exists an integer p so that $[pv] = [\mathbf{1}]$. If p is the least integer with that property then p is the *order* of v, $p = \mathrm{ord}(v)$, and $[\mathbf{0}] < [v] < [2v] \ldots < [pv]$. Recall that for an integer k, the notation kv is an abbreviation for the expression $(v \oplus \cdots \oplus v)$ k-times.

For each element $v \in L$, $v \notin [\mathbf{0}]$, we introduce the following FL-functions and call them *interval functions*:

$$I_k^v(x) = \begin{cases} \mathbf{1}, & \text{if } [kv] \leq [x] < [(k+1)v], \\ \mathbf{0}, & \text{otherwise} \end{cases} \tag{5.13}$$

where $0 \leq k \leq p-1$, $p = \mathrm{ord}(v)$, and

$$I_p^v(x) = \begin{cases} \mathbf{1}, & \text{if } x \in [\mathbf{1}], \\ \mathbf{0}, & \text{otherwise.} \end{cases}$$

Clearly,

$$\bigvee_{k=0}^{p} I_k^v(x) \equiv \mathbf{1}_L(x).$$

As can be seen, each function I_k^v, $0 \leq k \leq p-1$, is a characteristic function of the subset $\{x \mid [kv] \leq [x] < [(k+1)v]\}$. Hence, we will use it as the first part of the elementary conjunction and its negation as the first part of the elementary disjunction in the respective normal forms.

DEFINITION 5.5

Let $f(x_1, \ldots, x_n)$ be an FL-function and v be an element from L such that $v \notin [\mathbf{0}]$ and $p = \mathrm{ord}(v)$. The formula over $\mathcal{P}_L$

$$\mathrm{DNF}(x_1, \ldots, x_n) = \bigvee_{k_1=0}^{p} \cdots \bigvee_{k_n=0}^{p} (I_{k_1}^v(x_1) \otimes \cdots \otimes I_{k_n}^v(x_n) \otimes f(k_1 v, \ldots, k_n v)) \tag{5.14}$$

is called the MV-disjunctive normal form for $f(x_1, \ldots, x_n)$ and

$$\mathrm{CNF}(x_1, \ldots, x_n) = \bigwedge_{k_1=0}^{p} \cdots \bigwedge_{k_n=0}^{p} (\neg I_{k_1}^v(x_1) \oplus \cdots \oplus \neg I_{k_n}^v(x_n) \oplus f(k_1 v, \ldots, k_n v)) \tag{5.15}$$

is called the MV-conjunctive normal form for $f(x_1, \dots, x_n)$. We will call the disjunctive component $I^v_{k_1}(x_1) \otimes \cdots \otimes I^v_{k_n}(x_n) \otimes f(k_1 v, \dots, k_n v)$ the elementary conjunction and the conjunctive component $\neg I^v_{k_1}(x_1) \oplus \cdots \oplus \neg I^v_{k_n}(x_n) \oplus f(k_1 v, \dots, k_n v)$ the elementary disjunction.

It is not difficult to see that for $L = \{\mathbf{0}, \mathbf{1}\}$ the expressions (5.14) and (5.15) transform into classical Boolean normal forms.

Let us apply the identity $\neg a \oplus b = a \rightarrow b$ to the expression (5.15) of CNF and thus derive the following equivalent form

$$\mathrm{CNF}(x_1, \dots, x_n) = \bigwedge_{k_1=0}^{p} \cdots \bigwedge_{k_n=0}^{p} ((I^v_{k_1}(x_1) \otimes \cdots \otimes I^v_{k_n}(x_n)) \rightarrow f(k_1 v, \dots, k_n v)). \tag{5.16}$$

REMARK 5.3
The expressions $\mathrm{DNF}(x_1, \dots, x_n)$ and $\mathrm{CNF}(x_1, \dots, x_n)$ are used to denote the corresponding formulas over $\mathcal{P}_L$. When dealing with FL-functions represented by DNF and CNF, we will use the notation $f_{\mathrm{DNF}}(x_1, \dots, x_n)$ and $f_{\mathrm{CNF}}(x_1, \dots, x_n)$, respectively where f refers to the original FL-function. The dependence on the parameter v will be indicated in words.

Two facts have to be noticed when comparing the normal forms in Boolean and MV-algebras of functions. First, contrary to the case of Boolean functions, the FL-functions $f_{\mathrm{DNF}}(x_1, \dots, x_n)$ and $f_{\mathrm{CNF}}(x_1, \dots, x_n)$ are not generally equal to their origin $f(x_1, \dots, x_n)$. Second, the expressions for the normal forms in MV-algebra of functions depend on the parameter v, which, as will be shown below, indicates a proximity between the original FL-function and that represented by certain normal form.

EXAMPLE 5.3
Let the underlying MV-algebra $\mathcal{L}$ be the Łukasiewicz algebra $\mathcal{L}_{\mathrm{L}}$. The induced MV-algebra of functions $\mathcal{P}_{[0,1]}$ is then based on the set $P_{[0,1]} = \{f(x_1, \dots, x_n) \mid f : [0,1]^n \longrightarrow [0,1], \quad n \geq 0\}$. We will show in this example how the normal forms DNF and CNF look like and moreover, how the functions represented by them can be characterized. For the demonstration, consider the case when functions depend on one variable.

First, we will take into account the peculiarity of $\mathcal{L}_{\mathrm{L}}$ namely, that the support $[0,1]$ is a linearly ordered set and there exists only one ultrafilter coinciding with $\{1\}$. The consequence is that for each element $a \in [0,1]$ it holds that $[a] = \{a\}$ and thus, the parameter v can be chosen as any one which differs from 0. Let $v = 1/m$ where m is a positive integer such that $m = \mathrm{ord}(v)$. In this case, the interval functions can be expressed as follows:

$$I^m_k(x) = \begin{cases} 1, & \text{if } \frac{k}{m} \leq x < \frac{(k+1)}{m}, \\ 0, & \text{otherwise} \end{cases} \tag{5.17}$$

where $0 \le k \le m-1$, and

$$I_m^m(x) = \begin{cases} 1, & \text{if } x = 1, \\ 0, & \text{otherwise.} \end{cases} \tag{5.18}$$

If the normal forms include one variable then their representation is a simple reduction of (5.14)–(5.15) such that

$$\mathrm{DNF}(x) = \bigvee_{k=0}^{m} \left(I_k^m(x) \otimes f\left(\frac{k}{m}\right)\right),$$

$$\mathrm{CNF}(x) = \bigwedge_{k=0}^{m} \left(\neg I_k^m(x) \oplus f\left(\frac{k}{m}\right)\right).$$

Both these forms represent the same function (for the proof see below)

$$f_{\mathrm{DNF}}(x) = f_{\mathrm{CNF}}(x) = \begin{cases} f(\frac{k}{m}), & \text{if } \frac{k}{m} \le x < \frac{k+1}{m}, \quad 0 \le k \le m-1, \\ f(1), & \text{if } x = 1. \end{cases}$$

□

In general, the relation between DNF and CNF is established in the following lemma.

LEMMA 5.6
Let the normal forms $\mathrm{DNF}(x_1, \dots, x_n)$ *and* $\mathrm{CNF}(x_1, \dots, x_n)$, $n \ge 1$, *be constructed for the same FL-function* $f(x_1, \dots, x_n)$ *and use the same parameter* v *of the order* p. *Then they represent the same function.*

PROOF: It is sufficient to restrict ourselves to the case when $n = 1$. Suppose that $\mathrm{DNF}(x)$ and $\mathrm{CNF}(x)$ are constructed for some common FL-function $f(x)$ and are based on $v \in L$ so that $v \notin [\mathbf{0}]$ and $p = \mathrm{ord}(v)$. Choose an arbitrary $x \in L$ and verify that $f_{\mathrm{DNF}}(x) = f_{\mathrm{CNF}}(x)$. If $x \notin [\mathbf{1}]$ then find a unique integer k, $0 \le k \le p-1$, such that $[kv] \le [x] < [(k+1)v]$, otherwise let $k = p$. For the chosen k, the corresponding interval function $I_k^v(x)$ is equal to $\mathbf{1}$ whereas the other interval functions are equal to $\mathbf{0}$. This implies that $I_k^v(x) \otimes f(kv) = f(kv)$ and $\neg I_k^v(x) \oplus f(kv) = f(kv)$ while the other elementary conjunctions are equal to $\mathbf{0}$ and elementary disjunctions are equal to $\mathbf{1}$. Hence, $f_{\mathrm{DNF}}(x) = f_{\mathrm{CNF}}(x) = f(kv)$. □

It has already been mentioned that in general, the normal form (disjunctive or conjunctive) for the function $f(x_1, \dots, x_n)$ does not represent the same function as f. We can only prove that there is an effect of approximation. In order to speak about it correctly we will use the notion which plays the role of a distance in MV-algebra. According to the definition of C. C. Chang in [12], the FL-function

$$d(x, y) = (\neg x \otimes y) \oplus (\neg y \otimes x) \tag{5.19}$$

can be taken as a distance. The following properties justify well this choice (see [12]).

LEMMA 5.7
The following holds true for every $x, y, z \in L$.

(a) $d(x, x) = \mathbf{0}$,

(b) *if* $d(x, y) = \mathbf{0}$, *then* $x = y$,

(c) $d(x, y) = d(y, x)$,

(d) $d(x, z) \leq d(x, y) \oplus d(y, z)$.

DEFINITION 5.6
An FL-function $f(x_1, \dots, x_n)$ is uniformly continuous on L if for any $u \in L$, $[\mathbf{0}] < [u] < [\mathbf{1}]$, there exists $v \in L$, $[v] > [\mathbf{0}]$, such that $[d(x_1, y_1)] < [v], \dots, [d(x_n, y_n)] < [v]$ implies $[d(f(x_1, \dots, x_n), f(y_1, \dots, y_n))] < [u]$.

Note, that in the case of $L = [0, 1]$, the distance $d(x, y) = |x - y|$ and thus, the notion of uniform continuity coincides with the classical one for real valued real functions.

DEFINITION 5.7
Let $u \in L$ and $[u] < [\mathbf{1}]$. We say that an FL-function $g(x_1, \dots, x_n)$ u-approximates the FL-function $f(x_1, \dots, x_n)$, or in other words $g(x_1, \dots, x_n)$ approximates $f(x_1, \dots, x_n)$ with the accuracy u, if the inequality

$$[d(g(x_1, \dots, x_n), f(x_1, \dots, x_n))] \leq [u]$$

holds for all $(x_1, \dots, x_n) \in L^n$.

Obviously, the u-approximation is symmetrical.

THEOREM 5.3
Let an FL-function $f(x_1, \dots, x_n)$ be uniformly continuous on L. Then for any $u \in L$, $[\mathbf{0}] < [u] < [\mathbf{1}]$, there exists $v \in L$, $[v] > [\mathbf{0}]$, such that the function $f_{\mathrm{DNF}}(x_1, \dots, x_n)$ based on the parameter v u-approximates $f(x_1, \dots, x_n)$.

PROOF: For simplicity, let us consider the case of one variable. Fix an element $u \in L$, $[\mathbf{0}] < [u] < [\mathbf{1}]$, and, taking into account the uniform continuity of $f(x)$, find an element $v \in L$, $v > [\mathbf{0}]$, such that $[d(x, y)] < [v]$ implies $[d(f(x), f(y))] < [u]$. On the basis of v let us construct the normal form

$$\mathrm{DNF}(x) = \bigvee_{k=0}^{p} (I_k^v(x) \otimes f(kv))$$

where $p = \mathrm{ord}(v)$. We will show that $f_{\mathrm{DNF}}(x)$ is the function we are looking for.

Take an arbitrary element $x \in L$ and find an integer k, $0 \leq k < p$, such that either $[kv] \leq [x] < [(k+1)v]$, or $k = p$. It is true that $[d(kv, x)] < [v]$. Now evaluate the desired distance and see that

$$[d(f_{\mathrm{DNF}}(x), f(x))] = [d(f(kv), f(x))] < [u].$$

□

On the basis of Lemma 5.6, the following theorem can be established.

THEOREM 5.4
Let an FL-function $f(x_1, \dots, x_n)$ be uniformly continuous on L. Then for any $u \in L$, $[\mathbf{0}] < [u] < [\mathbf{1}]$, there exists $v \in L$, $[v] > [\mathbf{0}]$, such that the function $f_{\mathrm{CNF}}(x_1, \dots, x_n)$ based on the parameter v u-approximates $f(x_1, \dots, x_n)$.

5.2.2 *Normal forms for functions represented by formulas*

In this subsection, we will restrict ourselves to $L = [0,1]$ and deal with the functional system related to fuzzy propositional logic, i.e. $[0,1]$- valued FL-functions represented by formulas. By Corollary 5.2 on page 188 (McNaughton theorem) this system is equal to the set of piecewise linear functions defined on $[0,1]$. The same problems as above will be considered, namely construction of normal forms and examination of what can be expressed by them.

Let us denote by $PL_{[0,1]}$ the set of all piecewise linear (continuous) functions defined on $[0,1]$. Then, the MV-algebra

$$\mathcal{PL}_{[0,1]} = \langle PL_{[0,1]}, \oplus, \otimes, \neg, \mathbf{1}_{[0,1]}, \mathbf{0}_{[0,1]} \rangle$$

is the subalgebra of $\mathcal{P}_{[0,1]}$. The normal forms, which we aim at constructing should be formulas over $PL_{[0,1]}$. This implies that those introduced in Definition 5.5 are useless because they include symbols I_k^v, which stand for the non-continuous functions $I_k^v(x)$. Therefore, we need to modify the previously introduced normal forms (5.14)–(5.16) in all parts containing those symbols. For $L = [0,1]$, these functions have been denoted by $I_k^m(x)$ and defined by (5.17). To be able to obtain normal forms over $PL_{[0,1]}$, we need to approximate functions $I_k^m(x)$ by piecewise linear (continuous) ones. The resulting expressions will be substituted into DNF and CNF for the symbols $I_k^v(x)$.

Let $m > 0$ and $0 \leq k \leq m-1$ be integers. We will consider closed intervals $[\frac{k}{m}, \frac{k+1}{m}]$ as special cases of the convex polyhedra. Then, in accordance with the methodology described in Theorem 5.2 on page 188 we can establish a correspondence between each closed interval $[\frac{k}{m}, \frac{k+1}{m}]$ and the FL-function $\hat{I}_k^m(x)$ given by

$$\hat{I}_k^m(x) = \neg\left(-x + \frac{k}{m}\right)^* \wedge \neg\left(x - \frac{k+1}{m}\right)^*,$$

where the operations '$-$' and '$+$' are ordinary operations on reals and '$*$' has been defined by (5.1). Evidently, the right-hand side of this expression is not a formula of the propositional fuzzy logic but it can be transformed into it using

Lemma 5.5. Therefore, $\hat{I}_k^m(x) \in PL_{[0,1]}$ by Theorem 5.1. It is easy to see that $\hat{I}_k^m(x) = 1$ iff $x \in [\frac{k}{m}, \frac{k+1}{m}]$, $0 \leq k \leq m-1$.

For each function $\hat{I}_k^m(x)$, $0 \leq k \leq m-1$, $m > 0$, and each integer $r \geq 1$ let us introduce the function

$$(\hat{I}_k^m(x))^{rm} = \left(\neg\left(-x + \frac{k}{m}\right)^* \wedge \neg(x - \frac{k+1}{m}\Big)^*\right)^{rm}. \tag{5.20}$$

Obviously, $(\hat{I}_k^m(x))^{rm} \in PL_{[0,1]}$.

LEMMA 5.8
Let function $(\hat{I}_k^m(x))^{rm}$ be given by (5.20) where $0 \leq k \leq m-1$, $m > 0$, and $r \geq 1$. Let $x \in [0,1]$ be an arbitrary fixed element. Then

$$(\hat{I}_k^m(x))^{rm} = \begin{cases} 1 & \text{iff } x \in [\frac{k}{m}, \frac{k+1}{m}], \text{ and } 0 \leq k \leq m-1, \\ a & \text{iff } x \in (\frac{k}{m} - \frac{1}{rm}, \frac{k}{m}) \cup (\frac{k+1}{m}, \frac{k+1}{m} + \frac{1}{rm}) \text{ and } 1 \leq k \leq m-2, \\ & \quad \text{or } x \in (\frac{k+1}{m}, \frac{k+1}{m} + \frac{1}{rm}) \text{ and } k = 0, \\ & \quad \text{or } x \in (\frac{k}{m} - \frac{1}{rm}, \frac{k}{m}) \text{ and } k = m-1, \\ 0 & \text{otherwise} \end{cases}$$

where $0 < a < 1$.

The proof of this lemma is a technical exercise and is omitted.

COROLLARY 5.4
For each function $\hat{I}_k^m(x)$, $0 \leq k \leq m-1$, $m > 0$, and each integer $r \geq 1$, the function $(\hat{I}_k^m(x))^{rm}$ represented by (5.20) coincides with $I_k^m(x)$ for all $x \in [0,1]$ except for those belonging to the set

$$E_k^m = \begin{cases} (\frac{k}{m} - \frac{1}{rm}, \frac{k}{m}) \cup [\frac{k+1}{m}, \frac{k+1}{m} + \frac{1}{rm}) & \text{if } 1 \leq k \leq m-2, \\ [\frac{k+1}{m}, \frac{k+1}{m} + \frac{1}{rm}) & \text{if } k = 0, \\ (\frac{k}{m} - \frac{1}{rm}, \frac{k}{m}) & \text{if } k = m-1. \end{cases}$$

By virtue of Lemma 5.8 and Corollary 5.4, the function $(\hat{I}_k^m(x))^{rm}$ approximates $I_k^m(x)$ in the sense that they differ on the set E_k^m, which consists of one or two intervals of arbitrary small length. Of course, such kind of approximation is different from the u-approximation introduced earlier, but this sufficiently corresponds to our purposes.

We are now able to give new definitions of the disjunctive and conjunctive normal forms for the piecewise linear functions from $PL_{[0,1]}$.

DEFINITION 5.8

Let $f(x_1, \dots, x_n)$ be a piecewise linear function from $PL_{[0,1]}$, $m > 0$ and $r \geq 2$ be integers. The following is the MV-disjunctive normal form for $f(x_1, \dots, x_n)$

$$\mathrm{DNF}(x_1, \dots, x_n) = \bigvee_{k_1=0}^{m-1} \cdots \bigvee_{k_n=0}^{m-1} \Big((\hat{I}^m_{k_1}(x_1))^{rm} \otimes \cdots \otimes (\hat{I}^m_{k_n}(x_n))^{rm} \otimes \otimes f\left(\frac{k_1}{m}, \dots, \frac{k_n}{m}\right)\Big), \quad (5.21)$$

and the MV-conjunctive normal form for $f(x_1, \dots, x_n)$

$$\mathrm{CNF}(x_1, \dots, x_n) = \bigwedge_{k_1=0}^{m-1} \cdots \bigwedge_{k_n=0}^{m-1} \Big(\neg(\hat{I}^m_{k_1}(x_1))^{rm} \oplus \cdots \oplus \neg(\hat{I}^m_{k_n}(x_n))^{rm} \oplus \oplus f\left(\frac{k_1}{m}, \dots, \frac{k_n}{m}\right)\Big). \quad (5.22)$$

Using the identity $\neg a \oplus b = a \rightarrow b$ we can equivalently obtain

$$\mathrm{CNF}(x_1, \dots, x_n) = \bigwedge_{k_1=0}^{m-1} \cdots \bigwedge_{k_n=0}^{m-1} \Big((\hat{I}^m_{k_1}(x_1))^{rm} \otimes \cdots \otimes (\hat{I}^m_{k_n}(x_n))^{rm} \rightarrow \rightarrow f\left(\frac{k_1}{m}, \dots, \frac{k_n}{m}\right)\Big). \quad (5.23)$$

It can be seen from the definition that both the disjunctive as well as the conjunctive normal forms depend on the parameters m and r. The first parameter m controls proximity between $f(x_1, \dots, x_n)$ and $f_{\mathrm{DNF}}(x_1, \dots, x_n)$ (or $f_{\mathrm{CNF}}(x_1, \dots, x_n)$) and the second parameter r controls local fitness between them. For the proof of the following approximation theorem, it is sufficient to put $r = 2$.

THEOREM 5.5

Let $f(x_1, \dots, x_n)$ be a piecewise linear function from $PL_{[0,1]}$. Then for any ε, $0 < \varepsilon < 1$, there exists integer $m > 0$ such that the function $f_{\mathrm{DNF}}(x_1, \dots, x_n)$ based on the parameters m and $r = 2$ approximates $f(x_1, \dots, x_n)$ with the accuracy ε.

PROOF: The proof is again done only for the case of one variable as it sufficiently illustrates the underlying idea. For a piecewise linear function $f(x)$ from $PL_{[0,1]}$ and any fixed ε, $0 < \varepsilon < 1$, find an integer $m > 0$ so that whenever $|x - y| < \frac{3}{2m}$ the inequality $|f(x) - f(y)| \leq \varepsilon$ holds true. The existence of such m follows from the fact that any piecewise linear function from $PL_{[0,1]}$ is uniformly continuous on a closed interval. The disjunctive normal form for $f(x)$ based on the parameters m and $r = 2$ is given by the expression

$$\mathrm{DNF}(x) = \bigvee_{k=0}^{m-1} \left((\hat{I}^m_k(x))^{2m} \otimes f\left(\frac{k}{m}\right)\right).$$

It remains to estimate the difference $|f(x) - f_{\text{DNF}}(x)|$ for different $x \in [0,1]$ and to check whether the desired inequality is valid.

(a) Let $x \in [0, \frac{1}{2m})$. By virtue of Lemma 5.8 $(\hat{I}_0^m(x))^{2m} = 1$ while other functions $(\hat{I}_k^m(x))^{2m} = 0$, $1 \leq k \leq m-1$. Hence,

$$(\hat{I}_0^m(x))^{2m} \otimes f(0) = f(0)$$

and consequently, $f_{\text{DNF}}(x) = f(0)$. Therefore,

$$|f(x) - f_{\text{DNF}}(x)| = |f(x) - f(0)| \leq \varepsilon.$$

(b) Let $x \in [1 - \frac{1}{2m}, 1]$. By virtue of Lemma 5.8 $(\hat{I}_{m-1}^m(x))^{2m} = 1$ while other functions $(\hat{I}_k^m(x))^{2m} = 0$, $0 \leq k \leq m-2$. Hence,

$$(\hat{I}_{m-1}^m(x))^{2m} \otimes f(\frac{m-1}{m}) = f(\frac{m-1}{m}),$$

which implies that $f_{\text{DNF}}(x) = f(\frac{m-1}{m})$. Further, if $x = 1$ then $(\hat{I}_{m-1}^m(1))^{2m} = 1$ and $f_{\text{DNF}}(1) = f(\frac{m-1}{m})$. Thus, for each $x \in [1 - \frac{1}{2m}, 1]$ it holds true that

$$|f(x) - f_{\text{DNF}}(x)| = |f(x) - f(\frac{m-1}{m})| \leq \varepsilon.$$

(c) Let $x \in [\frac{k}{m} - \frac{1}{2m}, \frac{k}{m})$ for some k, $1 \leq k \leq m-1$. Taking into account Lemma 5.8 we infer $(\hat{I}_{k-1}^m(x))^{2m} = 1$ and $(\hat{I}_k^m(x))^{2m} \geq 0$ so that

$$(\hat{I}_{k-1}^m(x))^{2m} \otimes f(\frac{k-1}{m}) = f(\frac{k-1}{m})$$

and

$$(\hat{I}_k^m(x))^{2m} \otimes f(\frac{k}{m}) \in [0, f(\frac{k}{m})).$$

Other functions $(\hat{I}_l^m(x))^{2m} = 0$, $l \neq k, k-1$. Hence, $f_{\text{DNF}}(x) = f(\frac{k-1}{m})$ if $f(\frac{k-1}{m}) \geq f(\frac{k}{m})$ or $f_{\text{DNF}}(x) \in [f(\frac{k-1}{m}), f(\frac{k}{m}))$ otherwise. In both cases, we have $|x - \frac{k-1}{m}| < \frac{1}{m}$ as well as $|x - \frac{k}{m}| < \frac{1}{m}$, which together with the observations about $f_{\text{DNF}}(x)$ implies $|f(x) - f_{\text{DNF}}(x)| \leq \varepsilon$.

(d) Let $x \in [\frac{k}{m}, \frac{k}{m} + \frac{1}{2m})$ for some k, $1 \leq k \leq m-1$. Similarly as above, only two functions, namely $(\hat{I}_{k-1}^m(x))^{2m}$ and $(\hat{I}_k^m(x))^{2m}$ are positive on x. In fact, $(\hat{I}_k^m(x))^{2m} = 1$ and $(\hat{I}_{k-1}^m(x))^{2m} > 0$ so that

$$(\hat{I}_k^m(x))^{2m} \otimes f(\frac{k}{m}) = f(\frac{k}{m})$$

and

$$\hat{I}_{k-1}^m(x))^{2m} \otimes f(\frac{k-1}{m}) \in (0, f(\frac{k-1}{m})].$$

Hence, $f_{\mathrm{DNF}}(x) = f(\frac{k}{m})$ if $f(\frac{k}{m}) \geq f(\frac{k-1}{m})$ or $f_{\mathrm{DNF}}(x) \in [f(\frac{k}{m}), f(\frac{k-1}{m})]$ otherwise. In both cases we have $|x - \frac{k-1}{m}| < \frac{3}{2m}$ as well as $|x - \frac{k}{m}| < \frac{1}{m}$, which together with the observations about $f_{\mathrm{DNF}}(x)$ implies $|f(x) - f_{\mathrm{DNF}}(x)| \leq \varepsilon$. □

As will be seen later, the functions f_{DNF} and f_{CNF} represented by (5.21)–(5.22) and based on the same parameters m and r are not generally equal. Nevertheless, applying the same technique as in the proof of the previous theorem one can prove the following.

THEOREM 5.6
Let $f(x_1, \dots, x_n)$ be a piecewise linear function from $PL_{[0,1]}$. Then for any ε, $0 < \varepsilon < 1$, there exists integer $m > 0$ such that the function $f_{\mathrm{CNF}}(x_1, \dots, x_n)$ based on the parameters m and $r = 2$ approximates $f(x_1, \dots, x_n)$ with the accuracy ε.

The following corollary concerns the problem of transformation of any formula of propositional fuzzy logic into one of the normal forms.

COROLLARY 5.5
For any formula of the propositional fuzzy logic and any ε, $0 < \varepsilon < 1$, there exists a formula over $PL_{[0,1]}$ of the form of DNF *(*CNF*) such that the functions represented by these formulas ε-approximate each other.*

The proof of this corollary is based on Theorem 5.5 (Theorem 5.6) and the fact that each formula of propositional fuzzy logic represents a piecewise linear function from $PL_{[0,1]}$.

In other words, this corollary states that each formula of propositional fuzzy logic can be transformed into the form of DNF (CNF) by means of the function represented by the former. However, during this transformation some precision is lost.

Finally, we will establish the relation between the functions f_{DNF} and f_{CNF} represented by normal forms based on the same parameters m and $r = 2$.

THEOREM 5.7
Let $f(x_1, \dots, x_n) \in PL_{[0,1]}$. Furthermore, suppose that $f_{\mathrm{DNF}}(x_1, \dots, x_n)$ and $f_{\mathrm{CNF}}(x_1, \dots, x_n)$ be functions represented by normal forms for f, which are given by the expressions (5.21),(5.23) and based on the parameters $m > 0$ and $r = 2$. Then for all $x \in [0, 1]$

$$f_{\mathrm{CNF}}(x) \leq f_{\mathrm{DNF}}(x)$$

and

$$f_{\mathrm{DNF}}(x) - f_{\mathrm{CNF}}(x) \leq \begin{cases} 0, & \text{if } x \in [0, \frac{1}{2m}) \cup [1 - \frac{1}{2m}, 1], \\ \left|f\left(\frac{k}{m}\right) - f\left(\frac{k-1}{m}\right)\right|, & \text{if } x \in [\frac{k}{m} - \frac{1}{2m}, \frac{k}{m} + \frac{1}{2m}) \\ & \text{where } 1 \leq k \leq m-1. \end{cases}$$

PROOF: For the convenience, we have chosen the second expression for CNF based on the implication. Let the conditions of the theorem be valid. Choose an arbitrary $x \in [0,1]$ and consider the same cases for it as in the proof of Theorem 5.5.

(a) If $x \in [0, \frac{1}{2m})$ then by virtue of Lemma 5.8 $(\hat{I}_0^m(x))^{2m} = 1$ so that $(\hat{I}_0^m(x))^{2m} \otimes f(0) = f(0)$ and $(\hat{I}_0^m(x))^{2m} \to f(0) = f(0)$. Hence,

$$f_{\mathrm{DNF}}(x) = f_{\mathrm{CNF}}(x) = f(0)$$

and the conclusion of the theorem is obviously valid.

(b) If $x \in [1 - \frac{1}{2m}, 1]$ then by virtue of Lemma 5.8 $(\hat{I}_{m-1}^m(x))^{2m} = 1$ so that $(\hat{I}_{m-1}^m(x))^{2m} \otimes f(\frac{m-1}{m}) = f(\frac{m-1}{m})$ and $(\hat{I}_{m-1}^m(x))^{2m} \to f(\frac{m-1}{m}) = f(\frac{m-1}{m})$. Hence,

$$f_{\mathrm{DNF}}(x) = f_{\mathrm{CNF}}(x) = f(\frac{m-1}{m}).$$

Furthermore, $(\hat{I}_{m-1}^m(1))^{2m} = 1$ then $f_{\mathrm{DNF}}(1) = f_{\mathrm{CNF}}(1) = f(\frac{m-1}{m})$ and, similarly as above, the conclusion of the theorem is valid.

(c) Let $x \in [\frac{k}{m} - \frac{1}{2m}, \frac{k}{m})$ for some k, $1 \le k \le m-1$. Again, taking into account Lemma 5.8 we can infer two facts: first, $(\hat{I}_{k-1}^m(x))^{2m} = 1$ so that $(\hat{I}_{k-1}^m(x))^{2m} \otimes f(\frac{k-1}{m}) = f(\frac{k-1}{m})$ and $(\hat{I}_{k-1}^m(x))^{2m} \to f(\frac{k-1}{m}) = f(\frac{k-1}{m})$, and second, $(\hat{I}_k^m(x))^{2m} \ge 0$ so that $(\hat{I}_k^m(x))^{2m} \otimes f(\frac{k}{m}) \in [0, f(\frac{k}{m}))$ and $(\hat{I}_k^m(x))^{2m} \to f(\frac{k}{m}) \in (f(\frac{k}{m}), 1]$. Hence, if $f(\frac{k-1}{m}) \ge f(\frac{k}{m})$ then

$$f_{\mathrm{DNF}}(x) = f(\frac{k-1}{m}) \quad \text{and} \quad f_{\mathrm{CNF}}(x) \in (f(\frac{k}{m}), f(\frac{k-1}{m})]$$

otherwise

$$f_{\mathrm{DNF}}(x) \in [f(\frac{k-1}{m}), f(\frac{k}{m})) \quad \text{and} \quad f_{\mathrm{CNF}}(x) = f(\frac{k-1}{m}).$$

In both cases we have $f_{\mathrm{CNF}}(x) \le f_{\mathrm{DNF}}(x)$ and $(f_{\mathrm{DNF}}(x) - f_{\mathrm{CNF}}(x)) < |f(\frac{k}{m}) - f(\frac{k-1}{m})|$.

(d) Let $x \in [\frac{k}{m}, \frac{k}{m} + \frac{1}{2m})$ for some k, $1 \le k \le m-1$. As above, only two functions namely, $(\hat{I}_{k-1}^m(x))^{2m}$ and $(\hat{I}_k^m(x))^{2m}$ are positive on x. In fact, $(\hat{I}_k^m(x))^{2m} = 1$ so that $(\hat{I}_k^m(x))^{2m} \otimes f(\frac{k}{m}) = f(\frac{k}{m})$ and $(\hat{I}_k^m(x))^{2m} \to f(\frac{k}{m}) = f(\frac{k}{m})$ while $(\hat{I}_{k-1}^m(x))^{2m} > 0$ so that $(\hat{I}_{k-1}^m(x))^{2m} \otimes f(\frac{k-1}{m}) \in (0, f(\frac{k-1}{m})]$ and $(\hat{I}_{k-1}^m(x))^{2m} \to f(\frac{k-1}{m}) \in [f(\frac{k-1}{m}), 1)$. Hence, if $f(\frac{k}{m}) \ge f(\frac{k-1}{m})$ then

$$f_{\mathrm{DNF}}(x) = f(\frac{k}{m}) \quad \text{and} \quad f_{\mathrm{CNF}}(x) \in [f(\frac{k-1}{m}), f(\frac{k}{m})]$$

otherwise

$$f_{\mathrm{DNF}}(x) \in [f(\frac{k}{m}), f(\frac{k-1}{m})] \quad \text{and} \quad f_{\mathrm{CNF}}(x) = f(\frac{k}{m})$$

In both cases we have $f_{\mathrm{CNF}}(x) \le f_{\mathrm{DNF}}(x)$ and $(f_{\mathrm{DNF}}(x) - f_{\mathrm{CNF}}(x)) \le |f(\frac{k}{m}) - f(\frac{k-1}{m})|$. □

5.3 FL-Relations and Their Connection with Formulas of Predicate Fuzzy Logic

What can be expressed by formulas of predicate fuzzy logic? It is widely declared that fuzzy logic is able to express different kinds of dependencies in e.g., control theory, operation research etc., especially when basic events are ill defined. Thus, on the theoretical level it is interesting to investigate what are the backgrounds for such assertions and in the case they are sound, what are the canonical expressions for description of the represented dependencies. Apparently, propositional fuzzy logic is not sufficient for this purpose. Therefore, we must pass to the predicate one where we again face the following two problems: first, what is the class of the represented dependencies and second, how is it possible to represent any element from this class using some standard (canonical or normal) formula. In this section, we will give a solution to both problems. As a consequence, we will obtain a non-direct transformation of any formula to another one in a normal form. As we will see, the latter is in some sense equivalent to the initial one.

Thus obtained two-way connection between formulas of predicate fuzzy logic and truth functions represented by them, together with the completeness theorem will give us an opportunity to judge the consistency of special theories. The most interesting consequence of such approach is presented in Theorem 5.12, which states that the analogy of classical consistency theorem is not valid in fuzzy logic.

Throughout this and subsequent sections, we will deal with some fixed language J of fuzzy predicate logic, which does not contain functional symbols. By formula we mean a formula of fuzzy predicate logic. Furthermore, by piecewise linear function we mean a function defined in Definition 5.4 on page 183.

5.3.1 FL-Relations and Their Representation by Formulas

We will prove the representation theorem for what we call FL-relations, which is based on Corollary 5.2 on page 188 (McNaughton theorem). At first, we recall what is understood by relations in connection with fuzzy logic (cf. also Definition 1.2 on page 9).

DEFINITION 5.9
Let D be a nonempty set of objects and L be a support of a complete MV-algebra. An n-ary fuzzy logic relation R on D or an FL-relation, is a fuzzy set $R \underset{\sim}{\subset} D^n$, $n \geq 0$, which is given by the membership function $R(x_1, \ldots, x_n)$ defined on the set D^n and taking values from L.

Note that according to this definition, we identify an FL-relation with its membership function to be able to work with functions in the sequel. Furthermore, we will consider the case $L = [0, 1]$ and the Łukasiewicz MV-algebra $\mathcal{L}_{\mathrm{L}}$.

An expected problem arises: how is it possible to represent an FL-relation and, are the formulas of fuzzy predicate logic suitable for this. To investigate this problem, the class of FL-relations, which consists of *relations represented by formulas* of fuzzy predicate logic will be introduced. This class forms a functional system related to fuzzy predicate logic. A precise definition of an FL-relation represented by a formula can be given using analogous inductive procedure as in Definition 5.2 on page 182. It is also based on the definition of an interpretation of a formula (to remember the latter concept, see Section 4.3).

DEFINITION 5.10
Let $A(x_1, \dots, x_n)$ be a formula of fuzzy predicate logic with free variables $x_1, \dots, x_n$ in a language J. Let $\mathcal{D}$ be a structure for J with a domain D and $e : FV(A) \longrightarrow D$ be an evaluation of its free variables. An n-ary FL-relation $R_A(x_1, \dots, x_n)$ on D is said to be represented by the formula $A(x_1, \dots, x_n)$ with respect to the structure $\mathcal{D}$ and all the possible evaluations e if it establishes a correspondence $R_A : D^n \longrightarrow [0,1]$ such that each n-tuple $(d_1, \dots, d_n) \in D^n$, $d_i = e(x_i)$, $1 \le i \le n$, is assigned a value

$$R_A(d_1, \dots, d_n) = \mathcal{D}(A_{x_1,\dots,x_n}[\mathbf{d}_1, \dots, \mathbf{d}_n]).$$

Since we work with the fixed language J of fuzzy predicate logic, in what following we will use the term "structure" instead of "structure for J".

Let us stress, that we obtain an FL-relation represented by a formula in connection with some structure. This means that all parts of a formula, except for its free variables, are uniquely interpreted.

The following theorem establishes a correspondence between FL-relations represented by formulas and the general ones. It describes the functional structure of the former and thus leads to the conjecture that the general class of FL-relations includes that of represented by formulas. We leave this problem unsolved in this book.

THEOREM 5.8

(a) *Let $A(x_1, \dots, x_n)$ be a quantifier-free formula of fuzzy predicate logic with free variables $x_1, \dots, x_n$. Let $B_1, \dots, B_m$ be a list of all pairwise different atomic subformulas of A not being logical constants. Without loss of generality we will suppose that none of the subformulas $B_1, \dots, B_m$ contains closed terms. Then there exists a piecewise linear function $l(p_1, \dots, p_m)$ defined on $[0,1]$ such that in any structure $\mathcal{D}$ the FL-relation R_A represented by the formula A is given by the expression*

$$R_A(x_1, \dots, x_n) = l(f_{B_1}(x_1^1, \dots, x_{n_1}^1), \dots, f_{B_m}(x_1^m, \dots, x_{n_m}^m)) \quad (5.24)$$

where $f_{B_j}(x_1^j, \dots, x_{n_j}^j)$, $1 \le n_j \le n$, denotes the membership function of an FL-relation, which is represented by the atomic subformula B_j, $1 \le j \le m$, w.r.t. to $\mathcal{D}$. Moreover, $\{x_1^1, \dots, x_{n_1}^1, \dots, x_1^m, \dots, x_{n_m}^m\} = \{x_1, \dots, x_n\}$.

(b) *Conversely, let some n-ary FL-relation $R(x_1, \dots, x_n)$ on D be represented by the expression (5.24) where $l(p_1, \dots, p_m) : [0,1]^m \longrightarrow [0,1]$ is a piecewise linear function and $f_{B_j}(x_1^j, \dots, x_{n_j}^j)$ are n_j-ary FL- relations on D, $n_j \ge 1$, $1 \le j \le m$. Then there exists a quantifier-free formula $A(x_1, \dots, x_n)$ of fuzzy predicate logic and a structure $\mathcal{D}$ with the support D such that $R(x_1, \dots, x_n)$ is represented by A w.r.t. to $\mathcal{D}$.*

PROOF: (a) Let $A(x_1, \ldots, x_n)$ and $B_1, \ldots, B_m$ be given in accordance with the assumptions and let $p_1, \ldots, p_m$ be propositional variables. Replace each subformula B_j in A, $1 \leq j \leq m$, by the propositional variable p_j. The result is a formula $A'(p_1, \ldots, p_m)$ of fuzzy propositional logic. In accordance with Theorem 5.1 it represents a piecewise linear function, which we denote by $l(p_1, \ldots, p_m)$.

Choose some structure $\mathcal{D}$ and find membership functions of the FL-relations $f_{B_1}(x_1^1, \ldots, x_{n_1}^1), \ldots, f_{B_m}(x_1^m, \ldots, x_{n_m}^m)$ represented by the subformulas $B_1, \ldots, B_m$, respectively in such a way that $\{x_1^1, \ldots, x_{n_1}^1, \ldots, x_1^m, \ldots, x_{n_m}^m\} = \{x_1, \ldots, x_n\}$. Observe that even if some functions are denoted by different symbols, they may be equal. This may happen if different subformulas contain the same predicate symbol. Replace each variable p_j in the function l by $f_{B_j}(x_1^j, \ldots, x_{n_j}^j)$, $1 \leq j \leq m$, and thus obtain the FL-relation R_A represented by the formula A.

(b) Now, let $R(x_1, \ldots, x_n) : D^n \longrightarrow [0, 1]$ be an n-ary FL-relation represented by (5.24). In accordance with Theorem 5.2, the piecewise linear function $l(p_1, \ldots, p_m)$ can be represented by some formula $A'(p_1, \ldots, p_m)$ of fuzzy propositional logic. Now, choose a structure $\mathcal{D}$ with the support D such that each FL-relation $f_{B_j}(x_1^j, \ldots, x_{n_j}^j)$ on D is assigned an n_j-ary predicate symbol P_j so that different relations are assigned different symbols. Replace each occurrence of a propositional variable p_j in the formula $A'(p_1, \ldots, p_m)$ by the subformula $P_j(x_1^j, \ldots, x_{n_j}^j)$ and thus, obtain a formula A of fuzzy predicate logic. It remains to observe that $R(x_1, \ldots, x_n)$ is represented by A w.r.t. to $\mathcal{D}$. □

REMARK 5.4

It is worth noticing that the piecewise linear function l, which corresponds to the formula A, does not depend on any structure.

On the basis of Theorem 5.8 we can speak about some properties of FL-relations represented by formulas with respect to the given structure.

COROLLARY 5.6

Suppose that D being a support of a structure $\mathcal{D}$ is supplied with a topology $\mathcal{T}$. Moreover, let all the predicate symbols in predicate fuzzy logic be assigned FL-relations on D with continuous membership functions w.r.t. $\mathcal{T}$. Then, any quantifier-free formula represents an FL-relation on D w.r.t. $\mathcal{D}$ with continuous membership function.

PROOF: This directly follows from the expression (5.24) for an FL-relation represented by a formula. □

It seems to be true that an opposite statement saying that any FL-relation with continuous membership function can be represented by some formula of predicate fuzzy logic is valid. We leave this statement as our second conjecture. In connection with this, it can be proved that any FL-relation with continuous membership function can be approximated by some FL-relation represented by a certain formula of fuzzy predicate logic.

5.3.2 *Normal Forms and Approximation Theorems*

In this subsection we introduce the disjunctive and conjunctive normal forms and prove approximation theorems for uniformly continuous FL-relations. In accordance with the suggested conception concerning the structure of the normal forms, in the definition below, we will preserve two parts in each elementary construction (conjunction or disjunction): first, the characterization (positive or negative) of a certain set, which is given by a conjunction of atomic subformulas, and second, the average value of the respective FL-relation, which is represented by a certain atomic formula.

DEFINITION 5.11
Let $P_1, \ldots, P_k$ be unary predicate symbols and $E_{i_1 \ldots i_n}$, $1 \le i_j \le k$, $1 \le j \le n$, $n \ge 1$, be some atomic formulas. The following formulas of fuzzy predicate logic are called the MV-disjunctive normal form

$$\mathrm{DNF}(x_1, \ldots, x_n) = \bigvee_{i_1=1}^{k} \cdots \bigvee_{i_n=1}^{k} (P_{i_1}(x_1) \,\&\, \cdots \,\&\, P_{i_n}(x_n) \,\&\, E_{i_1 \ldots i_n}) \tag{5.25}$$

and the MV-conjunctive normal form

$$\mathrm{CNF}(x_1, \ldots, x_n) = \bigwedge_{i_1=1}^{k} \cdots \bigwedge_{i_n=1}^{k} (\neg P_{i_1}(x_1) \nabla \cdots \nabla \neg P_{i_n}(x_n) \nabla E_{i_1 \ldots i_n}). \tag{5.26}$$

As in the previous section, the expression (5.26) can be rewritten into the following one based on implication, namely

$$\mathrm{CNF}(x_1, \ldots, x_n) = \bigwedge_{i_1=1}^{k} \cdots \bigwedge_{i_n=1}^{k} (P_{i_1}(x_1) \,\&\, \cdots \,\&\, P_{i_n}(x_n) \Rightarrow E_{i_1 \ldots i_n}). \tag{5.27}$$

In the sequel, we will use the shorts DNF and CNF. Contrary to the case of propositional fuzzy logic where we suggested the expressions for the normal forms as algebraic formulas, the expressions (5.25)–(5.27) are formulas of fuzzy predicate logic. This explains the occurrence of atomic subformulas $E_{i_1 \ldots i_n}$ instead of the values of the membership functions of FL-relations, which one could expect by the analogy with propositional case (cf. (5.21)– (5.22)). Here, we aim at proving that any uniformly continuous FL-relation can be approximated by each of the normal forms (5.25) or (5.26), respectively. In what follows we suppose a support D of a structure $\mathcal{D}$ to be a compact uniform topological space. Let ρ be a pseudometrics, which belongs to the gage of the uniformity on D, i.e. to the family of all pseudometrics which are uniformly continuous on $D \times D$ (cf. [53]).

DEFINITION 5.12
An n-ary FL-relation $R \mathrel{\underset{\sim}{\subset}} D^n$ is uniformly continuous on D^n if for any ε, $0 < \varepsilon < 1$, there exists a positive number r, such that if $\rho(x_{11}, x_{21}) < r, \ldots, \rho(x_{1n}, x_{2n}) < r$ then $|R(x_{11}, \ldots, x_{1n}) - R(x_{21}, \ldots, x_{2n})| < \varepsilon$.

Let us stress that this notion coincides with the analogous classical notion for real valued functions defined on a compact uniform space.

DEFINITION 5.13
An FL-relation $S(x_1, \dots, x_n)$ on D is said to ε- approximate another FL-relation $R(x_1, \dots, x_n)$ on D if

$$|S(x_1, \dots, x_n) - R(x_1, \dots, x_n)| \leq \varepsilon$$

is true for all $(x_1, \dots, x_n) \in D^n$.

THEOREM 5.9
Let an FL-relation $R \underset{\sim}{\subset} D^n$ be uniformly continuous on D^n. Then for any ε, $0 < \varepsilon < 1$, there exist a formula DNF$(x_1, \dots, x_n)$ *and a structure $\mathcal{D}$ with the support D such that the FL-relation represented by* DNF *w.r.t. $\mathcal{D}$ ε-approximates $R(x_1, \dots, x_n)$.*

PROOF: Choose ρ from the gage of the uniformity on D and for some ε, $0 < \varepsilon < 1$, find a positive number r such that $\rho(x_{11}, x_{21}) < r, \dots, \rho(x_{1n}, x_{2n}) < r$ implies $|R(x_{11}, \dots, x_{1n}) - R(x_{21}, \dots, x_{2n})| < \varepsilon$. Since D is a compact uniform space it is covered by a finite union $\bigcup_i^k S_i$ of open balls with the diameter smaller than r. Suppose that each ball S_i has its center at the point a_i, $1 \leq i \leq k$.

In order to construct DNF given by (5.25) we only need to specify the atomic subformulas $E_{i_1 \dots i_n}$. Let $E_{i_1 \dots i_n}$ be a logical constant corresponding to the truth value $R(a_{i_i}, \dots a_{i_n})$ where $1 \leq i_j \leq k$, $1 \leq j \leq n$.

Choose a structure $\mathcal{D}$ with the support D where each predicate symbol P_i, $1 \leq i \leq k$, from (5.25) is assigned an FL-relation $P_{i,D} \underset{\sim}{\subset} D$ defined by

$$P_{i,D}(x) = \begin{cases} 1, & \text{if } x \in S_j, \\ 0, & \text{otherwise.} \end{cases}$$

Thus, we come to the FL-relation $R_{\text{DNF}}(x_1, \dots, x_n)$ represented by (5.25) w.r.t. $\mathcal{D}$, which has the explicit form

$$R_{\text{DNF}}(x_1, \dots, x_n) = \bigvee_{i_1=1}^{k} \cdots \bigvee_{i_n=1}^{k} (P_{i_1,D}(x_1) \otimes \cdots \otimes P_{i_n,D}(x_n) \otimes R(a_{i_i}, \dots a_{i_n})). \tag{5.28}$$

To verify the ε-approximation property, choose an arbitrary point $(d_1, \dots d_n) \in D^n$ and evaluate the distance $|R(d_1, \dots, d_n) - R_{\text{DNF}}(d_1, \dots, d_n)|$. At first, observe that $R_{\text{DNF}}(d_1, \dots, d_n) = R(a_{l_1}, \dots a_{l_n})$ where

$$\begin{aligned} R(a_{l_1}, \dots a_{l_n}) &= \max_{(i_1, \dots, i_n)} \{R(a_{i_1}, \dots a_{i_n}) \mid P_{i_1,D}(d_1) = 1, \dots, P_{i_n,D}(d_n) = 1\} = \\ &= \max_{(i_1, \dots, i_n)} \{R(a_{i_1}, \dots a_{i_n}) \mid d_1 \in S_{i_1}, \dots, d_n \in S_{i_n}\}. \end{aligned} \tag{5.29}$$

Therefore,

$$|R(d_1, \dots, d_n) - R_{\text{DNF}}(d_1, \dots, d_n)| = |R(d_1, \dots, d_n) - R(a_{l_1}, \dots a_{l_n})|.$$

Since $d_i \in S_i$ and thus $\rho(d_i, a_{l_i}) < r$, $1 \le i \le n$, it follows that

$$|R(d_1, \dots, d_n) - \mathrm{DNF}(d_1, \dots, d_n)| < \varepsilon,$$

which has been required. □

In predicate fuzzy logic, the formulas $\mathrm{DNF}(x_1, \dots, x_n)$ and $\mathrm{CNF}(x_1, \dots, x_n)$ do not necessarily represent the same FL-relations, even in the same structure. Thus, the approximation property for FL-relations, which are represented by $\mathrm{CNF}(x_1, \dots, x_n)$, must be proved as well.

THEOREM 5.10
Let an FL-relation $R \underset{\sim}{\subseteq} D^n$ be uniformly continuous on D^n. Then for any ε, $0 < \varepsilon < 1$, there exist a formula $\mathrm{CNF}(x_1, \dots, x_n)$ and a structure $\mathcal{D}$ with the support D such that the FL-relation represented by CNF w.r.t. $\mathcal{D}$ ε-approximates $R(x_1, \dots, x_n)$.

PROOF: For the given R and ε, $0 < \varepsilon < 1$, we will use the same covering $\bigcup_i^k S_i$ of D by open balls as in the previous proof. Choose the second expression (5.27) for CNF based on implication and specify the atomic subformulas $E_{i_1 \dots i_n}$ to be logical constants for the truth values $R(a_{i_i}, \dots a_{i_n})$, respectively where $1 \le i_j \le k$ and $1 \le j \le n$. Furthermore, choose the same structure $\mathcal{D}$ as in the proof of Theorem 5.9. Thus, we obtain the FL-relation $R_{\mathrm{CNF}}(x_1, \dots, x_n)$ represented by (5.27) w.r.t. $\mathcal{D}$, which has the explicit form

$$R_{\mathrm{CNF}}(x_1, \dots, x_n) = \bigwedge_{i_1=1}^{k} \cdots \bigwedge_{i_n=1}^{k} (P_{i_1,D}(x_1) \otimes \cdots \otimes P_{i_n,D}(x_n) \to R(a_{i_i}, \dots a_{i_n})). \tag{5.30}$$

It remains to check the property of the ε-approximation. Choose an arbitrary point $(d_1, \dots d_n) \in D^n$ and evaluate the distance

$$|R(d_1, \dots, d_n) - R_{\mathrm{CNF}}(d_1, \dots, d_n)|.$$

As a first step, we obtain that $R_{\mathrm{CNF}}(d_1, \dots, d_n) = R(a_{l_1}, \dots a_{l_n})$ where

$$\begin{aligned} R(a_{l_1}, \dots a_{l_n}) &= \min_{(i_1, \dots, i_n)} \{R(a_{i_1}, \dots a_{i_n}) \mid P_{i_1,D}(d_1) = 1, \dots, P_{i_n,D}(d_n) = 1\} = \\ &= \min_{(i_1, \dots, i_n)} \{R(a_{i_1}, \dots a_{i_n}) \mid d_1 \in S_{i_1}, \dots, d_n \in S_{i_n}\}. \end{aligned} \tag{5.31}$$

Therefore,

$$|R(d_1, \dots, d_n) - R_{\mathrm{CNF}}(d_1, \dots, d_n)| = |R(d_1, \dots, d_n) - R(a_{l_1}, \dots a_{l_n})|.$$

Since $d_i \in S_i$ and thus $\rho(d_i, a_{l_i}) < r$, $1 \le i \le n$, it follows that

$$|R(d_1, \dots, d_n) - \mathrm{CNF}(d_1, \dots, d_n)| < \varepsilon,$$

which has been required. □

COROLLARY 5.7
Let an FL-relation $R \subseteq D^n$ be uniformly continuous on D^n. Then for any ε, $0 < \varepsilon < 1$, there exist formulas DNF *and* CNF *and a structure $\mathcal{D}$ with the support D such that each of FL-relations $R_{\mathrm{DNF}}(x_1, \dots, x_n)$ and $R_{\mathrm{CNF}}(x_1, \dots, x_n)$ represented by* DNF *and* CNF, *respectively w.r.t. $\mathcal{D}$ ε-approximates $R(x_1, \dots, x_n)$. Furthermore, the latter 2ε-approximates the former. Additionally, the inequality*

$$R_{\mathrm{CNF}}(x_1, \dots, x_n) \leq R_{\mathrm{DNF}}(x_1, \dots, x_n) \tag{5.32}$$

between the FL-relations w.r.t. $\mathcal{D}$ is valid for all $(x_1, \dots, x_n)$.

PROOF: For the given R and ε, $0 < \varepsilon < 1$, let us construct DNF, CNF and the structure $\mathcal{D}$ as in the proofs of Theorems 5.9 and 5.10. This leads to the FL-relations $R_{\mathrm{DNF}}(x_1, \dots, x_n)$ and $R_{\mathrm{CNF}}(x_1, \dots, x_n)$ given by (5.28) and (5.30), respectively. To verify the property of 2ε-approximation we choose an arbitrary point $(d_1, \dots d_n) \in D^n$ and evaluate the distance

$$|R_{\mathrm{DNF}}(d_1, \dots, d_n) - R_{\mathrm{CNF}}(d_1, \dots, d_n)|.$$

By virtue of Theorems 5.9 and 5.10 the initial FL-relation R is ε-approximated by R_{DNF} and R_{CNF}. Thus,

$$\begin{aligned} |R_{\mathrm{DNF}}(d_1, \dots, d_n) - R_{\mathrm{CNF}}(d_1, \dots, d_n)| \leq \\ \leq |R_{\mathrm{DNF}}(d_1, \dots, d_n) - R(d_1, \dots, d_n)| + \\ + |R(d_1, \dots, d_n) - R_{\mathrm{CNF}}(d_1, \dots, d_n)| \leq 2\varepsilon. \end{aligned}$$

To prove the inequality (5.32), it is sufficient to analyze the expression (5.29) for $R_{\mathrm{DNF}}(d_1, \dots, d_n)$ and the expression (5.31) for $R_{\mathrm{CNF}}(d_1, \dots, d_n)$. □

COROLLARY 5.8
Let D be a compact uniform topological space. Then for any quantifier- free formula A and any ε, $0 < \varepsilon < 1$, there exist DNF (CNF) *and a structure $\mathcal{D}$ with the support D such that the FL-relations R_A and R_{DNF} (R_{CNF}) represented by the respective formulas w.r.t. $\mathcal{D}$ ε-approximate each other.*

PROOF: We will construct the structure $\mathcal{D}$ in two steps. First, assign each predicate symbol in A a continuous membership function defined on D. Moreover, different predicate symbols are assigned different functions. If A contains constants, we assign them arbitrary elements from D. Thus, the structure $\mathcal{D}$ is sufficiently determined to interpret A. Then, by virtue of Corollary 5.6, the membership function of the FL- relation R_A represented by A w.r.t. $\mathcal{D}$ is continuous. Hence, it is uniformly continuous on D.

For thus obtained R_A and ε, $0 < \varepsilon < 1$, let us construct DNF (CNF) following the proof of Theorem 5.9 (5.10). If necessary, rename the predicate symbols in DNF (CNF) to have them different from those occurring in A. As a second step in the construction of $\mathcal{D}$ assign the predicate symbols and logical constants from DNF (CNF) membership functions and truth values respectively, in accordance with the proof of Theorem 5.9 (5.10). □

This corollary states that the problem of transformation of any quantifier-free formula into the form of DNF or CNF can be solved by means of the relations represented by them with respect to the certain structure. This kind of transformation can be called *approximation* since it proceeds using relations, which approximate each other. The same situation we saw in propositional fuzzy logic. The explanation of this is based on the fact that there is no elaborated technique for equivalent transformation of formulas in both calculi.

5.3.3 Representation of FL-relations and Consistency of Fuzzy Theories

In Theorem 5.8, it has been proved that any FL-relation represented by a quantifier-free formula of fuzzy predicate logic can be represented as a composition of functions (5.24) with a piecewise linear external one. This fact will be used below when speaking about truth evaluation of formulas. Together with the completeness theorem it will give us an opportunity to judge the consistency of special theories. The most interesting consequent of such approach is that the analogy of the classical consistency theorem is not valid in fuzzy logic (see Theorem 5.12).

We begin with proving of the basic theorem concerning the truth evaluation of a quantifier-free formula in fuzzy predicate calculus.

THEOREM 5.11
Let T be fuzzy predicate calculus and $A(x_1, \dots, x_n)$ be a quantifier-free formula. Then $T \models_a A$ iff

$$a = \min_{(p_1,\dots,p_m)\in[0,1]^m} l_A(p_1, \dots, p_m)$$

where $l_A(p_1, \dots, p_m)$ is a piecewise linear function referring to A in accordance with Theorem 5.8.

PROOF: The inequality $\min_{(p_1,\dots,p_m)\in[0,1]^m} l_A(p_1, \dots, p_m) \leq a$ immediately follows from the definition of the semantic consequence (4.10) on page 102 and the fact that l_A is a piecewise linear function.

To prove the converse inequality, let us consider the point $(p_1^0, \dots, p_m^0) \in [0,1]^m$ at which l_A is minimal. Our task is to find a structure $\mathcal{D}$ such that $\mathcal{D}(P_i(t_{i1}, \dots, t_{in_i})) = p_i^0$, $i = 1, \dots, m$, where $P_1, \dots, P_m$ be a set of all occurrences of predicate symbols in A and $P_i(t_{i1}, \dots, t_{in_i})$ is an instance of an atomic subformula of A.

Let us fix some set D of elements such that their number is not smaller than the number of different terms occurring in A. We assign each term t_{ij} an element $\mathcal{D}(t_{i1}) \in D$ so that different terms are assigned different elements. Take an instance $P_i(t_{i1}, \dots, t_{in_i})$ constructed on the basis of a predicate symbol P_i. The desired structure $\mathcal{D}$ is then obtained if we assign each P_i a membership function $f_{P_i} : D^n \longrightarrow [0,1]$ of the corresponding fuzzy relation in such a way that the equality

$$\mathcal{D}(P_i(t_{i1},\dots,t_{in_i})) = f_{P_i}(\mathcal{D}(t_{i1}),\dots,\mathcal{D}(t_{in_i})) = p_i^0$$

holds true. □

The theorem below plays the central role in this subsection and demonstrates that there is a difference between classical and fuzzy logics concerning characterization of the consistency.

THEOREM 5.12
Let T be a fuzzy predicate calculus. Then there exist quantifier-free evaluated formulas $a_1/A_1(x),\dots,\ a_n/A_n(x)$ where $0 < a_1 \otimes \dots \otimes a_n < 1$ and $A_1(x),\dots,A_n(x) \in F_{J(T)}$ such that

$$T' = T \cup \{\, a_1/A_1(x),\dots,\ a_n/A_n(x)\}$$

is contradictory and

$$T'' = T \cup \{\, a_1 \otimes \dots \otimes a_n/A_1(x) \,\&\, \dots \,\&\, A_n(x)\}$$

is consistent.

PROOF: We will demonstrate the proof only for the case of $n = 2$. Its extension to arbitrary (finite) n is straightforward.

Choose some element $c \in (0,1)$ and consider two piecewise linear functions $l_1(p), l_2(p)$ on $[0,1]$ such that $0 < l_1(p) < 1$, $0 < l_2(p) < 1$ and $l_1(p) \otimes l_2(p) = c > 0$ for all $p \in [0,1]$. In accordance with Theorem 5.2 the functions $l_1(p), l_2(p)$ can be represented by some formulas $A_1(p), A_2(p)$ of fuzzy propositional logic, respectively.

Let D be a non-empty domain and $r(x) : D \longrightarrow [0,1]$ be an FL-relation with the range equal to $[0,1]$. Substitute $r(x)$ into $l_1(p), l_2(p)$ instead of each occurrence of p and thus, obtain the FL-relations $l_1(r(x))$ and $l_2(r(x))$ on D.

Choose a structure $\mathcal{D}$ with the support D such that the FL- relation $r(x)$ is assigned to some unary predicate symbol $R \in F_{J(T)}$. Replace each occurrence of the propositional variable p in the formulas $A_1(p), A_2(p)$ by the subformula $R(x)$ and thus, obtain formulas $A_1(x), A_2(x)$ of fuzzy predicate logic. It remains to find evaluations a_1, a_2, which fulfil the conditions of the theorem.

Since $l_1(p)$ is piecewise linear, there exists $p_0 \in [0,1]$ such that $\max_p l_1(p) = l_1(p_0)$. Put $a_1 = l_1(p_0) + \varepsilon$ and $a_2 = l_2(p_0) - \varepsilon$ where $\varepsilon \in (0,1)$ is chosen so that $0 < a_1, a_2 < 1$. Observe that

$$a_1 \otimes a_2 = l_1(p_0) + \varepsilon + l_2(p_0) - \varepsilon - 1 = l_1(p_0) \otimes l_2(p_0) = c.$$

Now, consider the formula $A_1(x) \,\&\, A_2(x)$. By Theorem 5.8, it corresponds to the piecewise linear function $l_1(p) \otimes l_2(p)$ whence by Theorem 5.11, $T \models_c A_1(x) \,\&\, A_2(x)$. Hence, in any model $\mathcal{D} \models T$ we have $\mathcal{D}(A_1(x) \,\&\, A_2(x)) \geq c = a_1 \otimes a_2$, which implies $\mathcal{D} \models T''$, i.e. T'' is consistent.

On the other hand, $T \models_{b_1} \neg A_1$. By Theorem 5.11, $b_1 = \min_p(1 - l_1(p)) = \neg \max_p l_1(p) = \neg l_1(p_0)$. Since $a_1 > l_1(p_0)$ then $b_1 > \neg a_1$ and consequently,

$$b_1 \otimes a_1 > \neg a_1 \otimes a_1 = 0.$$

But this means that, by completeness, $T \vdash_{b_1} \neg A_1$. Therefore, $T' \vdash_d \neg A_1$ for some $d \geq b_1$ and $T' \vdash_e A_1$ for some $e \geq a_1$. We conclude that T' is contradictory. □

5.4 Approximation of Continuous Functions by Fuzzy Logic Normal Forms

In the previous section we have considered a class of FL-relations represented by normal forms (disjunctive or conjunctive) of predicate fuzzy logic and proved that any FL-relation with continuous membership function can be approximated by some element from this class. In this section, we will focus on analogous problem of approximation of any continuous function on a compact set in the same class.

First, we will apply the above mentioned results. However, this is possible only if we take a function as a particular case of an FL-relation with continuous membership function. We will then obtain a non-direct solution of the problem. The second solution will be provided by a direct algorithmic construction of an FL-relation represented by a normal form, which approximates the given function.

Another type of approximation of continuous functions will be suggested by applying a defuzzification operation to the approximating FL-relation. In this case, we are able to prove the best approximation property.

For simplicity, we will confine ourselves to real valued real functions. Note, that the general case requires to consider many sorted relations, which might lead to unnecessary complexity. Moreover, throughout this section when speaking about a formula we will suppose some quantifier-free one of predicate fuzzy logic.

5.4.1 Approximation of continuous functions through FL- relations

Let us fix some continuous real valued real function $f(x_1, \ldots, x_n)$, $n \geq 1$, defined on a compact set $D^n \in \mathbb{R}^n$. Obviously, $f(x_1, \ldots, x_n)$ is also uniformly continuous on D^n, i.e. for any $\varepsilon > 0$ there exists $\delta > 0$ such that

$$|f(x_{11}, \ldots, x_{1n}) - f(x_{21}, \ldots, x_{2n})| < \varepsilon \tag{5.33}$$

whenever $|x_{1j} - x_{2j}| < \delta$, $1 \leq j \leq n$, and $(x_{11}, \ldots, x_{1n}), (x_{21}, \ldots, x_{2n}) \in D^n$. Moreover, $f(D^n)$ is a compact set, as well.

Let $\varepsilon > 0$ be an arbitrary number, and let $\delta > 0$ be the number, which provides the inequality (5.33). Since D is a compact set there exists a finite covering of it by open δ-intervals $I_1, \ldots, I_k$, $k \geq 1$, respectively. It follows that the set D^n is also covered by all the direct products of those intervals. Since f is uniformly continuous, the set $f(D^n)$ will be covered by intervals

$f(I_{i_1} \times \cdots \times I_{i_n})$, $1 \le i_j \le k$, $1 \le j \le n$, whose lengths are not greater than ε. Their centers will be denoted by $y_{i_1 \dots i_n}$.

For each δ-interval I_i, $1 \le i \le k$, choose any closed subinterval $IC_i \subset I_i$ and denote the closure of I_i by CI_i. Furthermore, consider open ε-intervals around the points $y_{i_1 \dots i_n}$, $1 \le i_j \le k$, $1 \le j \le n$, and denote them by $H_{i_1 \dots i_n}$. Observe that $f(I_{i_1} \times \cdots \times I_{i_n}) \subseteq H_{i_1 \dots i_n}$. Choose any closed subinterval $HC_{i_1 \dots i_n} \subset H_{i_1 \dots i_n}$ containing the point $y_{i_1 \dots i_n}$. By virtue of Urysohn lemma (see [53]), for each i, $1 \le i \le k$, there exists a continuous function $g_i(x)$ on D taking values from $[0, 1]$, which is equal to one on IC_i and equal to zero on $D \setminus I_i$. At the same time, for each n-tuple $(i_1 \dots i_n)$, $1 \le i_j \le k$, $1 \le j \le n$, there exists a continuous function $h_{i_1 \dots i_n}(y)$ on $f(D^n)$ such that it takes values from $[0, 1]$, it is equal to one on $HC_{i_1 \dots i_n}$ and it is equal to zero on $f(D^n) \setminus H_{i_1 \dots i_n}$.

Now, for the given function f let us construct an $(n + 1)$-ary FL-relation $R_f \mathrel{\underset{\sim}{\subseteq}} D^n \times f(D^n)$

$$R_f(x_1, \dots, x_n, y) = \bigvee_{i_1}^{k} \dots \bigvee_{i_n}^{k} (g_{i_1}(x_1) \cdots g_{i_n}(x_n) \cdot h_{i_1 \dots i_n}(y)), \tag{5.34}$$

which approximates f in a sense explained below. Observe, that '$\cdot$' denotes the ordinary product and thus, the expression (5.34) is not a formula of fuzzy predicate logic.

DEFINITION 5.14
Let $f(x_1, \dots, x_n)$, $n \ge 1$, be a continuous real valued real function defined on a compact set $D^n \in \mathbb{R}^n$ and $Q(x_1, \dots, x_n, y) \mathrel{\underset{\sim}{\subseteq}} D^n \times f(D^n)$ be an FL-relation. We say that Q ε-approximates f if the condition $Q(x_1, \dots, x_n, y) > 0$ implies the condition $|y - f(x_1, \dots, x_n)| \le \varepsilon$.

LEMMA 5.9
The FL-relation $R_f(x_1, \dots, x_n, y)$ given by (5.34) is uniformly continuous on $D^n \times f(D^n)$.

PROOF: This evidently follows from the fact that the right-hand side of (5.34) is a composition of continuous functions. □

LEMMA 5.10
Let $f(x_1, \dots, x_n)$, $n \ge 1$, be a continuous real valued real function defined on a compact set $D^n \in \mathbb{R}^n$. Furthermore, let $\varepsilon > 0$ be an arbitrary number and $\delta > 0$ be the number, which provides the inequality (5.33). If $R_f(x_1, \dots, x_n, y)$ is the FL-relation given by (5.34) with respect to the function f and values ε, δ, then R_f 2ε-approximates f.

PROOF: Let $R_f(x_1, \dots, x_n, y) > 0$. Then there are subscripts $i_1, \dots, i_n$ such that $g_{i_1}(x_1) \cdots g_{i_n}(x_n) \cdot h_{i_1 \dots i_n}(y) > 0$. Hence, $g_{i_1}(x_1) > 0, \dots, g_{i_n}(x_n) > 0$, $h_{i_1 \dots i_n}(y) > 0$, which means that $x_1 \in I_{i_1}, \dots, x_n \in I_{i_n}$ and $y \in H_{i_1 \dots i_n}$. Remember that intervals $H_{i_1 \dots i_n}$ have their centers at points $y_{i_1 \dots i_n}$ and $y_{i_1 \dots i_n} \in f(I_{i_1} \times \cdots \times I_{i_n})$. Due to the continuity of the function $f(x_1, \dots, x_n)$ there exists

$(c_{i_1}, \ldots, c_{i_n}) \in D^n$ such that $c_{i_1} \in CI_{i_1}, \ldots, c_{i_n} \in CI_{i_n}$ and $f(c_{i_1}, \ldots, c_{i_n}) = y_{i_1 \ldots i_n}$. It follows that

$$|y - f(x_1, \ldots, x_n)| \leq |y - y_{i_1 \ldots i_n}| + |f(c_{i_1}, \ldots, c_{i_n}) - f(x_1, \ldots, x_n)| \leq 2\varepsilon.$$

□

In order to have the desired result about the approximation of continuous functions, it remains to combine the result of this lemma with the result of Theorem 5.9 on page 205.

THEOREM 5.13
Let $f(x_1, \ldots, x_n)$, $n \geq 1$, be a continuous real valued real function defined on a compact set $D^n \in \mathbb{R}^n$, $\varepsilon > 0$ be an arbitrary number and $\delta > 0$ be the number, which provides the inequality (5.33). Then there exists an FL-relation represented by a formula, which 2ε-approximates f.

PROOF: For the given function f and $\varepsilon > 0$ we will keep the construction of the FL-relation $R_f(x_1, \ldots, x_n, y)$, except for the choice of $h_{i_1 \ldots i_n}(y)$, $1 \leq i_j \leq k$, $1 \leq j \leq n$, whose range is assumed to be $[0, \varepsilon']$ instead of $[0, 1]$ and $0 < \varepsilon' < \varepsilon$. It follows that the range of $R_f(x_1, \ldots, x_n, y)$ will be also $[0, \varepsilon']$. By Lemma 5.10, R_f 2ε-approximates f. By Lemma 5.9, $R_f(x_1, \ldots, x_n, y)$ is uniformly continuous on the compact set $D^n \times f(D^n)$. Hence, by Theorem 5.9 on page 205 there exists an FL-relation represented by DNF given by (5.25), which ε-approximates R_f. Therefore, to prove the theorem it is sufficient to construct a structure $\mathcal{D}$ such that the FL-relation $Q(x_1, \ldots, x_n, y)$ say, represented by DNF w.r.t. $\mathcal{D}$ ε-approximates $R_f(x_1, \ldots, x_n, y)$ and the condition $Q(x_1, \ldots, x_n, y) > 0$ implies that there are subscripts $i_1, \ldots, i_n$ such that $g_{i_1}(x_1) \cdots g_{i_n}(x_n) \cdot h_{i_1 \ldots i_n}(y) > 0$. We leave this to the reader as a technical exercise. □

The following constructive procedure gives the precise expression for the FL-relation, which ε-approximates f and which can be represented by a formula in the disjunctive normal form. In comparison with the FL-relation $R_f(x_1, \ldots, x_n, y)$ given by (5.34), the one suggested below has not a continuous membership function.

- Let $f(x_1, \ldots, x_n)$, $n \geq 1$, be a continuous real valued real function defined on a compact set $D^n \in \mathbb{R}^n$ and let $\varepsilon > 0$ be an arbitrary number. Choose $\delta > 0$ such that the inequality (5.33) holds and find a finite covering of D by open δ-intervals $I_1, \ldots, I_k$. As a result, the set D^n will be also covered by all the direct products of those intervals.

- Based on the uniform continuity of f on D^n find the corresponding finite covering of $f(D^n)$ by intervals $f(I_{i_1} \times \cdots \times I_{i_n})$, $1 \leq i_j \leq k$, $1 \leq j \leq n$, whose lengths are not greater than ε. Denote their centers by $y_{i_1 \ldots i_n}$ and consider open ε-intervals $H_{i_1 \ldots i_n}$ around the points $y_{i_1 \ldots i_n}$, $1 \leq i_j \leq k$, $1 \leq j \leq n$.

- Define functions $\bar{g}_1(x), \dots, \bar{g}_k(x)$ on D and $\bar{h}_{i_1 \dots i_n}(y)$ on $f(D^n)$, so that

$$\bar{g}_i(x) = \begin{cases} 1, & \text{if } x \in I_i, \\ 0, & \text{otherwise} \end{cases}$$

  and

$$\bar{h}_{i_1 \dots i_n}(y) = \begin{cases} 1, & \text{if } y \in H_{i_1 \dots i_n} \\ 0, & \text{otherwise,} \end{cases}$$

  where $1 \leq i \leq k$, $1 \leq i_j \leq k$, $1 \leq j \leq n$.

- Form the FL-relation

$$\bar{R}_f(x_1, \dots, x_n, y) = \bigvee_{i_1}^{k} \dots \bigvee_{i_n}^{k} (\bar{g}_{i_1}(x_1) \otimes \dots \otimes \bar{g}_{i_n}(x_n) \otimes \bar{h}_{i_1 \dots i_n}(y)). \quad (5.35)$$

It remains to verify that this FL-relation ε-approximates f and can be represented by a formula in the disjunctive normal form w.r.t to some structure. The first directly follows from the proof of Lemma 5.10. To prove the second, let us consider the respective normal form obtained from the general expression (5.25) by specifying atomic subformulas $E_{i_1 \dots i_n}$ using unary predicate symbols $P_{i_1 \dots i_n}$

$$\mathrm{DNF}(x_1, \dots, x_n, y) = \bigvee_{i_1=1}^{k} \cdots \bigvee_{i_n=1}^{k} (P_{i_1}(x_1) \,\&\, \cdots \,\&\, P_{i_n}(x_n) \,\&\, P_{i_1 \dots i_n}(y)). \quad (5.36)$$

Let us form the structure $\mathcal{D}$ for an interpretation of (5.36). To do this, we take the set D as a support of the structure and assign unary predicate symbols P_{i_j} and $P_{i_1 \dots i_n}$ fuzzy sets $P_{i_j,D} \underset{\sim}{\subset} D$ and $P_{i_1 \dots i_n, D}(y)$ defined by

$$P_{i_j,D}(x_j) = \bar{g}_{i_j}(x_j),$$
$$P_{i_1 \dots i_n, D}(y) = \bar{h}_{i_1 \dots i_n}(y)$$

where $1 \leq i_j \leq k$ for all $1 \leq j \leq n$. It is not difficult to see that the formula (5.36) represents the FL-relation $\bar{R}_f$ (5.35) with respect to the structure $\mathcal{D}$.

Observe, that the described constructive procedure can be evidently rearranged for the case of the conjunctive normal form. For this, it is sufficient to rewrite the expression (5.35) into the form following the structure of CNF.

5.4.2 *Approximation of continuous functions determined by a defuzzification operation*

As seen, fuzzy logic formulas represent a special type of relations (called FL-relations), which can, in particular, approximately represent functions. This representation has an advantage consisting in a finite character of the description of a continuous function by a fuzzy logic formula. Moreover, it would have a real use if the approximating function could be extracted from the formula representing it. This is done by applying the defuzzification procedure.

Strictly speaking, defuzzification belongs to a class of generalized operations, which correspond an object with a set. The specificity of the defuzzification is that it corresponds an *object* with a *fuzzy set.* Therefore, we can regard the defuzzification as a *generalized fuzzy operation* on a set of elements.

DEFINITION 5.15
A defuzzification of a fuzzy set $A \underset{\sim}{\subseteq} X$ *given by its membership function* $A : X \longrightarrow [0,1]$ *is a mapping* $\Theta : \mathcal{F}(X) \longrightarrow X$ *such that*

$$A(\Theta(A)) > 0. \tag{5.37}$$

The set of all defuzzification operations on $\mathcal{F}(X)$ will be denoted by $\mathrm{DEF}_{\mathcal{F}(X)}$.

Let X be a compact set and $A(x) \not\equiv 0$ be a continuous membership function of a fuzzy set $A \underset{\sim}{\subseteq} X$. Put $x^* = \Theta(A)$ where $\Theta \in \mathrm{DEF}_{\mathcal{F}(X)}$. The following specifications of the defuzzification Θ are used most frequently:

$$x^* = \inf\{x \mid A(x) = \sup_{u\in X} A(u)\}, \qquad \text{(Least of Maxima)} \tag{5.38}$$

$$x^* = \frac{x_L + x_G}{2}, \qquad \text{(Mean of Maxima)} \tag{5.39}$$

where $x_L = \inf\{x \mid A(x) = \sup_{u\in X} A(u)\}$, $x_G = \sup\{x \mid A(x) = \sup_{u\in X} A(u)\}$,

$$x^* = \frac{\int_X xA(x)dx}{\int_X A(x)dx}, \qquad \text{(Center of Gravity)} \tag{5.40}$$

In particular, if $X = \{x_1, \dots, x_n\}$ then

$$x^* = \frac{1}{k}\sum_{j=1}^{k} x_{ij}, \quad \text{where} \quad A(x_{ij}) = \max_{u\in X} A(u),$$

$$x^* = \frac{\sum_{i=1}^{n} x_i A(x_i)}{\sum_{i=1}^{n} A(x_i)}.$$

REMARK 5.5
In the examples of the defuzzification given above the continuity of $A(x)$ on the compact set X guarantees validity of (5.37). If $A(x)$ is not continuous then the defuzzification may be also realized in accordance with (5.38)–(5.40), but the inequality (5.37) should be verified subsequently. In the sequel, we will not assume the membership functions of the defuzzified fuzzy sets to be necessarily continuous.

Now, given a certain defuzzification we can uniquely correspond a function to an FL-relation. Denote the latter by $R(x_1, \dots, x_n, y)$ and suppose that it is defined on a set $X_1 \times \ldots \times X_n \times Y$. Choose a defuzzification $\Theta \in \mathrm{DEF}_{\mathcal{F}(Y)}$ and define the function $f_{R,\Theta}(x_1, \ldots, x_n) : X_1 \times \ldots \times X_n \longrightarrow Y$ by putting

$$f_{R,\Theta}(x_1, \ldots, x_n) = \Theta(R(x_1, \ldots, x_n, y)). \tag{5.41}$$

The function $f_{R,\Theta}$ defined by (5.41) is said to be adjoined to an FL-relation R by means of the chosen defuzzification Θ .

The following theorem shows the relation between functions f and $f_{R,\Theta}$ where R is an FL-relation, which approximates f.

THEOREM 5.14
Let $f(x_1, \ldots, x_n)$, $n \geq 1$, be a continuous real valued real function defined on a compact set $D^n \in \mathbb{R}^n$ and $\varepsilon > 0$ be an arbitrary number. Furthermore, let an FL-relation $R(x_1, \ldots, x_n, y)$ ε-approximate f. Then for any defuzzification $\Theta \in \mathrm{DEF}_{\mathcal{F}(Y)}$ the function $f_{R,\Theta}$ defined by (5.41) also ε-approximates f as well, i.e. for any $(x_1, \ldots, x_n) \in D^n$

$$|f(x_1, \ldots, x_n) - f_{R,\Theta}(x_1, \ldots, x_n)| \leq \varepsilon.$$

PROOF: By Definition 5.14, the inequality $R(x_1, \ldots, x_n, y) > 0$ implies $|y - f(x_1, \ldots, x_n)| \leq \varepsilon$. Let $f_{R,\Theta}(x_1, \ldots, x_n)$ be defined by (5.41). Then it remains to observe that $R(x_1, \ldots, x_n, f_{R,\Theta}(x_1, \ldots, x_n)) > 0$. □

At a first glance, Theorem 5.14 convinces us that it makes no difference, which operation of defuzzification is needed to be chosen for the satisfactory approximation. This could be so if we do not take into consideration any kind of measuring the goodness of an approximation. In order to provide the reader with the extended investigation of the problem, we will consider the problem of the best approximation in accordance with a certain criterion. To simplify the description, we will confine ourselves to the case of functions with one variable.

DEFINITION 5.16
Let $f(x)$ be a continuous real valued function defined on the closed interval $[a, b]$ and let G be a set of approximation, which consists of continuous real valued functions defined also on $[a, b]$. Moreover, let d determine a distance on the set of all continuous real valued functions defined on $[a, b]$. Following [108] we say that $g^*(x) \in G$ is the best approximation from G to f (with respect to d) if the condition

$$d(f, g^*) \leq d(f, g) \tag{5.42}$$

holds for all $g(x) \in G$.

In the sequel, we will consider a set of approximation consisting of a set of real valued real functions, which are adjoined to some special FL-relations represented by formulas by means of some defuzzification operation. We will be interested in finding the best approximation to the given continuous real valued real function $f(x)$ in a sense close to the above defined one.

On the account of results from the previous subsection we may regard those special FL-relations represented by normal forms. In particular, they are given by

$$\mathrm{DNF}(x,y) = \bigvee_{i=1}^{k} (P_i(x) \mathbin{\&} Q_i(y)) \tag{5.43}$$

and

$$\mathrm{CNF}(x,y) = \bigwedge_{i=1}^{k} (P_i(x) \Rightarrow Q_i(y)) \tag{5.44}$$

where P_i and Q_i, $1 \le i \le k$, are unary predicate symbols.

Let $f(x)$ be a continuous real valued function defined on the closed interval $[a,b]$ so that $f : [a,b] \longrightarrow [c,d]$. In order to obtain FL-relations $R_{\mathrm{DNF}}(x,y)$ and $R_{\mathrm{CNF}}(x,y)$, which will be used in the construction of the set of approximation for f we will describe a structure $\mathcal{D}$ dependent on f.

Choose an arbitrary $\varepsilon > 0$ and find $\delta > 0$ so that the inequality $|f(x') - f(x'')| < \varepsilon$ holds true whenever $|x' - x''| < \delta$ and $x', x'' \in [a,b]$. Consider a finite covering of $[a,b]$ by the following half-open δ-intervals, with the exception of the last closed one

$$[a,b] = \bigcup_{i=1}^{k-1} [x_i, x_{i+1}) \cup [x_k, x_{k+1}]$$

where for the simplicity suppose that $k = (b-a)/\delta$ and thus, $x_1 = a$, $x_{k+1} = b$, $x_{i+1} = a + i\delta$, $1 \le i \le k-1$. Denote $I_i = [x_i, x_{i+1})$, $1 \le i \le k-1$, and $I_k = [x_k, x_{k+1}]$. Furthermore, denote by $\bar{f}[I_i] = [f_i, f_{i+1}]$ the closed interval corresponding to I_i, $1 \le i \le k$. Observe that the length of $\bar{f}[I_i]$ is not greater than ε.

Now, put $D = [a,b] \cup [c,d]$ as a support of the structure $\mathcal{D}$ and for each $1 \le i \le k$ assign unary predicate symbols P_i and Q_i fuzzy sets $P_{i,D} \mathrel{\underset{\sim}{\subset}} D$ and $Q_{i,D} \mathrel{\underset{\sim}{\subset}} D$ given by

$$P_{i,D}(x) = \begin{cases} 1, & \text{if } x \in I_i, \\ 0, & \text{otherwise} \end{cases}$$

and

$$Q_{i,D}(y) = \begin{cases} 1, & \text{if } y \in \bar{f}[I_i], \\ 0, & \text{otherwise.} \end{cases}$$

This completes the construction of the structure $\mathcal{D}$ sufficient for the interpretation of the normal forms (5.43) and (5.44). It is easy to see that the FL-relations $R_{\mathrm{DNF}}(x,y)$ and $R_{\mathrm{CNF}}(x,y)$ represented by these normal forms w.r.t.

$\mathcal{D}$ are equal. We will therefore omit the subscript and denote each of them by $R(x, y)$. The precise expression is

$$R(x,y) = \begin{cases} 1, & \text{if } (\exists i)(1 \le i \le k \quad \text{and} \quad x \in I_i, y \in \bar{f}[I_i]), \\ 0, & \text{otherwise.} \end{cases} \tag{5.45}$$

Finally, define the set of approximation $\mathcal{G}_f$ for f as the set of all functions adjoined to $R(x, y)$ given by (5.45) by means of the defuzzifications $\Theta \in \mathrm{DEF}_{\mathcal{F}(D)}$

$$\mathcal{G}_f = \{g(x) \mid g(x) = f_{R,\Theta}(x) = \Theta(R(x,y)),\ \Theta \in \mathrm{DEF}_{\mathcal{F}(D)}\}. \tag{5.46}$$

In what follows we will fix the structure $\mathcal{D}$, which lead to $R(x, y)$ and $\mathcal{G}_f$ and preserve all the intermediate notations. Remember that in accordance with Definition 5.15, the defuzzification is a generalized fuzzy operation, which maps a fuzzy subset onto an element belonging to its support. Therefore, it is easy to see that the following characterization lemma holds true.

LEMMA 5.11
Let $f(x) : [a, b] \longrightarrow [c, d]$ be a continuous function and $\mathcal{G}_f$ be the set of approximation for f given by (5.46). Then, using the notation introduced above, each function $g(x)$ from $\mathcal{G}_f$ can be characterized as follows:

$$g(x) = g_i \quad \textit{iff} \quad x \in I_i \tag{5.47}$$

where $g_i \in \bar{f}[I_i]$, $1 \le i \le k$.

Moreover, for each k-tuple $(g_1, \dots, g_k) \in \bar{f}[I_1] \times \cdots \times \bar{f}[I_k]$ there exists a function $g(x) \in \mathcal{G}_f$, which can be characterized by (5.47).

By this lemma, the set of approximation $\mathcal{G}_f$ can be regarded as a subset of piecewise polynomials of zero degree. Below, we will see, what is the explicit form of some functions from $\mathcal{G}_f$ determined by the specifications of the defuzzification operations listed above.

LEMMA 5.12
Let $f(x) : [a, b] \longrightarrow [c, d]$ be a continuous function and $\mathcal{G}_f$ be the set of approximation for f given by (5.46). Furthermore, let $g_{LOM}(x), g_{MOM}(x), g_{COG}(x) \in \mathcal{G}_f$ be functions specified by the defuzzifications (5.38)–(5.40), respectively. Then using the notation introduced above they can be characterized by

$$g_{LOM}(x) = f_i \quad \textit{iff} \quad x \in I_i, \tag{5.48}$$

$$g_{MOM}(x) = g_{COG}(x) = \frac{f_i + f_{i+1}}{2} \quad \textit{iff} \quad x \in I_i \tag{5.49}$$

where $1 \le i \le k$.

In what follows, we wish to analyze, whether the functions given by (5.48)–(5.49) can be somehow distinguished in $\mathcal{G}_f$. We will show that $g_{MOM}(x)$ is in some sense the best approximation to the given continuous function $f(x)$. The words "in some sense" are used here because the functions from $\mathcal{G}_f$ are not, in general, continuous. Therefore, we will speak about the best approximation on each closed interval $CI_i = [x_i, x_{i+1}]$, $1 \le i \le k$, separately.

Let us introduce the set $\mathcal{G}_{f,i}$ of continuous restrictions of functions from $\mathcal{G}_f$ on CI_i as the set of approximation for $f|CI_i$, $1 \le i \le k$. By Lemma 5.11, for each element $g_i \in \bar{f}[I_i]$, $1 \le i \le k$, there exist a function $g(x) \in \mathcal{G}_f$ such that $g|I_i = g_i$. Thus, $\mathcal{G}_{f,i}$ is simply the set of constant functions defined on CI_i so that $\mathcal{G}_{f,i} = \{g(x) : CI_i \longrightarrow \bar{f}[I_i] \mid g(x) \equiv g_i,\ g_i \in \bar{f}[I_i]\}$.

THEOREM 5.15
Let $f(x) : [a,b] \longrightarrow [c,d]$ be a continuous function and $\mathcal{G}_f$ be the set of approximation for f given by (5.46). Furthermore, let $g_{MOM}(x) \in \mathcal{G}_f$ be the function specified by the defuzzification (5.39). Then for each closed interval $CI_i = [x_i, x_{i+1}]$, $1 \le i \le k$, the continuous restriction $g_{MOM}|CI_i(x)$ is the best approximation from $\mathcal{G}_{f,i}$ to $f|CI_i(x)$ with respect to the distance d given by

$$d(u(x), v(x)) = \max_{CI_i} |u(x) - v(x)|. \tag{5.50}$$

PROOF: Let us fix some index i, $1 \le i \le k$. Remember that in accordance with the given notation, $\bar{f}[CI_i] = [f_i, f_{i+1}]$ and $g_{MOM}|CI_i(x) \equiv (f_i + f_{i+1})/2$. Thence, the theorem follows from the general criterion for a function to be the best approximation with respect to the distance d given by (5.50) (see [108]). We reformulate this criterion using the introduced notation. Let $g^*(x) \in \mathcal{G}_{f,i}$ and $M_i \subseteq CI_i$ be the set of points, at which the function $|f|CI_i(x) - g^*(x)|$ takes its maximum value. The function $g^*(x)$ is the best approximation from $\mathcal{G}_{f,i}$ to $f|CI_i(x)$ with respect to d iff there is no element $g(x) \in \mathcal{G}_{f,i}$, which satisfies the condition

$$(f|CI_i(x) - g^*(x)) \cdot g(x) > 0, \quad x \in M_i.$$

□

5.5 Representation of Continuous Functions by the Conjunctive Normal Form

We are interested here in the precise representation of continuous functions by FL-relations in such a way that given a function find the FL-relation, to which it can be adjoined by means of some defuzzification. In addition, the FL-relation is supposed to be represented by one of the normal forms.

After a thorough analysis of the methodology of approximation described in the previous section we deduce that the problem of representation can be solved if a correspondence between function $f(x)$ and the FL-relation $R(x,y)$ adjoined to it, is more rigorous. We will suppose that the latter is built up in such a way that

$$y = f(x) \quad \text{implies} \quad R(x,y) = 1. \tag{5.51}$$

Moreover, the defuzzification is supposed to be given by (5.38). As above, we will work with real valued real functions with one variable.

Till now, both kinds of normal forms have been interchangeable. In this section, we will see that the conjunctive normal form is more suitable for the solution of our problem. The justification of this follows from the subsequent lemma where the functions adjoined to FL-relations represented by DNF are shown to be very specific. Throughout this section we will not use a language of the first order logic when dealing with the normal forms. For our purposes it is sufficient to work with the algebraic expressions, which inherit the structure of the respective logical formulas. However, we will call these expressions by normal forms as before.

LEMMA 5.13
Let $f(x)$ be a real valued function defined on $[a,b]$, which is adjoined to the FL-relation $R(x,y) \underset{\sim}{\subset} [a,b] \times f([a,b])$ by means of the defuzzification (5.38) so that (5.51) holds. Suppose that $R(x,y)$ is expressed in the disjunctive normal form

$$R(x,y) = \bigvee_{i=1}^{k} (P_i(x) \otimes Q_i(y))$$

where $P_i(x), Q_i(y)$, $1 \le i \le k$, are membership functions of normal fuzzy sets defined on intervals $[a,b]$ and $f([a,b])$, respectively. Moreover, suppose that those membership functions are continuous and their a-cuts are closed intervals. Then $f(x)$ is a piecewise constant with finite number of steps (some steps may degenerate into points).

PROOF: Let $f(x)$, $R(x,y)$ fulfil the assumptions. Then for each $x \in [a,b]$ there exists at least one $y \in f([a,b])$ such that $R(x,y) = 1$. Since $R(x,y)$ can be represented in the disjunctive normal form, for each $x \in [a,b]$ there exist i, $1 \le i \le k$, and $y \in f([a,b])$ such that $P_i(x) = 1$ and $Q_i(y) = 1$. Therefore, taking into account that the 1-cuts of $P_i(x)$, $1 \le i \le k$, are closed intervals deduce that $[a,b]$ is covered by closed subintervals $I_1, \dots, I_k$ such that $x \in I_i$ iff $P_i(x) = 1$. Observe, that some of those subintervals may be simply points.

Let us consider the partition of $[a,b]$ given by all the non-empty constituents of $I_1, \dots, I_k$ and denote elements of this partition by $J_1, \dots, J_l$, $k \le l \le 2^k$. We will prove that $f(x)$ is equal to a constant on each subinterval J_j, $1 \le j \le l$. For this purpose, let us fix an arbitrary $x \in J_j$, $1 \le j \le l$, and suppose that $J_j = I_1^{\sigma_1} \cap \dots \cap I_k^{\sigma_k}$ where $\sigma_i \in \{0,1\}$ and

$$I_i^{\sigma_i} = \begin{cases} I_i, & \text{if } \sigma_i = 1, \\ [a,b] \setminus I_i, & \text{if } \sigma_i = 0, \end{cases}$$

$1 \le i \le k$. Consider the k-tuple $(\sigma_1, \dots, \sigma_k)$ and choose all those subscripts $\{i_1, \dots, i_r\}$, which correspond to the unit elements. Denote the 1-cuts of $Q_{i_1}(y), \dots Q_{i_r}(y)$ by $C_{i_1}, \dots, C_{i_r}$, respectively. Thus, for the chosen x obtain that $R(x,y) = 1$ iff $y \in (C_{i_1} \cup \dots \cup C_{i_r})$. In accordance with the chosen defuzzification the function f maps x onto $\check{y} = \inf(C_{i_1} \cup \dots \cup C_{i_r})$. Observe that $\check{y}$ does not depend on the choice of x from J_j. Therefore, $f(x) = \check{y}$ for any $x \in J_j$, $1 \leq j \leq l$. □

It is worth noticing that neither the defuzzification given by (5.38) nor the other one given by (5.40) can be used in Lemma 5.13. This is true because the condition (5.51), which assumed as necessary may not be fulfilled in those cases.

Now, we will show that the same assumptions as in Lemma 5.13 except the choice of the normal form lead to the wider class of real valued functions, which can be expressed.

LEMMA 5.14
Let $f(x)$ be a continuous and strictly monotonous real valued function defined on $[a, b]$. Then there exists an FL-relation $R(x, y)$ defined on $[a, b] \times [f(a), f(b)]$ and expressed in the conjunctive normal form

$$R(x, y) = \bigwedge_{i=1}^{2} (P_i(x) \to Q_i(y)) \tag{5.52}$$

so that (5.51) holds, and $f(x)$ is the function adjoined to $R(x, y)$ by means of the defuzzification (5.38).

PROOF: Let $f(x)$ fulfils the assumptions. To obtain the proof, it is sufficient to construct the membership functions $P_i(x), Q_i(y)$, $i = 1, 2$, defined on $[a, b]$ and $[f(a), f(b)]$, respectively and then verify the conclusions.

Since $f(x)$ is continuous and monotonous on $[a, b]$ it defines a one-to-one correspondence between $[a, b]$ and $[f(a), f(b)]$. Therefore, the inverse function $f^{-1}(y)$ exists. For the certainty, assume that $f(x)$ monotonously increases. Put the above mentioned membership functions as follows:

$$P_1(x) = 1 - \frac{x-a}{b-a}, \qquad P_2(x) = \frac{x-a}{b-a}, \quad x \in [a, b],$$
$$Q_1(y) = 1 - \frac{f^{-1}(y)-a}{b-a}, \quad Q_2(y) = \frac{f^{-1}(y)-a}{b-a}, \quad y \in [f(a), f(b)].$$

Let $x^0 \in [a, b]$ be an arbitrary element. We will show that

$$f(x^0) = \inf\{y \mid R(x^0, y) = 1\} \quad \text{and} \quad R(x^0, f(x^0)) = 1.$$

Indeed,

$$\begin{aligned}
P_1(x^0) \to Q_1(y) = 1 \quad &\text{iff} \quad P_1(x^0) \leq Q_1(y) \\
&\text{iff} \quad 1 - \frac{x^0-a}{b-a} \leq 1 - \frac{f^{-1}(y)-a}{b-a} \quad \text{iff} \quad y \leq f(x^0), \\
P_2(x^0) \to Q_2(y) = 1 \quad &\text{iff} \quad P_2(x^0) \leq Q_2(y) \quad \text{iff} \quad \frac{x^0-a}{b-a} \leq \frac{f^{-1}(y)-a}{b-a} \\
&\text{iff} \quad f(x^0) \leq y.
\end{aligned}$$

Thus,

$$R(x^0, y) = 1 \quad \text{iff} \quad f(x^0) = y$$

what has been necessary. □

The same technique can be applied for the representation of the more general class of real valued functions which are constituted of strictly monotonous restrictions. To be precise, we introduce the following definition.

DEFINITION 5.17
By a piecewise monotonous real valued function $f(x)$ defined on $[a, b]$ we mean a function, for which there exists a finite partition of $[a, b]$ into subintervals such that the restriction of f to each subinterval from the partition is strictly monotonous.

THEOREM 5.16
Let $f(x)$ be a continuous and piecewise monotonous real valued function defined on $[a, b]$. Then there exist a number N, and an FL-relation $R(x, y)$ defined on $[a, b] \times f([a, b])$ and expressed in the conjunctive normal form

$$R(x, y) = \bigwedge_{i=1}^{N} \bigwedge_{j=1}^{2} (P_{ij}(x) \rightarrow Q_{ij}(y)) \tag{5.53}$$

so that (5.51) holds and $f(x)$ is the function adjoined to $R(x, y)$ by means of the defuzzification (5.38).

PROOF: Since $f(x)$ is continuous and piecewise monotonous on $[a, b]$, there exists a finite partition of $[a, b]$ into a finite number of subintervals $I_1, \dots, I_N$ such that the restriction $f|I_i$, $1 \leq i \leq N$, is strictly monotonous. Suppose that $I_1, \dots, I_N$ are closed subintervals. In accordance with Lemma 5.14, for each $f|I_i$ there exists an FL-relation $R_i(x, y)$ defined on $I_i \times f(I_i)$ and given by the expression

$$R_i(x, y) = \bigwedge_{j=1}^{2} (P_{ij}(x) \rightarrow Q_{ij}(y))$$

so that (5.51) holds and $f|I_i(x)$ is the function adjoined to $R_i(x, y)$ by means of the defuzzification (5.38).

We extend the domains of the membership functions $P_{ij}(x), Q_{ij}(y)$ from I_i and $f(I_i)$ to $[a, b]$ and $f([a, b])$ respectively, by putting them equal to 0 outside I_i and $f(I_i)$, $1 \leq i \leq N$, $1 \leq j \leq 2$. Thus, each FL-relation $R_i(x, y)$ will be extended to $[a, b] \times f([a, b])$. Now, consider the fuzzy relation $R(x, y)$ defined on $[a, b] \times f([a, b])$ and given by

$$R(x, y) = \bigwedge_{i=1}^{N} R_i(x, y).$$

Evidently, $R(x,y)$ is given by (5.53) as well. Therefore, it remains to check for it the assertion of the theorem. Let $x^0 \in [a,b]$ be an arbitrary element, then either it belongs to one subinterval I_i, $1 \leq i \leq N$, or it is an internal boundary of two neighboring subintervals I_i and I_{i+1}, $1 \leq i \leq N-1$. In any case, it follows that not more than two membership functions $P_{i1}(x), P_{i2}(x)$ or $P_{i2}(x), P_{i+1,1}(x)$ can differ from 0 in x^0. Thus, from the continuity of f

$$x^0 \in I_i \quad \text{implies} \quad R(x^0,y) = R_i(x^0,y), \; y \in f(I_i).$$

It remains to refer Lemma 5.14 to complete the proof. □

Both Lemma 5.14 and Theorem 5.16 have been based on the certain construction of the FL-relation $R(x,y)$ given by (5.52) and (5.53). However, the latter is completely determined by the choice of the membership functions. The natural question arises related to the possibility of a generalization the suggested method of construction the membership functions. The way of a generalization is described in the following lemma where for the simplicity the character of monotonicity is supposed to be fixed.

LEMMA 5.15
Let $f(x)$ be a continuous and strictly monotonously increasing real valued function defined on $[a,b]$. Moreover, let $e(x) : [a,b] \longrightarrow [0,1]$ also be a continuous and strictly monotonously increasing function. Then the FL-relation $R(x,y)$ defined on $[a,b] \times [f(a), f(b)]$ and expressed in the conjunctive normal form

$$R(x,y) = \bigwedge_{i=1}^{2} (P_i(x) \rightarrow Q_i(y))$$

where $P_1(x) = 1 - e(x)$, $P_2(x) = e(x)$, $Q_1(y) = 1 - e(f^{-1}(y))$, $Q_2(y) = e(f^{-1}(y))$, satisfies the conclusion of Lemma 5.14.

PROOF: The proof reduces to examination of (5.51) and to verification that $f(x)$ is the function adjoined to $R(x,y)$ by means of the defuzzification (5.38). Both conditions are proved by the direct checking using the same technique as in the proof of Lemma 5.14. □

It is easy to see that the method of construction the FL-relation $R(x,y)$ given in the proof of Theorem 5.16 can be generalized in the same way as above.

Let us stress that the way, how Theorem 5.16 was proved, opens interesting combination of fuzzy technique with ordinary numerical methods. In the case where the dependence between inputs and outputs is given by point-to-point correspondence, any approximation function obtained using numerical methods is easily transformed into a formula in the conjunctive normal form.

Concerning the problem of how many representations exist for the given continuous function, we see from the text that not only one way is possible. In this case the search of the most efficient (in some sense) way of representation is worth to be done.

6 FUZZY LOGIC IN BROADER SENSE

In Chapter 4, fuzzy logic in narrow sense has been presented at length. A lot of arguments have been given in favour of this logic, among which the most important is its role in providing tools for modelling of the vagueness phenomenon via graded approach. Consequently, FLn should be a theoretical basis standing beyond methods and techniques of fuzzy logic. The mediator between the latter and FLn is natural language. To be more specific, most of the applications of fuzzy logic are based on the generalized modus ponens mentioned in Chapter 1, which is a model of one of the fundamental principles of human reasoning. Recall that it typically uses fuzzy IF-THEN rules, which are conditional statements containing expressions of natural language and interpreted using fuzzy sets.

Since the success of applications of fuzzy logic is convincing, it is important to elaborate good formalization of natural language and to employ it in the model of the commonsense human reasoning. As a contribution to this task we bring forward the concept of *fuzzy logic in broader sense* (FLb) in this chapter. Let us remark that there is a gap between the applications of natural language and the used means interpreting it. The applications are quite often based only on indirect interpretation and the conclusions are derived without trying to penetrate more deeply into understanding of its semantics. This chapter is an attempt to fill in this gap and to provide an explicit expression of the meaning of the natural language statements so that the expected conclusions could be better justified.

Of course, our formalization concerns only a small fragment of natural language, namely that which is used in evaluation of the behaviour of dynamic systems and various decision situations. Furthermore, because of the limited space, we focus only on two fundamental kinds of the reasoning schemes: a simple deduction based on modus ponens, and a more specific reasoning the aim of which is approximation either of a crisp relation, or of a function. On the other hand, we try to give some general principles for extension of FLb, which could be applied when introducing other types of reasoning and other types of linguistic expressions not included so far (of course, provided this turns out to be necessary). For example, we see a significant potential in the use of linguistic quantifiers (most, some, often, etc.), which seem not to be satisfactorily elaborated so far.

From the formal point of view, FLb can be seen as an extension of FLn. The expressions of natural language (syntagms) are in FLb translated into sets of evaluated instances of open formulas of FLn (we call them multiformulas). The translation follows both the principles of FLn as well as the results of the theory of linguistic semantics.

We expect that FLb might help in answering questions raised in fuzzy control theory, for example the aspect of controllability, and it could also be a theoretical basis for the development of fuzzy expert systems. According to the main result of this chapter, formulas used in the generalized modus ponens, both in implication as well as classical Mamdani's forms, lead to the best possible truth values, provided that we confine ourselves to the specific kinds of the linguistic expressions (syntagms). The concept of FLb has been initiated by V. Novák in [90, 91, 96] and in some aspects further elaborated also by I. Perfilieva (cf. [100, 101]).

6.1 Partial Formalization of Natural Language

In this section, we define a special class of linguistic syntagms (expressions) which are of interest for us. The stress will be put on the so called evaluating syntagms since they play a crucial role in the applications of fuzzy logic. Furthermore, we define translation rules into the language of FLn. This is fundamental for further elaboration of their meaning. We formalize the concepts of intension and extension, and show, how the intension of the basic syntagms "small, medium" and "big", as well as of the linguistic modifiers can be constructed.

6.1.1 Evaluating and simple conditional linguistic syntagms

The fragment of natural language we will work with in the sequel consists of a set $\mathcal{S}$ of linguistic syntagms specified below. Our specification, however, should not be taken as fixed. We suppose $\mathcal{S}$ to be modified whenever necessary.

We will use the term *syntagm* which can be understood as any part of a sentence (casually the whole) constructed according to grammatical rules and having a specific meaning[1]. For example, the following are syntagms: "big number, young girl, very small distance, a letter is important, John goes to the cinema", etc. On the other hand, the following are not syntagms: "pen is, I would, is interesting, roughly table", etc.

The syntagms we will work with will be constructed from the following classes of words.

(i) *Nouns*, which mostly will be taken as primitives without studying their structure and formally represented simply by variables denoting objects.

(ii) *Adjectives* where we will confine to few proper ones, mostly forming an evaluation linguistic trichotomy of the type "small", "medium" and "big". These words can be taken as prototypical for other similar linguistic trichotomies, such as "weak, medium strong, strong", "clever, average, dull", etc.

(iii) *Linguistic hedges*, which form a special kind of the so called intensifying adverbs, such as "very, slightly, roughly", etc. We distinguish linguistic hedges with narrowing effect, such as "very, significantly", etc. and those with extending effect, such as "roughly, more or less", etc.

(iv) *Connectives* "and", "or" and the *conditional clause* "if ... then ...".

Further class to be added in the future are linguistic quantifiers. For the time being, however, we will not consider them.

In the construction of the set $\mathcal{S}$, we employ the so called *evaluating syntagms*, i.e., linguistic expressions which characterize a position on a bounded ordered scale (usually an interval of some numbers (real, rational, natural)). These are syntagms such as "small, medium, big, very small, roughly large, extremely high", etc. We will also include into $\mathcal{S}$ some combinations of evaluating syntagms using connectives[2], and conditionals.

The symbol $\langle \dots \rangle$ used below denotes a metavariable for the kind of a word given inside the angle brackets.

DEFINITION 6.1
An evaluating syntagm is one of the following linguistic expressions:

(i) Atomic evaluating syntagm, which is any of the following:

- *An adjective "small", "medium", or "big",*
- *A fuzzy quantity "approximately z", which is a linguistic expression characterizing some quantity z from an ordered set.*

(ii) Negative evaluating syntagm, which is an expression

$$\mathsf{not}\ \langle \textit{atomic evaluating syntagm} \rangle.$$

[1]It is not easy to give a more precise formal definition. In [89, 119] it is defined as any connected part of a tectogrammatical tree, which corresponds to an expression of natural language on the surface level.

[2]In the linguistic theory we speak about *close coordination*.

(iii) *Simple evaluating syntagm, which is an expression*

$$\langle \textit{linguistic hedge}\rangle\langle \textit{atomic evaluating syntagm}\rangle$$

(iv) *Compound evaluating syntagms constructed as follows: if* $\mathcal{A}, \mathcal{B}$ *are evaluating syntagms then '*$\mathcal{A}$ and $\mathcal{B}$*', '*$\mathcal{A}$ or $\mathcal{B}$*' are compound evaluating syntagms.*

EXAMPLE 6.1
Atomic evaluating syntagms are *small, medium, big.* Fuzzy quantities are, e.g. *twenty five, the value z*, etc. Simple evaluating syntagms are *very small, more or less medium, roughly big, about twenty five, approximately z*, etc. Compound evaluating syntagms are *roughly small or medium, quite roughly medium and not big*, etc. □

The "fuzzy quantity" is a linguistic characterization of some quantity — usually a number. This means that every linguistic characterization of a number is understood imprecisely. This imprecision may be increased using some linguistic hedge, for example, "very roughly one thousand". Specific is the expression "*precisely z*" whose meaning is the precise quantity (number) z. We will take the form "approximately z" as canonical.

Evaluating syntagms are essential in the definition of the evaluating predications, which are used in the description of dynamic systems.

DEFINITION 6.2
An evaluating predication is one of the following syntagms.

(i) *Let* $\mathcal{A}$ *be an evaluating syntagm. Then the syntagm*

$$\langle \textit{noun}\rangle \text{ is } \mathcal{A} \tag{6.1}$$

is an evaluating predication. In special case, if $\mathcal{A}$ *is a simple evaluating syntagm then (6.1) is a a simple evaluating predication*[3].

(ii) *Let* $\mathcal{A}, \mathcal{B}$ *be evaluating predications. Then '*$\mathcal{A}$ and $\mathcal{B}$*', '*$\mathcal{A}$ or $\mathcal{B}$*' are compound evaluating predications.*

Note that from the point of view of the linguistic theory, evaluating predications are also syntagms.

EXAMPLE 6.2
Evaluating predications are, e.g. "*temperature is very high*" (here "high" is taken as synonym for "big"), "*pressure is not small*", "*frequency is small or*

[3]At first sight, the other possible form, namely "$\mathcal{A}$ ⟨noun⟩" seems to be synonymous with our definition of the evaluating predication (for example, a "small boy" is sometimes taken as synonymous to a "boy is small"). However, this is not the case since the latter uses a verb phrase which has a more complicated semantic structure than the former. In this book, however, we are using the "is" only as an assignment without considering its verb structure.

medium", "*cost is not small and not big*", "*income is roughly three million*", etc. Compound evaluating predication is, e.g. "*temperature is high and pressure is very high*". □

The following definition characterizes the set of linguistic expressions which will be employed in FLb.

DEFINITION 6.3
The set $\mathcal{S}$ of syntagms consists of evaluating syntagms, evaluating predications and the conditional clause

$$\text{IF } \mathcal{A} \text{ THEN } \mathcal{B} \tag{6.2}$$

where $\mathcal{A}$, $\mathcal{B}$ are linguistic terms.

To simplify our explanation, we will speak about *simple syntagm* when having on mind a simple evaluating syntagm or a simple evaluating predication, and *complex* one otherwise.

Recall that all three definitions should be taken as temporary and can be modified when a satisfactory theory of other kind words and expressions is developed (this concerns especially introduction of linguistic quantifiers).

6.1.2 *Translation of evaluating syntagms and predications into fuzzy logic*

Language and basic assignments. This subsection is devoted to a possible way, how the syntagms introduced in the previous subsection can be translated into the language of FLn. For this purpose, we fix some many-sorted language J of the predicate FLn with desirable properties and look for a reasonable assignment of components of syntagms from $\mathcal{S}$ to elements of J. Recall from Remark 4.3 on page 150 that many-sorted fuzzy predicate logic is also complete. We will use this fact in this chapter.

First of all, we must realize that the syntagms $\mathcal{A} \in \mathcal{S}$ are certain names of properties of objects. Thus in general, we will assign them predicate formulas. In the center of our interest, however, lay evaluating predications, which contain nouns, i.e. words denoting objects. Though we will not be explicitly interested in objects, it is necessary to characterize relations among them. A crucial role is played by an ordering relation but other relations might be also possible. Therefore, we will assume that we are given a certain *background fuzzy theory* BT, which will play a significant role in characterization of the meaning of the syntagms $\mathcal{A}$ in concern. Furthermore, we choose some model $\mathcal{K} \models BT$ and fix it as a canonical one.

In addition to the general definition of the language of the predicate FLn, the language J we are fixing will be supposed to have the following properties.

(i) J contains all the symbols of the language $J(BT)$ of the background fuzzy theory.

(ii) J has a finite number of sorts ι, $\iota = 1, \dots, p$. For each sort, J contains a nonempty set K_ι of constants being names of objects from the support of the canonical model $\mathcal{K}$ of the background theory BT.

(iii) J contains a finite set of additional logically fitting unary connectives $\Theta = \{\triangleleft_1, \dots, \triangleleft_p\}$.

The language J will be occasionally extended by some additional logically fitting conjunction connectives $\hat{\&}_1, \dots, \hat{\&}_p$ and disjunction ones $\hat{\nabla}_1, \dots, \hat{\nabla}_q$. In general, various connectives interpreted by the t-norms and t-conorms may be suitable for local cases of the mentioned close coordination (cf. our discussion of the conjunction in Section 4.2). Since this part of the theory is still not satisfactory developed, for the present we will mostly use only classical logical connectives, namely $\wedge, \&, \vee, \nabla$.

By $M_{J\iota}$ we denote a *set of closed terms* of the sort ι of the language J (the subscript J is often omitted). Recall that the structure for J is

$$\mathcal{D} = \langle D_1, \dots, D_p, P_D, \dots, f_D, \dots \rangle$$

where D_ι are sets corresponding to the respective sorts, $P_D, \dots$ are fuzzy relations and $f_D, \dots$ are functions defined on them.

The following definition provides fundamental rules for translation of syntagms from $\mathcal{S}$ into fuzzy logic.

DEFINITION 6.4

(i) A ⟨noun⟩ is assigned a variable $x \in J$ of some sort ι.

(ii) An ⟨atomic evaluating syntagm⟩ is assigned a unary predicate symbol $P \in J$.

(iii) The negation ‘ not ’ is assigned the connective $\neg$.

(iv) A ⟨linguistic hedge⟩ is assigned a unary connective $\triangleleft \in \Theta$.

(v) The connective ‘ and ’ is, dependingly on the local case, assigned some of the conjunction connectives $\hat{\&}_1, \dots, \hat{\&}_p$.

(vi) The connective ‘ or ’ is, dependingly on the local case, assigned some of the disjunction connectives $\hat{\nabla}_1, \dots, \hat{\nabla}_q$.

(vii) The conditional clause ‘ IF ... THEN ... ’ is assigned the implication connective $\Rightarrow$.

In the case (iv) we suppose that each linguistic hedge is uniquely assigned a specific connective $\triangleleft$. However, because of the locality of the close coordination, we cannot be precise in the items (v) and (vi). We will mostly assume that ‘ and ’ is assigned the connective $\wedge$ or at most $\&$, and ‘ or ’ the connective $\vee$ (exclusively ∇).

Intension and extension of syntagms. In the logical analysis of natural language, three concepts play a distinguished role, namely intension, extension and possible world.

A possible world is, for example by C. A. Anderson in [1], a "way of things it could be in order for the expression to be true". D. Gallin in [27] speaks about "particular state of affairs" which determine truth of a sentence. The celebrated example by W. V. O. Quine is "the morning star is the evening star"[4], which is true in our world but may not be true in another world. From the formal point of view, possible worlds are certain formal objects (sets, etc.) whose structure is unimportant.

Intension of a linguistic syntagm, sentence, or of a concept, can be identified with the property denoted it. Intension may lead to different truth values in various possible worlds but it is invariant with respect to them. In intensional logic, intensions are functions assigning truth values to objects in each possible world.

Extension is a set of elements determined by an intension, which fall into the meaning of a syntagm in a given possible world. Thus, it depends on the particular context of use (see [27]). In the formalism of intensional logic, extension is the result of applying a function, which is an intension, to the world in question. It is accepted that extension is a function of the intensions of the semantically relevant parts of a syntagm in the given possible world (i.e. extension can be constructed from them).

There are many arguments supporting the idea that extensions are not sufficient to characterize the meaning of a syntagm. Most of them are based on the characterization of an identity between two differently specified objects. For example, to identify the meaning of the syntagm "morning star" with the set of objects denoted by it (i.e. Venus), and the same with the syntagm "evening star", is apparently insufficient since these syntagms take the meaning in all possible worlds (contexts) and thus, it is not sufficient simply to give a set of the corresponding objects.

The extensional approach is evident in all applications of the fuzzy set theory to the meaning of syntagms. For example, the meaning of "small" is usually identified with some fuzzy set of real numbers $A_{small} \underset{\sim}{\subset} \mathbb{R}$. When concerning people, we then say that small people form the fuzzy set

$$A_{small}(x) = \begin{cases} 1, & \text{if } x \leq 165 \text{ (cm)}, \\ \max\{0, \frac{1-(x-165)}{15}\} & \text{otherwise.} \end{cases}$$

However, we cannot use the same definition when speaking about small people 200 years ago, or in Guliver's Laputa. Moreover, the word "small" can also be used to characterize height of other things than people. On the other hand, it is important to observe that the shape of the fuzzy set assigned to "small" is always the same. Hence, there should exist a common characterization of the meaning of this word (and, of course, also of other ones), which does not depend on the extension. The mentioned intensional approach to the meaning of syntagms of natural language in connection with fuzzy logic seems to be a clue.

For further explanation, we will need the following definition.

[4]Venus.

DEFINITION 6.5
Let $A(x_1, \ldots, x_n) \in F_J$ be a formula with the free variables $x_1, \ldots, x_n$ of the respective sorts $\iota_1, \ldots, \iota_n$. Then the set

$$\mathbf{A}_{\langle x_1,\ldots,x_n\rangle} = \left\{ {}^{a_{t_1,\ldots,t_n}}/A_{x_1,\ldots,x_n}[t_1, \ldots, t_n] \,\middle|\, t_1 \in M_{\iota_1}, \ldots, t_n \in M_{\iota_n} \right\} \quad (6.3)$$

of evaluated formulas being closed instances of $A(x_1, \ldots, x_n)$ is called a multiformula.

We may also say that the formula $A(x_1, \ldots, x_n)$ generates the multiformula $\mathbf{A}_{\langle x_1,\ldots,x_n\rangle}$. Obviously, a multiformula $\mathbf{A}$ can be at the same time viewed as a *fuzzy set* $\mathbf{A} \underset{\sim}{\subset} F_J$ of *closed instances* of the formula $A(x_1, \ldots, x_n)$.

Though the free variables in $A(x_1, \ldots, x_n)$ are replaced by closed terms in each instance $A_{x_1,\ldots,x_n}[t_1, \ldots, t_n]$, it is sometimes important to notify them explicitly in the symbol $\mathbf{A}$. This is the reason why they have been put into subscript in (6.3). If the variables $x_1, \ldots, x_n$ are clear from the context then we will write simply $\mathbf{A}$ instead of $\mathbf{A}_{\langle x_1,\ldots,x_n\rangle}$.

If $\mathcal{D}$ is a structure for J then the interpretation $\mathcal{D}(\mathbf{A}_{\langle x_1,\ldots,x_n\rangle})$ is a fuzzy relation $R_{D,A} \underset{\sim}{\subset} D_1 \times \cdots \times D_n$ assigned to A in a way described in Section 4.3.

Recall that syntagms $\mathcal{A} \in \mathcal{S}$ are names of properties of objects. The latter can be represented in a formal language of logic using formulas $A \in F_J$. At the same time, vagueness of syntagms must be taken into account and thus, a simple assignment of A to $\mathcal{A}$ is not sufficient. The following definition offers a solution.

DEFINITION 6.6
Let $\mathcal{A} \in \mathcal{S}$ be a syntagm and $A(x_1, \ldots, x_n) \in F_J$ be a formula assigned to it. Then the multiformula $\mathbf{A}_{\langle x_1,\ldots,x_n\rangle}$ is the intension of $\mathcal{A}$. A structure $\mathcal{D}$ for J is a possible world and the extension of $\mathcal{A}$ in a given possible world $\mathcal{D}$ is an interpretation $\mathcal{D}(\mathbf{A}_{\langle x_1,\ldots,x_n\rangle})$.

We may explicitly write an extension of the syntagm $\mathcal{A} \in \mathcal{S}$ as

$$R_{D,A} = \mathcal{D}(\mathbf{A}) = \Big\{ \mathcal{D}(A_{x_1,\ldots,x_n}[t_1, \ldots, t_n]) / \langle \mathcal{D}(t_1), \ldots, \mathcal{D}(t_n)\rangle \Big| \quad t_i \in M_{\iota_i}, i = 1, \ldots, n \Big\}$$

where $\mathcal{D}(t_1) \in D_1, \ldots, \mathcal{D}(t_n) \in D_n$ are interpretations of the terms $t_1, \ldots, t_n$, respectively in the structure $\mathcal{D}$. It is clear that one intension $\mathbf{A}$ may lead to (infinitely) many extensions $\mathcal{D}(\mathbf{A})$ with different truth evaluations.

EXAMPLE 6.3
Let $\mathcal{A} :=$ *Young*. For simplicity, we will consider the background theory BT to be the theory of natural numbers and let $M_J = \{t_0, \ldots, t_{100}, \ldots\}$ be a set of terms representing years.

We choose a formula $Y(x) \in F_J$ and assign it to *Young*. Then the following multiformula is the intension of *Young*:

$$\mathbf{Y}_{\langle x\rangle} = \{ {}^{1}/Y_x[t_0], \ldots, {}^{1}/Y_x[t_{20}], \ldots, {}^{0.6}/Y_x[t_{30}], \ldots, {}^{0.2}/Y_x[t_{45}], \ldots, {}^{0}/Y_x[t_{60}]\}. \quad (6.4)$$

Let $\mathcal{D}$ be a structure with $D \subset \mathbb{N}$. Then a possible extension of *Young* can be, for example, the following.

$$\mathcal{D}(\mathbf{Y}_{\langle x\rangle}) = \{\, {}^{1}/1, \ldots, {}^{1}/20, \ldots, {}^{0.6}/30, \ldots, {}^{0.2}/45, \ldots, {}^{0}/60\} \tag{6.5}$$

where $\mathcal{D}(t_0) = 1, \ldots, \mathcal{D}(t_{20}) = 20, \ldots, \mathcal{D}(t_{30}) = 30, \ldots, \mathcal{D}(t_{45}) = 45, \ldots, \mathcal{D}(t_{60}) = 60$ are interpretations of the terms from M_J when representing age of people.

Other possible extension $\mathcal{D}'(\mathbf{Y})$ can be obtained when representing ages of dogs, i.e.

$$\mathcal{D}'(\mathbf{Y}_{\langle x\rangle}) = \{\, {}^{1}/0.1, \ldots, {}^{1}/4, \ldots, {}^{0.7}/6, \ldots, {}^{0.3}/8, \ldots, {}^{0}/14\} \tag{6.6}$$

where $\mathcal{D}(t_0) = 0.1, \ldots, \mathcal{D}(t_{20}) = 4, \ldots, \mathcal{D}(t_{30}) = 6, \ldots, \mathcal{D}(t_{45}) = 8, \ldots, \mathcal{D}(t_{60}) = 14$. Recall in these examples that if ${}^{a}/A$ is an evaluated formula then only the inequality $\mathcal{D}(A) \geq a$ should be fulfilled. This is illustrated in (6.6).

The fuzzy sets of the from (6.5) and (6.6) are introduced in various examples in the literature on fuzzy set theory as characterization of the meaning of *Young*. From our point of view, the authors always considered some extension of it. □

REMARK 6.1
Note that the role of the background theory is important since this enables us to characterize precisely the intension of the syntagm in concern. In the above example, this was the theory of natural numbers where we needed ordering and the successor. If we wanted to characterize the age more precisely using real numbers then the background theory would have to be the theory of fields.

6.1.3 The meaning of simple evaluating syntagms

We will now focus on the model of the simple evaluating syntagms. A question arises, what is their linguistic meaning and how it can be expressed in fuzzy logic. We will give a possible answer in this subsection.

Finite numbers and the meaning of the basic evaluation linguistic trichotomy. The deliberation leading to our model of the meaning of the words "small", "medium" and "big", i.e. the words forming the basic linguistic trichotomy of evaluating syntagms stems from the concept of finite numbers which is motivated by the ideas of the Alternative Set Theory (AST) presented by P. Vopěnka in [129, 130] and elaborated from the point of view of linguistics and fuzzy logic in [89]. These ideas have already been noticed in Chapter 1, namely in Example 1.1 on page 3, and also on page 10 while explaining the solution of the sorites paradox in fuzzy logic.

Unlike classically infinite sets, whose infinity is actual, i.e. there is no "world" continuing beyond them and therefore there is no possibility to cross over it, the AST infinity is natural; in some sense it is a synonym for non-transparency. The infinity is given by the *horizon*, which *can be crossed over*; the world is not finished by it and continues beyond. The horizon and the world before are represented by a class of finite numbers which are all "transparent" numbers, i.e. those whose infiniteness is easy to be verified[5].

There is a strong parallel between the concepts of finite numbers in the above sense and the small ones. Therefore, it seems reasonable to rely fuzzy logic definition of the meaning of "small" on the ideas presented above. Moreover, also the meaning of the other two members of the evaluation linguistic trichotomy, i.e. "medium" and "big", can be based on the same ideas and so, we can get a unified theory.

We begin with the observation that there is a tendency of people to classify three positions on any ordered scale, namely "the leftmost", "the rightmost", and "in the middle". The horizon of *small* lays "somewhere to the right from the leftmost position". Small numbers are numbers which everybody easily understands and is able to verify that they are small. On the other hand, there is no last small number, i.e. small numbers form a class but not set and we may encounter only the horizon of small numbers running "somewhere to big ones". Thus, small numbers behave similarly as finite ones. In the same way, the horizon of *big* lays "somewhere to the left from the rightmost position" and there is no first big number. In the case of *medium*, the horizon is spread both to the left as well as to the right from the middle.

This reasoning is perceptible also from the accepted form of the membership functions modelling the meaning of "small", "medium" and "big" as depicted on Figure 6.1. Below, we will give a formal setting of the above concept of

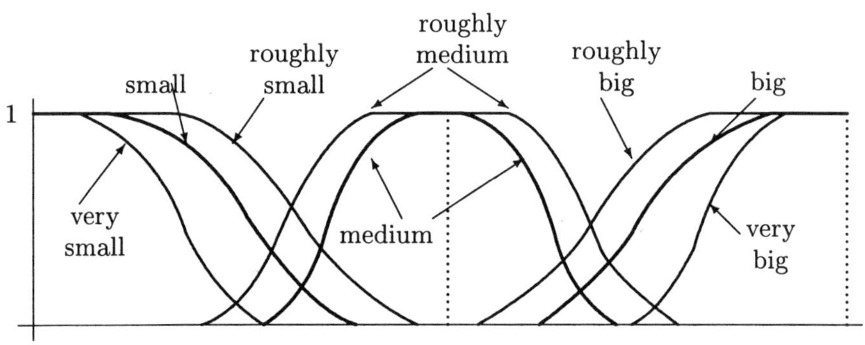

Figure 6.1. Typical membership functions of simple evaluating syntagms.

horizon and the meaning of the basic evaluation linguistic trichotomy as well as the other simple evaluating syntagms.

[5]For example, the number 10^{50}, though classically finite should be better taken as infinite since there is no real way (e.g. by counting lines) how to verify that this number is indeed finite. For the detailed discussion of these ideas see especially [130]

Canonical multiformula for the horizon on natural numbers. The meaning of the evaluation linguistic trichotomy can be characterized using the fuzzy logic formalization of the concept of horizon in the form provided by Theorem 4.25 on page 150. The horizon is there spread to the right from 0. Since we need it spread to both sides, we will introduce two special multiformulas. We will confine to the background fuzzy theory $\mathbb{N}$ of natural numbers without axiom of induction.

REMARK 6.2
A possible generalization to real numbers, i.e. when the background theory is the theory of fields, would have to be based on the following modification of axioms of the theory T in Theorem 4.25

$$T = \{\, {}^{1}/\mathbb{HR}_x[0],\ {}^{1-\varepsilon}/(\forall x)(\forall y)((\mathbf{0} \leq y - x)\,\&(y - x \leq \mathbf{1}) \Rightarrow \\ \Rightarrow (\varepsilon \Leftrightarrow (\mathbb{HR}(x)\,\&\,\neg\mathbb{HR}(y)))\,\&(\mathbb{HR}(x) \Rightarrow \mathbb{HR}(y)))\ {}^{1}/(\exists x)\neg\mathbb{HR}(x)\}$$

where the inequality predicate is taken as crisp and all the axioms of the background theory of fields are evaluated formulas in the degree 1. When confining to $\mathbb{N}$, the ideas are not basically changed but the explanation is significantly simplified, and so we will consider it in the sequel.

A term whose interpretation is a natural number will be called a *numeral*. In the following definition, '$*$' denotes the truncation operation (5.1) on page 5.1.

DEFINITION 6.7
Let $\mathbb{HR}$ be a predicate of horizon and m, n be numerals. Then we define the following multiformulas.

(i) Right horizon

$$\boldsymbol{RHR}_x(m) = \left\{ {}^{1}/\mathbb{HR}_x[m], \ldots, {}^{(1-n\varepsilon)^*}/\mathbb{HR}_x[m+n], \ldots \right\}. \tag{6.7}$$

(ii) Left horizon

$$\boldsymbol{LHR}_x(m) = \left\{ {}^{1}/\mathbb{HR}_x[m], \ldots, {}^{(1-n\varepsilon)^*}/\mathbb{HR}_x[m-n], \ldots \right\}. \tag{6.8}$$

These multiformulas are fundamental for our definition of the intension of the words forming the basic linguistic trichotomy.

Linguistic hedges. The model of the meaning of the linguistic hedges belongs to the important contributions of the fuzzy set theory. L. A. Zadeh proposed to model their meaning using certain transformations of the fuzzy sets, which represent the meaning of the words standing with the linguistic hedges.

The first studied linguistic hedge was "very". L. A. Zadeh proposed to assign it the operation $CON(a) = a^2$, $a \in [0, 1]$, using which the transformation of the corresponding fuzzy set is realized. For example, let us consider the simple evaluating syntagm "very small" and let the atomic evaluating syntagm "small" be assigned a fuzzy set $A_{small} \underset{\sim}{\subseteq} \mathbb{R}$. Then "very small" is assigned the fuzzy set $A_{very\ small} \underset{\sim}{\subseteq} \mathbb{R}$ given by the membership function

$$A_{very\ small}(x) = CON(A_{small}(x)) = A^2_{small}(x), \qquad x \in \mathbb{R}.$$

A similar model of the meaning of other linguistic hedges has been presented in [138, 141] or [87].

As can be immediately seen, the membership function of the fuzzy set A_{small} is steepened but its kernel remains unchanged because $1^2 = 1$. G. Lakoff in [61] pointed out that such model does not fit well the linguistic meaning because according to it, the set of surely small numbers (i.e. those with the membership degree equal to 1) is equal to the set of surely very small ones. However, there should exist small numbers not being very small. Consequently, the hedge *very* and other similar ones should steepen the membership function as well as shift it. Furthermore, the shifting depends on the kind of the atomic syntagm so that there is a difference between modification of the syntamgs "small" and "big" (see Figure 6.1). Therefore, another model of linguistic hedges must be searched for. A theory stemming from the mentioned concept of horizon has been rendered by V. Novák and I. Perfilieva in [101]. The main idea is as follows.

Let us focus on the linguistic hedge "very". If it is joined with the syntagm "small numbers" then the the meaning of the resulting evaluating syntagms is more accurate than the meaning of the former. Because "very small numbers" are small but there are small numbers which are not very small, a very small number is closer to 0; it is more transparent in the sense that it is easier to verify its finiteness. In other words, the horizon of very small numbers is nearer than that of the small ones. Hence, the main principle of the model of hedges consists in *shifting* of the horizon.

Note that in comparison with the linguistic hedge "very", an opposite effect is encountered with linguistic hedges such as "more or less, roughly", etc. , i.e. the horizon is there shifted farther. In general, we speak about the linguistic hedges with *narrowing effect* (very, highly, etc.) and those with *widening effect* (more or less, roughly, etc.). The effect of both kinds of hedges when applied to the three atomic syntagms "small", "medium", and "big" is depicted on Figure 6.1. Note that there is no linguistic hedge with narrowing effect to be applied to "medium".

In this book, the theory presented in [101] is slightly modified. The resulting membership function is there obtained using the information about element taken into consideration. In this book, we will model linguistic hedges by introducing special unary connectives into the language of FLb.

A special class of unary connectives. The importance of additional connectives in fuzzy logic has already been stressed. A suitable class is formed by unary connectives interpreted by functions monotonously increasing or decreasing truth degrees. We will call them hedges and denote by $\lhd$.

Hedges will be interpreted by functions $\lhd : L \longrightarrow L$ fulfilling the following conditions.

(i) The function $\lhd(x)$ is *logically fitting* in the sense of Definition 4.12 on page 113. If $L = [0,1]$ then in accordance with Theorem 4.3(c), the function $\lhd(x)$ is Lipschitz continuous on $[0,1]$.

(ii) There are points $a, b \in L$, $a < b$ such that the function $\lhd(x)$ is strictly increasing on (a,b), $\lhd(x) = 0$ if $x \in [0,a]$ and $\lhd(x) = 1$ if $x \in [b,1]$.

Hedges can be naturally ordered on the basis of ordering of the functions interpreting them. We will distinguish them using subscripts assigned in accordance with their ordering, i.e.

$$\alpha \leq \beta \quad \text{iff} \quad \lhd_\alpha \leq \lhd_\beta \quad \text{iff} \quad \lhd_\alpha(x) \leq \lhd_\beta(x), \qquad x \in L. \tag{6.9}$$

There are a lot of functions which may serve us as hedges. For our purposes, the following class of functions seem to be convenient.

$$\lhd(x) = (\lambda \sin(2.05x - \nu))^* \tag{6.10}$$

where λ and ν are special parameters. It can be verified that this function fulfils the requirements specified above.

Intensions of simple evaluating syntagms. We are now ready to define intensions of the class of simple evaluating syntagms due to Definition 6.1(i) and (iii) based on the basic linguistic trichotomy consisting of the syntagms "small", "medium", and "big".

Recall that we will work in the background fuzzy theory BT, which is the theory of natural numbers where the inequality predicate $\leq$ is taken as crisp and all the axioms are evaluated formulas in the degree 1. Let ϑ be a numeral representing some number, which will be taken as maximal, and δ a numeral such that $BT \vdash (0 < \delta) \,\&\, (\delta < \vartheta)$, i.e. it represents some chosen number between 0 and the maximal number.

Our definition of the intensions of the simple evaluating terms is based on the idea of modification of the horizon. For this purpose, the language J is supposed to contain a set of unary hedges Θ. Among them, one hedge is taken as central and all of them are ordered in the sense of (6.9) as follows:

$$\Theta = \{\lhd_{-k}, \ldots, \lhd_{-1}, \lhd_0, \lhd_1, \ldots, \lhd_p\}.$$

The mentioned central hedge is $\lhd_0$. Then the hedges $\lhd_{-k}, \ldots, \lhd_{-1}$ will be called the *hedges with narrowing effect* and $\lhd_1, \ldots, \lhd_p$ the *hedges with widening effect*.

DEFINITION 6.8

Let ϑ, δ be the above considered numerals representing the maximal number and some number between 0 and the maximal number, respectively.

(i) *The following multiformulas are intensions of the respective atomic evaluating syntagms "small", "medium" and "big":*

$$\mathbf{Sm}_x := \lhd_0 \, R\mathbb{HR}_x(0), \tag{6.11}$$
$$\mathbf{Me}_x := \lhd_0 \, L\mathbb{HR}_x(\delta) \cup \lhd_0 \, R\mathbb{HR}_x(\delta), \tag{6.12}$$
$$\mathbf{Bi}_x := \lhd_0 \, L\mathbb{HR}_x(\vartheta). \tag{6.13}$$

(ii) *Let $\mathcal{C}$:= '⟨linguistic hedge⟩$\mathcal{A}$' be a simple evaluating syntagm where the ⟨linguistic hedge⟩ is assigned the hedge $\lhd_\alpha \in \Theta \setminus \{\lhd_0\}$ and the intension* **A** *of its atomic evaluating syntagm $\mathcal{A}$ is one of (6.11)–(6.13). Then the intension of $\mathcal{C}$ is* **C** *obtained from* **A** *by replacing the hedge $\lhd_0$ by $\lhd_\alpha$.*

If ⟨linguistic hedge⟩ has narrowing effect then it is assigned the hedge $\lhd_\alpha \in \Theta \setminus \{\lhd_0\}$ where $\alpha \in \{-k, \dots, -1\}$, and if it has widening effect then it is assigned the hedge $\lhd_\alpha \in \Theta \setminus \{\lhd_0\}$ where $\alpha \in \{1, \dots, p\}$.

The main outcome of this definition is that we have obtained a unified model of the meaning of all the simple evaluating syntagms.

We can be more specific if we suppose that the hedges are interpreted by the functions $\lhd(x)$ of the form (6.10) for various settings of the parameters λ, ν. Then the following values give satisfactory results:

Linguistic hedge	*Hedge*	λ	ν
Extremely	$\lhd_{-3}$	1.45	1.26
Significantly	$\lhd_{-2}$	1.42	1.18
Very	$\lhd_{-1}$	1.3	1.02
—	$\lhd_0$	1.12	0.74
More or less	$\lhd_1$	1.2	0.58
Roughly	$\lhd_2$	1.44	0.62

Fuzzy quantities. To characterize intensions of the fuzzy quantities due to Definition 6.1(i), we must assume the existence of the fuzzy equality predicate $\doteq$ in the background fuzzy theory BT. Then we define the intension of the fuzzy quantity "*approximately* t_0" to be the multiformula

$$\mathbf{Q}[t_0]_{\langle x \rangle} = \left\{ \left. {}^{a_t}/(t \doteq t_0) \right| t \in M_\iota \text{ and } BT \vdash_{a_t} t \doteq t_0 \right\}. \tag{6.14}$$

Note that $a_{t_0} = \mathbf{1}$. We furthermore suppose that $a_t < \mathbf{1}$ for all $t \neq t_0$. As a special case, the intension of "*precisely* t_0" is the multiformula $\{ {}^{\mathbf{1}}/(t = t_0) \mid t \in M_\iota\}$. In general, intensions of the different fuzzy quantities can be defined using different fuzzy equalities. However, if not otherwise necessary, we will suppose one fuzzy equality only.

We say that fuzzy quantities with the respective intensions $\mathbf{Q}[t_0]_{\langle x \rangle}$ and $\mathbf{Q}[s_0]_{\langle y \rangle}$ are *disjoint* if $BT \vdash_{\mathbf{0}} (t \doteq t_0) \,\&\, (s \doteq s_0)$ for all terms $t, s \in M_\iota$. Since $BT \vdash t_0 \doteq t_0$ and $BT \vdash s_0 \doteq s_0$, it follows from this definition that $BT \vdash_{\mathbf{0}} t_0 \doteq s_0$.

Construction of intension of syntagms. On the basis of Definition 6.4, we will now define the intension of more complex syntagms from $\mathcal{S}$.

DEFINITION 6.9
Let $\mathcal{A}$ and $\mathcal{B}$ be syntagms with the intensions $\mathbf{A}_{\langle x_1,\dots,x_n\rangle}$ and $\mathbf{B}_{\langle y_1,\dots,y_p\rangle}$, respectively.

(i) Let $\mathcal{C}$:= ' not $\mathcal{A}$' be a negative syntagm. Then its intension is

$$\mathbf{C} = \neg\mathbf{A}_{\langle x_1,\dots,x_n\rangle} = \{\ \neg a_{t_1,\dots,t_n}/\neg A_{x_1,\dots,x_n}[t_1,\dots,t_n] \mid t_i \in M_{\iota_i}, i = 1,\dots,n\}.$$

(ii) Let $\mathcal{C}$:= '$\mathcal{A}$ and $\mathcal{B}$' be a compound syntagm. Then its intension is

$$\mathbf{C}_{\langle x_1,\dots,x_n,y_1,\dots,y_p\rangle} = \mathbf{A}_{\langle x_1,\dots,x_n\rangle} \mathbin{\hat{\&}} \mathbf{B}_{\langle y_1,\dots,y_p\rangle} = \\ = \Big\{\ a_{t_1,\dots,t_n}\ \mathbf{t}\ b_{s_1,\dots,s_p}/A_{x_1,\dots,x_n}[t_1,\dots,t_n] \mathbin{\hat{\&}} B_{y_1,\dots,y_p}[s_1,\dots,s_p] \Big| \\ t_i \in M_{\iota_i}, i = 1,\dots,n, s_k \in M_{\iota_k}, k = 1,\dots,p\} \tag{6.15}$$

where $\mathbf{t}$ is a t-norm interpreting the connective $\hat{\&}$.

(iii) Let $\mathcal{C}$:= '$\mathcal{A}$ or $\mathcal{B}$' be a compound syntagm. Then its intension is

$$\mathbf{C}_{\langle x_1,\dots,x_n,y_1,\dots,y_p\rangle} = \mathbf{A}_{\langle x_1,\dots,x_n\rangle} \hat{\nabla} \mathbf{B}_{\langle y_1,\dots,y_p\rangle} = \\ = \Big\{\ a_{t_1,\dots,t_n}\ \mathbf{s}\ b_{s_1,\dots,s_p}/A_{x_1,\dots,x_n}[t_1,\dots,t_n] \hat{\nabla} B_{y_1,\dots,y_p}[s_1,\dots,s_p] \Big| \\ t_i \in M_{\iota_i}, i = 1,\dots,n, s_k \in M_{\iota_k}, k = 1,\dots,p\} \tag{6.16}$$

where $\mathbf{s}$ is a t-conorm interpreting the connective $\hat{\nabla}$.

Recall from the remark on page 228 beyond Definition 6.4 that we will mostly consider the basic connectives $\wedge$, $\&$, $\vee$, ∇.

Construction of intension of evaluating predications. Definitions 6.8 and 6.9 give us rules how to construct intensions of the evaluating syntagms including the compound ones. We must now define the intension of the evaluating predications (see Definition 6.2).

Evaluating syntagms characterize in a linguistic way a property of objects. On the one hand, we may understand them without closer specification of the objects in concern. For example, we understand what does it mean "small" without closer specification of the objects which are spoken about to be small. This is reflected in the above definitions of their intensions, which are multi-formulas defined using some terms of the language J.

On the other hand, only the evaluating predications have for us concrete meaning. The problem lays in the meaning of ⟨noun⟩, which is a name of some objects. However, in the syntax of our logic, we have no means how to specify the objects concretely; this can be done only on the level of semantics where we assign the terms of the language J specific objects which then correspond to ⟨noun⟩. Thus, there are no means in our logic how the intension of the latter could be represented. This observation stands behind Definition 6.4, according to which ⟨noun⟩ is assigned a variable x. Remind also Example 6.3 where $\mathbf{Y}_{\langle x\rangle}$ can be understood as the intension of the evaluating predication " noun is young" where noun may be "man" as well as "dog". What noun is indeed taken into account can be seen only from the interpretation (i.e. the extensions (6.5) or (6.6). We see that the meaning of the evaluating predication is fully determined by the evaluating syntagm.

Before we give the definition, let us still observe the following. The evaluating syntagm contained in the evaluating predication expresses a property of a certain kind of objects. Therefore, we will consider its intension dependent on one variable only, i.e. of the form $\mathbf{A}_{\langle x\rangle}$.

DEFINITION 6.10

(i) Let $\mathcal{C}$:= '⟨noun⟩ is $\mathcal{A}$' be an evaluating predication, x be a variable assigned to ⟨noun⟩ and the intension of $\mathcal{A}$ be $\mathbf{A}_{\langle x\rangle}$. Then the intension of $\mathcal{C}$ is

$$\mathbf{C}_{\langle x\rangle} := \mathbf{A}_{\langle x\rangle}.$$

(ii) Let $\mathcal{A}$ and $\mathcal{B}$ be evaluating predications with the intensions $\mathbf{A}_{\langle x\rangle}$ and $\mathbf{B}_{\langle y\rangle}$, respectively. Then the intension of the compound evaluating predication $\mathcal{C}$:= '$\mathcal{A}$ and $\mathcal{B}$' is

$$\mathbf{C}_{\langle x,y\rangle} = \mathbf{A}_{\langle x\rangle} \mathbin{\hat{\&}} \mathbf{B}_{\langle y\rangle} = \left\{ {}^{a_t\,\mathbf{t}\,b_s}/A_x[t] \mathbin{\hat{\&}} B_y[s] \,\middle|\, t \in M_{\iota_1}, s \in M_{\iota_2} \right\} \tag{6.17}$$

where $\mathbf{t}$ is a t-norm interpreting the connective $\hat{\&}$. Similarly we define the intension of the compound evaluating predication containing the connective 'or', which is assigned $\hat{\nabla}$ interpreted by a t-conorm $\mathbf{s}$.

(iii) Let $\mathcal{A}$ and $\mathcal{B}$ be evaluating predications with the intensions $\mathbf{A}_{\langle x\rangle}$ and $\mathbf{B}_{\langle y\rangle}$, respectively. If $\mathcal{C}$:= ' IF $\mathcal{A}$ THEN $\mathcal{B}$' is a conditional clause then its intension is

$$\mathbf{C}_{\langle x,y\rangle} = \mathbf{A}_{\langle x\rangle} \Rightarrow \mathbf{B}_{\langle y\rangle} = \left\{ {}^{a_t \to b_s}/A_x[t] \Rightarrow B_y[s] \,\middle|\, t \in M_{\iota_1}, s \in M_{\iota_2} \right\}. \tag{6.18}$$

Let us stress that such interpretation of the evaluating predications is possible because the main constituent of them is the evaluating syntagm. From the linguistic point of view, our notion of evaluating predication is a special case of the syntagm formed by an adjective joined with a noun by means of the *general dependency relationship*. Construction of the meaning of such syntagms is in more general way elaborated in [89].

The above definition makes possible to construct intension of the evaluating predications consisting of compound evaluating syntagms as well as compound evaluating predications.

EXAMPLE 6.4
Let us consider the evaluating syntagm "very small" with the intension $\mathbf{VeSm}_x = \lhd_{-1} R\mathbb{HR}_x(0)$. Then $\mathbf{VeSm}_x \vee \mathbf{Me}_x$ is the intension of the evaluating syntagms "very small or medium" as well as of the predication "temperature is very small or medium" is $\mathbf{VeSm}_x \vee \mathbf{Me}_x$, provided that the noun "temperature" is assigned the variable x . On the other hand, the intension of "temperature is very small or pressure is medium" is $\mathbf{VeSm}_x \vee \mathbf{Me}_y$, provided that the noun "pressure" is assigned the variable y.

The intension of the conditional clause "IF temperature is very small THEN pressure is medium" is $\mathbf{VeSm}_x \Rightarrow \mathbf{Me}_y$. □

Linguistic variable. We have arrived to the slightly modified concept of the linguistic variable in comparison with that presented in Chapter 1, page 12. The *linguistic variable* is given by a set $T(\mathcal{X}) \subset \mathcal{S}$ of evaluating predications (6.1) *containing the same* ⟨noun⟩, which is the *name* $\mathcal{X}$ *of the linguistic variable.* The syntactic rule is a rule according to which the evaluating syntagms are constructed (e.g. given by our Definition 6.2). The significant difference lays in the semantical part since the definition on page 12 is purely extensional considering some universe U. From our point of view, U should be replaced by the set M_ι of closed terms of the sort ι on the basis of which the intensions of $\mathcal{A} \in T(\mathcal{X})$ are formed. Furthermore, the concept of the linguistic variable should be completed by a set of possible worlds (models) in which extensions of $\mathcal{A}$ are defined. Hence, we may define the linguistic variable as follows.

DEFINITION 6.11
The linguistic variable is a tuple

$$\langle \mathcal{X} := \langle noun \rangle, T(\mathcal{X}), G, M_\iota, \mathcal{M}, \mathcal{P} \rangle$$

where $T(\mathcal{X}) \subset \mathcal{S}$, G, M_ι have been defined above, $\mathcal{M}$ is a semantical rule assigning to each evaluating predication $\mathcal{A} \in T(\mathcal{X})$ its intension

$$\mathbf{A}_{\langle x \rangle} = \{ {}^{a_t}/A_x[t] \mid t \in M_\iota \}$$

and

$$\mathcal{P} = \{\mathcal{D} \mid \mathcal{D} \textit{is a structure for } J\}$$

is a set of possible worlds for $\mathcal{X}$.

6.2 Formal Scheme of FLb

Few basic concepts. The way how syntagms of natural language are formalized within fuzzy logic makes it possible to design a formal system of FLb analogously as any other formal logical system (cf. Sections 3.3 or 4.1). Namely, the syntagms from $\mathcal{S}$ take a role of linguistically expressed special axioms. Thus, we may introduce the following definition.

DEFINITION 6.12
Let $\mathcal{A}_i \in \mathcal{S}$, $i = 1, \ldots, m$ be syntagms with the respective intensions $\mathbf{A}_i$. A formal theory of FLb is given by

$$\mathcal{T} = \{\mathcal{A}_0[\mathbf{A}_0], \ldots, \mathcal{A}_m[\mathbf{A}_m]\}. \tag{6.19}$$

Since intensions are multiformulas, i.e. sets of evaluated formulas, the theory $\mathcal{T}$ in (6.19) is adjoined a fuzzy theory T of FLn

$$T = BT \cup \mathbf{A}_0 \cup \cdots \cup \mathbf{A}_m \tag{6.20}$$

where BT is the background fuzzy theory. Hence, all essential manipulations in FLb can be transformed into FLn.

In the sequel, we will denote the theory of FLb by the script letter $\mathcal{T}$ and the adjoint theory of FLn by the corresponding italic letter T. Moreover, the syntagms from $\mathcal{S}$ will usually be considered in the initial stage when constructing the intensions and the other formal procedures will proceed only with the latter. Therefore, the syntagms $\mathcal{A}$ themselves will be included in the explicit formalism (as is the case, e.g. of (6.19)) only when necessary and omitted otherwise.

DEFINITION 6.13
An inference rule of FLb is the scheme

$$R : \frac{\mathbf{A}_1, \ldots, \mathbf{A}_n}{\mathbf{B}} \tag{6.21}$$

where the multiformulas $\mathbf{A}_1, \ldots, \mathbf{A}_n$ are premises and the multiformula $\mathbf{B}$ is a conclusion of R.

Inference rules are usually based on some corresponding rule of FLn. Then the rule (6.21) takes a more specific form determined on the basis of Theorem 4.1 on page 100.

DEFINITION 6.14
The following are inference rules of FLb (recall that due to our agreement, we write one variable only at each intension).

(i) Modus ponens

$$R_{MP} : \frac{\mathbf{A}'_{\langle x \rangle}, \mathbf{A}_{\langle x \rangle} \Rightarrow \mathbf{B}_{\langle y \rangle}}{\mathbf{B}'_{\langle y \rangle}} \tag{6.22}$$

where $\mathbf{A}'_{\langle x \rangle}$ may differ from $\mathbf{A}_{\langle x \rangle}$ in the evaluations of the corresponding members. The resulting multiformula $\mathbf{B}'_{\langle y \rangle}$ has the form

$$\mathbf{B}'_{\langle y \rangle} = \left\{ \bigvee_{t \in M_{\iota_1}} (a_t \otimes c_{ts}) / B_y[s] \,\middle|\, s \in M_{\iota_2} \right\}.$$

(ii) Modus ponens with hedges

$$R_{MPH}: \frac{\lhd_\alpha \mathbf{A}_{\langle x\rangle}, \lhd_\beta \mathbf{A}_{\langle x\rangle} \Rightarrow \lhd_\gamma \mathbf{B}_{\langle y\rangle}}{\lhd_\delta \mathbf{B}'_{\langle y\rangle}} \tag{6.23}$$

where $\alpha \leq \beta$ and $\gamma \leq \delta$. The resulting multiformula $\lhd_\delta \mathbf{B}'_{\langle y\rangle}$ has the form

$$\lhd_\delta \mathbf{B}'_{\langle y\rangle} = \left\{ \bigvee_{t\in M_{\iota_1}} (a_t \otimes c_{ts}) / \lhd_\delta B_y[s] \,\middle|\, s \in M_{\iota_2} \right\}.$$

A *proof* in FLb is a sequence of multiformulas

$$\mathbf{B}_1, \ldots, \mathbf{B}_n \tag{6.24}$$

each of which is either an intension of a linguistically formulated axiom, or it is derived using some inference rule.

It is obvious that, given a theory $\mathcal{T}$ of FLb, (6.24) can be viewed as a multiple proof in the fuzzy theory T adjoint to $\mathcal{T}$. Thence, the concepts of syntactic and semantic consequence can be formulated in a straightforward way in accordance with FLn.

The general scheme of FLb is therefore the following. We form a fuzzy theory $\mathcal{T}$ in (6.19) using natural language (more precisely, using syntagms from $\mathcal{S}$). The theory $\mathcal{T}$ is assigned a fuzzy theory T of FLn. Then we proceed multiple reasoning within the latter. The result is a multiformula $\mathbf{B}$ which can be considered as a more precise intension of some syntagm $\mathcal{B} \in \mathcal{S}$ being the conclusion of the reasoning (deduction) if FLb.

A natural question arises, whether it is possible to find $\mathcal{B}$, given an intension $\mathbf{B}$. In the fuzzy set literature, this task is called a linguistic approximation and has been first formulated by L. A. Zadeh in [141]. A general algorithm solving this problem has not been given till now. For the description of some algorithms of the linguistic approximation, see, e.g. [87].

Fuzzy IF-THEN rules and the linguistic description. An outstanding role in fuzzy logic is played by the conditional clauses (6.2), which are implications characterized using natural language and used for description of some dynamic process or decision situation. A set of these statements is called the *linguistic description.*

DEFINITION 6.15
Let $\mathcal{A}_j, \mathcal{B}_j \in \mathcal{S}$, $j = 1, \ldots, m$ be evaluating predications with the respective intensions $\mathbf{A}_j$, $\mathbf{B}_j$ due to Definition 6.10. Then the linguistic description in FLb is either a finite set $\mathcal{LD}^I$ or a finite set $\mathcal{LD}^A$ of linguistic statements as follows.

(i) $\mathcal{LD}^I = \{\mathcal{R}_1^I, \ldots, \mathcal{R}_m^I\}$ where

$$\mathcal{R}_j^I = \textsf{IF } \mathcal{A}_j \textsf{ THEN } \mathcal{B}_j, \tag{6.25}$$

$j = 1, \ldots, m$, are conditional clauses with the intension (6.18).

(ii) $\mathcal{LD}^A = \{\mathcal{R}_1^A, \dots, \mathcal{R}_m^A\}$ *where*

$$\mathcal{R}_j^A = \mathcal{A}_j \text{ and } \mathcal{B}_j, \tag{6.26}$$

$j = 1, \dots, m$, *are compound evaluating predications with the intension (6.17).*

REMARK 6.3
The linguistic description can be itself understood as a complex syntagm. Then, depending on the form of the syntagms $\mathcal{R}_j$, we obtain the following two special linguistic descriptions:

$$\widehat{\mathcal{LD}^I} = \mathcal{R}_1^I \text{ and } \cdots \text{ and } \mathcal{R}_m^I \tag{6.27}$$

if $\mathcal{R}_j^I$ are conditional clauses (6.25) and

$$\widehat{\mathcal{LD}^A} = \mathcal{R}_1^A \text{ or } \cdots \text{ or } \mathcal{R}_m^A \tag{6.28}$$

if $\mathcal{R}_j^A$ are compound evaluating predications (6.26). Of course, we may then speak about intensions of the respective linguistic descriptions (6.27) and (6.28). Writing the detailed shape of the corresponding multiformulas is a technical task and it is left to the reader.

If all the evaluating predications in (6.25) or (6.26) are simple (see Definition 6.2(i)) then $\mathcal{LD}^I$ and $\mathcal{LD}^A$ are called ***simple linguistic descriptions***.

In correspondence with (6.20), the linguistic description defines a theory of FLb. Note that from the logical point of view, all (known to the authors) applications of fuzzy logic which use fuzzy IF-THEN rules, for instance in fuzzy control, can be understood as application of formal proving in FLb.

There are some misunderstandings concerning the meaning of fuzzy IF-THEN rules since in the applications, they are written in the if-then form but interpreted as (6.26), and sets of them are interpreted as (6.28). Consequently, erroneous terms such as "Mamdani's implication" interpreted by a simple symmetric operation (e.g. infimum $\wedge$) appeared. Though the latter term should be rejected, it has a good sense to call the compound evaluating predications (6.26) the fuzzy IF-THEN rules. The reason is that the linguistic description $\mathcal{LD}^A$ can be understood as an imprecise and partial characterization of some classical function using several couples of argument-value. We will analyze this question in more details in the next section.

6.3 Special Theories in FLb

In this section, we will be interested in specific kinds of theories of FLb based on the concept of the linguistic description. First, we will introduce the concept of independent formulas, which enables us to construct a structure in which the truth evaluation of the given set of independent formulas can be specified. Then, we will analyze two kinds of simple linguistic descriptions and show that in a special case, it is possible to attain the provability degrees.

6.3.1 *Independence of formulas*

DEFINITION 6.16

(i) Two formulas A and B are independent if no variant or instance of one is a subformula of the other one.

(ii) Let $F_0 \subset F_J$ be a set of evaluated formulas, which fulfils the following conditions:

(a) If $^a\!/A$, $^b\!/B \in F_0$ then A, B are independent.

(b) To each A there is at most one $a > \mathbf{0}$ such that $^a\!/A \in F_0$.

(c) If A is a logical axiom of FLn then $^a\!/A \in F_0$ implies $a = \mathrm{LAx}(A)$.

We will call F_0 the set of independent evaluated formulas.

Note that if $A(x), B(y)$ are independent then also all their respective instances are independent.

The proof of the following lemma is based on the technique developed by E. Turunen in [128].

LEMMA 6.1

Let F_0 be a set of independent evaluated formulas. We define a fuzzy theory $T = \{\, ^a\!/A \mid {}^a\!/A \in F_0\}$. Then there is a model $\mathcal{D} \models T$ such that

$$\mathcal{D}(A) = a \tag{6.29}$$

holds for all $^a\!/A \in F_0$.

PROOF: We construct a Henkin extension T_H of the theory T and the Lindenbaum algebra $\mathcal{L}(T_H)$. By Theorem 4.20(b) on page 143, $\mathcal{L}(T_H)$ is a residuated lattice. Let us denote its elements by $[\cdot]$.

Now, we construct an algebra Q being the smallest MV-algebra containing the set

$$Q_0 = \{[A] \mid {}^a\!/A \in F_0\} \cup \{[\bot]\}$$

so that if $[A], [B] \in Q$ then $[A] \to [B] = [A \Rightarrow B] \in Q$ is fulfilled. It can be verified that Q is a subalgebra of $\mathcal{L}(T_H)$.

Let us now define a function $f : Q \longrightarrow L$ as follows.

(a) $f([A]) = a$ if $^a\!/A \in F_0$,

(b) $f([\bot]) = \mathbf{0}$,

(c) $f([A] \to [B]) = f([A]) \to f([B])$.

Since F_0 is a set of independent formulas, the function f exists and it is a homomorphism. By Theorem 2.9(b) on page 41, the lattice of truth values L in consideration is injective and thus, f can be extended to a homomorphism

$$g : \mathcal{L}(T_H) \longrightarrow L.$$

Finally, we define a truth evaluation $G : F_J \longrightarrow L$ by $G(A) = g([A])$. Obviously, $G(A) = a$ for every $^a\!/A \in F_0$.

We will also show that $G((\forall x)B) = \bigwedge_{t\in M_\iota} H(B_x[t])$. Since T_H is Henkin and G is a homomorphism, we have $G((\forall x)B) = G(B_x[\mathbf{r}])$ where $\mathbf{r}$ is a special constant for $(\forall x)B$, both of the same sort ι.

At the same time, $G((\forall x)B) \leq G(B_x[t])$ holds for every term t of the sort ι. Furthermore, let $G(C) \leq G(B_x[t])$ hold for all terms t where C is some formula. Then $G(C) \leq G(B_x[\mathbf{r}]) = G((\forall x)B)$ as a special case, i.e. $G((\forall x)B)$ is infimum of all the truth evaluations $G(B_x[t])$, $t \in M_\iota$. Analogously we proceed for suprema, using negation.

Hence, using G, we can construct a canonical structure $\mathcal{D}$ (cf. Definition 4.18 on page 119), which is a model of the theory T_H and has the property (6.29). But then $\mathcal{D} \models T$ follows from the fact that T_H is a conservative extension of T. □

This lemma plays an important role in proving some theorems about approximate reasoning below. Using it and the completeness theorem we can prove the following two lemmas. However, first we give a corollary whose proof is immediate.

COROLLARY 6.1

Let Θ be a set of hedges, $J' = J \cup \Theta$ be an extended language, and let F_0, T and $\mathcal{D}$ be the same as in Lemma 6.1. Furthermore, let us extend $\mathcal{D}$ into a structure $\mathcal{D}'$ for J' by putting $\mathcal{D}'(A) = \mathcal{D}(A)$ for every formula $A \in F_J$ and $\mathcal{D}'(\lhd A) = \lhd(a)$ for every evaluated formula $a/A \in F_0$. Then $\mathcal{D}'$ is a model of the fuzzy theory

$$T' = T \cup \left\{ \left. {}^{\lhd(a)}/\lhd A \right| \; {}^{a}/A \in G \right\}.$$

for every set $G \subseteq F_0$ of evaluated formulas.

The following lemma characterizes conditions under which we can obtain maximal provability degrees when dealing with a finite set of implications between independent formulas.

LEMMA 6.2

Let $A_1(x), \ldots, A_m(x)$ be independent formulas where x is a variable of the sort ι_1 and similarly, $B_1(y), \ldots, B_m(y)$ be independent formulas where y is a variable of the sort ι_2. Let

$$T = \left\{ {}^{a_{k,t}}/A_{k,x}[t],\ {}^{c_{j,ts}}/A_{j,x}[t] \Rightarrow B_{j,y}[s] \,\middle|\, t \in M_{\iota_1}, s \in M_{\iota_2}, j = 1, \ldots, m \right\}$$

be a fuzzy theory where $1 \leq k \leq m$. Then T is consistent and

$$T \vdash_{b_{k,s}} B_{k,y}[s], \quad b_{k,s} = \bigvee_{t\in M_{\iota_1}} (a_{k,t} \otimes c_{k,ts}), \qquad s \in M_{\iota_2}. \tag{6.30}$$

PROOF: Put

$$F_0 = \left\{ {}^{a_{j,t}}/A_{j,x}[t],\ {}^{b_{j,s} = \bigvee_{t\in M_{\iota_1}} (a_{j,t} \otimes c_{j,ts})}/B_{j,y}[s] \,\middle|\, t \in M_{\iota_1}, s \in M_{\iota_2}, j = 1, \ldots m \right\}.$$

It follows from the assumptions that F_0 is a set of independent evaluated formulas. By Lemma 6.1, there exists a structure $\mathcal{D}$ such that

$$\mathcal{D}(A_{j,x}[t]) = a_{j,t}, \qquad \mathcal{D}(B_{j,y}[s]) = b_{j,s},$$

for all $j = 1, \ldots, m$. We will show that $\mathcal{D} \models T$.

Obviously,

$$a_{j,t} \otimes c_{j,ts} \leq \bigvee_{t \in M_{\iota_1}} (a_{j,t} \otimes c_{j,ts})$$

holds for all $t \in M_{\iota_1}$ and $s \in M_{\iota_2}$ and j. By the adjunction, we obtain

$$c_{j,ts} \leq a_{j,t} \to \bigvee_{t \in M_{\iota_1}} (a_{j,t} \otimes c_{j,ts}) = \mathcal{D}(A_{j,x}[t]) \to \mathcal{D}(B_{j,y}[s]) = \\ = \mathcal{D}(A_{j,x}[t] \Rightarrow B_{j,y}[s])$$

for all $j = 1, \ldots m$, and thus, $\mathcal{D} \models T$ is a model of T, which implies that T is consistent.

Consider the proofs

$$w_{k,ts} := a_{k,t}/A_{k,x}[t] \; \{spec.\ axiom\}, \; c_{k,ts}/A_{j,x}[t] \Rightarrow B_{j,y}[s] \; \{spec.\ axiom\}, \\ a_{k,t} \otimes c_{k,ts}/B_{k,y}[s] \; \{r_{MP}\},$$

$t \in M_{\iota_1}$, $s \in M_{\iota_2}$. Then

$$b_{k,s} \geq \bigvee_{t \in M_{\iota_1}} \mathrm{Val}_T(w_{k,ts}) = \bigvee_{t \in M_{\iota_1}} (a_{k,t} \otimes c_{k,ts}).$$

But at the same time, $b_{k,s} \leq \mathcal{D}(B_{k,y}[s]) = \bigvee_{t \in M_{\iota_1}} (a_{k,t} \otimes c_{k,ts})$, which gives (6.30) by the completeness. □

The final lemma of this subsection characterizes the provability degree when some relation R is approximated from below by a disjunction of conjunctions of independent formulas.

LEMMA 6.3
Let $A_1(x), \ldots, A_m(x)$ be independent formulas where x is a variable of the sort ι_1, $B_1(y), \ldots, B_m(y)$ be independent formulas where y is a variable of the sort ι_2, and $R(x, y)$ be an atomic formula. Furthermore, let $A'(x)$ be a formula either independent on $A_1, \ldots, A_n$ or one of them. Put

$$T = \Big\{ a_{j,t} \wedge b_{j,s}/A_{j,x}[t] \wedge B_{j,y}[s], \; \mathbf{1}/((A_{j,x}[t] \wedge B_{j,y}[s]) \Rightarrow R_{x,y}[t,s]), \\ a'_t/A'_x[t] \,\Big|\, t \in M_{\iota_1}, s \in M_{\iota_2}, j = 1, \ldots, m \Big\}. \quad (6.31)$$

Finally put $B'(y) := (\exists x)(A'(x) \wedge R(x, y))$. If $T \vdash_{a'_t} A'_x[t]$ for all $t \in M_{\iota_1}$ then $T \vdash_{b'_s} B'_y[s]$, $s \in M_{\iota_2}$ where

$$b'_s = \bigvee_{t \in M_{\iota_1}} \left(a'_t \wedge \bigvee_{j=1}^{m} (a_{j,t} \wedge b_{j,s}) \right). \quad (6.32)$$

PROOF: Using the tautology (P11) and modus ponens, we obtain

$$T \vdash_{d_{ts}} R_{x,y}[t,s], \qquad t \in M_{\iota_1}, s \in M_{\iota_2}$$

where $d_{ts} \geq \bigvee_{j=1}^{m}(a_{j,t} \wedge b_{j,s})$. Then there is a set of proofs in T

$$\begin{aligned} w_{j,t} := \ & a'_t / A'_x[t] \ \{assumption\}, \dots, \ a_{j,t} \wedge b_{j,s} / R_{x,y}[t,s] \ \{modus\ ponens\}, \\ & \dots, \ a'_t \wedge a_{j,t} \wedge b_{j,s} / A'_x[t] \wedge R_{x,y}[t,s], \dots \\ & a'_t \wedge (a_{j,t} \wedge b_{j,s}) / (\exists x)(A'(x) \wedge R(x)_y[s]) \ \{(Q8),\ r_{MP}\}, \end{aligned}$$

$t \in M_{\iota_1}$, $j = 1, \dots, m$. From it follows that

$$\bigvee_{\substack{t \in M_{\iota_1} \\ j=1,\dots,m}} \mathrm{Val}_T(w_{j,t}) = \bigvee_{t \in M_{\iota_1}} \bigvee_{j=1}^{m} (a'_t \wedge (a_{j,t} \wedge b_{j,s})) = \bigvee_{t \in M_{\iota_1}} \left(a'_t \wedge \bigvee_{j=1}^{m} (a_{j,t} \wedge b_{j,s}) \right)$$

which gives $T \vdash_{c_s} B'_y[s]$ where

$$c_s \geq \bigvee_{t \in M_{\iota_1}} \left(a'_t \wedge \bigvee_{j=1}^{m} (a_{j,t} \wedge b_{j,s}) \right). \tag{6.33}$$

Since the formulas A_j, B_j, $j = 1, \dots, m$, and A' are independent, by Lemma 6.1 there exists a structure $\mathcal{D}'$ such that

$$\mathcal{D}'(A_{j,x}[t] \wedge B_{j,y}[s]) = a_{j,t} \wedge b_{j,s}, \qquad \mathcal{D}'(A'_x[t]) = a'_t$$

for all $j = 1, \dots m$, $t \in M_{\iota_1}$, $s \in M_{\iota_2}$.

Since the formula $R(x,y)$ is independent on the above formulas, we can construct a structure $\mathcal{D}$ as follows. We put $D = D'$ and, furthermore,

$$\mathcal{D}(A_{j,x}[t] \wedge B_{j,y}[s]) = \mathcal{D}'(A_{j,x}[t] \wedge B_{j,y}[s]),$$

$$\mathcal{D}(R_{x,y}[t,s]) = r_{ts} \geq \bigvee_{j=1}^{m} (a_{j,t} \wedge b_{j,s}) \tag{6.34}$$

where $t \in M_{\iota_1}$, $s \in M_{\iota_2}$, $j = 1, \dots m$ and $r_{ts} \in L$ is an appropriate value. Then

$$\mathcal{D}(((A_{j,x}[t] \wedge B_{j,y}[s]) \Rightarrow R_{x,y}[t,s])) = \bigwedge_{t \in M_{\iota_1}, s \in M_{\iota_2}} ((a_{j,t} \wedge b_{j,s}) \to r_{ts}) = \mathbf{1}$$

by the definition of $\mathcal{D}$ and consequently, $\mathcal{D} \models T$.

Finally, we may, as a special case, take a model $\mathcal{D}$ such that the equality holds in (6.34) and therefore,

$$c_s \leq \mathcal{D}(B_y[s]) = \mathcal{D}((\exists x)(A'(x) \wedge R(x)_y[s])) =$$

$$= \bigvee_{t \in M_{\iota_1}} \mathcal{D}(A'_x[t] \wedge R_{x,y}[t,s]) = \bigvee_{t \in M_{\iota_1}} \left(a'_t \wedge \bigvee_{j=1}^{m} (a_{j,t} \wedge b_{j,s}) \right),$$

which together with (6.33) gives (6.32). □

6.3.2 *Deduction on simple linguistic descriptions*

The concept of independent formulas is useful for the characterization of the behaviour of linguistic descriptions containing at most simple evaluating predications due to Section 6.1. Intensions of these syntagms are given by multiformulas generated by formulas which, when using variables of different sorts, become independent. In this subsection, we present a theorem characterizing the basic behavior of the linguistic description $\mathcal{LD}^I$, which consists of the IF-THEN rules being linguistically characterized logical implications.

Recall that on page 227, the predicate language J has been fixed containing, besides others, also linguistic hedges. We will modify it into the language $\hat{J}$ as follows.

We will remove from J all the hedges and instead of them, add to it new predicate symbols $H, \ldots$ corresponding to all the possible simple evaluating predications. Then, with respect to Definition 6.10, we put:

- Let $\mathcal{C}$:= '⟨noun⟩ is ⟨linguistic hedge⟩$\mathcal{A}$' be a simple evaluating predication with the intension

$$\mathbf{C} = \lhd \mathbf{A}_{\langle x \rangle} = \left\{ \lhd(a_t) / \lhd A_x[t] \middle| t \in M_\iota \right\}.$$

  Then

$$\mathbf{C} := \mathbf{H}_{\langle x \rangle} = \left\{ h_t = \lhd(a_t) / H_x[t] \middle| t \in M_\iota \right\}$$

  where $H \in \hat{J}$ is the corresponding new predicate symbol.

This trick enables us to use Lemma 6.1.

We will now consider a simple linguistic description $\mathcal{LD}^I$ consisting of m IF-THEN rules and an evaluating predication $\mathcal{A}_k$ occurring in the antecedent of some rule from $\mathcal{LD}^I$. Intensions of all the evaluating predications are supposed to be translated into the language $\hat{J}$ in the above described way. Let $\mathcal{A}'_k$ be a possible modification of $\mathcal{A}_k$ such that the intension $\mathbf{A}'_{k,\langle x \rangle}$ may differ from $\mathbf{A}_{k,\langle x \rangle}$ in some (possibly all) syntactic truth evaluations.

THEOREM 6.1
Let

$$\mathcal{T} = \{\mathcal{A}'_k, \mathcal{LD}^I\},$$

be a theory of FLb where $\mathcal{LD}^I$ is a simple linguistic description and $\mathcal{A}'_k$ the evaluating predication described above. Then we may derive a conclusion $\mathcal{B}'_k$ with the intension

$$\mathbf{B}'_{k,\langle y \rangle} = \left\{ b'_{k,s} = \bigvee\nolimits_{t \in M_{\iota_1}} (a'_{k,t} \otimes c_{k,ts}) / B_{k,y}[s] \middle| s \in M_{\iota_2} \right\} \tag{6.35}$$

such that all $b'_{k,s}$ for $s \in M_{\iota_2}$ in the multiformula $\mathbf{B}'_{k,\langle y \rangle}$ are maximal.

PROOF: The theory $\mathcal{T}$ determines the fuzzy theory

$$T = \{\mathbf{A}'_{k,\langle x\rangle}, \mathbf{A}_{j,\langle x\rangle} \Rightarrow \mathbf{B}_{j,\langle y\rangle} \mid j = 1, \dots, m\} =$$
$$= \left\{ {}^{a'_{k,t}}/A_{k,x}[t],\ {}^{c_{j,ts}}/A_{j,x}[t] \Rightarrow B_{j,y}[s] \middle| t \in M_{\iota_1}, s \in M_{\iota_2}, j = 1, \dots, m \right\}$$

where $c_{j,ts} = a_{j,t} \to b_{j,s}$ due to Definition 6.10(iii). Because the IF-THEN rules from $\mathcal{LD}^I$ consist of the simple evaluating predications, we may translate them into the language $\hat{J}$. The theorem then follows from Lemma 6.2. □

This theorem explicitly states the following: if we interpret IF-THEN rules as logical implications formed of the simple evaluating predications then the deduction based on them leads to the conclusion, which is the best possible one — in the sense of maximization of the provability degrees within the fuzzy theory determined by them.

In fuzzy control, a certain precise value u_0 is usually at disposal, which is obtained by measuring of some characteristics of the controlled process. From our point of view, this value is an interpretation of some closed term t_0 in the language J. Given a linguistic description $\mathcal{LD}^I$, a question now arises whether it is possible to cope with it in such a way that Theorem 6.1 can be applied. To find an answer, let us consider the following example.

EXAMPLE 6.5
Given a linguistic description consisting of two rules

$$\mathcal{R}_1 := \ \mathsf{IF}\ x\ \mathsf{is\ small}\ \ \mathsf{THEN}\ y\ \mathsf{is\ big}$$
$$\mathcal{R}_2 := \ \mathsf{IF}\ x\ \mathsf{is\ big}\ \ \mathsf{THEN}\ y\ \mathsf{is\ small}\ .$$

These rules are interpreted by the respective multiformulas $\mathbf{Sm}_{\langle x\rangle} \Rightarrow \mathbf{Bi}_{\langle y\rangle}$ and $\mathbf{Bi}_{\langle x\rangle} \Rightarrow \mathbf{Sm}_{\langle y\rangle}$.

To be specific, let us consider a model $\mathcal{D}_0$ with the support $U = V = [0, 1]$. Then one may agree that small values should lay around 0.3 (or smaller) and big ones around 0.7 (or bigger). Of course, given the measured value, say $x = 0.3$, we expect the result being "big" due to the rule $\mathcal{R}_1$. Similarly, for $x = 0.75$ we expect the result being "small" due to the rule $\mathcal{R}_2$. With respect to our theory, this means that the value 0.3 is an interpretation of some term t_0 in $\mathcal{D}_0$, i.e. $\mathcal{D}_0(t_0) = 0.3$ (and the like for 0.7). Then we have a specific intension of the form $\mathbf{Sm}'_{\langle x\rangle} = \{\ {}^{\mathbf{1}}/Sm_x[t_0]\}$ and when applying Theorem 6.1, we expect the result to be a multiformula $\mathbf{Bi}'_{\langle y\rangle}$. □

The term t_0 considered in this example can be understood as a *typical example of the evaluating predication* "⟨noun⟩ is small". In general, given an evaluating predication $\mathcal{A}$ with the intension $\mathbf{A}_{\langle x\rangle}$, we derive a specific syntagm $Ex(\mathcal{A})$:="typical example of $\mathcal{A}$" with the intension

$$Ex(\mathbf{A}_{\langle x\rangle}) = \{\ {}^{\mathbf{1}}/A_x[t_0]\}\,. \tag{6.36}$$

Determination of the term t_0, however, is out of logic. There are algorithms for solution of this task, implemented, e.g. in the software system LFLC [95]. The described situation is mathematically characterized in the following corollary of Theorem 6.1.

COROLLARY 6.2
Let

$$\mathcal{T} = \{Ex(\mathcal{A}_k), \mathcal{LD}^I\}$$

be a theory of FLb where $\mathcal{LD}^I$ *is a simple linguistic description and* $Ex(\mathcal{A}_k)$ *is the syntagm "typical example of* $\mathcal{A}_k$*" with the intension (6.36). Then we may derive a conclusion* $\mathcal{B}'_k$ *with the intension*

$$\mathbf{B}'_{k,\langle y\rangle} = \left\{ \left. {}^{b'_{k,s}} = c_{k,t_0 s}/B_{k,y}[s] \right| s \in M_{\iota_2} \right\} \tag{6.37}$$

and all $b'_{k,s}$ *for* $s \in M_{\iota_2}$ *in the multiformula* $\mathbf{B}'_{k,\langle y\rangle}$ *are maximal.*

Since the intension of k-th rule of $\mathcal{LD}^I$ has the form $\mathbf{A}_{k,\langle x\rangle} \Rightarrow \mathbf{B}_{k,\langle y\rangle}$, the $b'_{k,s}$ are equal to $b'_{k,s} = a_{k,t_0} \rightarrow b_{k,s}$.

Let us remark that the concept of a "typical example of $\mathcal{A}$" appeared to be fruitful. It plays a fundamental role in the implementation of the algorithm (in the mentioned system LFLC), which learns the linguistic description $\mathcal{LD}^I$ on the basis of the data obtained from monitoring of the successful control. The goal was to control the given process using the learned description. The results are very promising (see [5]).

6.3.3 Fuzzy approximation based on simple linguistic descriptions

In this section we formulate the result stating that the standard Mamdani's Max-Min fuzzy approximation rule can be obtained as a consequence of special axioms in which we consider a predicate $R(x, y)$ representing some relation to be approximated by the linguistic description consisting of disjunction of conjunctions. The linguistic description is simple, i.e. consisting of exactly those syntagms considered in the known practical applications. Furthermore, the approximation rule gives the best possible result in the sense of the provability degree. In other words, the formulas used in the generalized modus ponens proposed already by the classics of fuzzy logic, namely L. A. Zadeh and E. H. Mamdani, give the maximal possible truth value, provided that we confine ourselves to linguistic syntagms of the special kind.

Fuzzy approximation of a relation. The theorem below is, analogously to Theorem 6.1, formulated on the basis of Lemma 6.3. First, we will introduce a special syntagm $\mathcal{R} :=$ "$\langle$noun$\rangle_1$ is in relation with $\langle$noun$\rangle_2$" with the intension

$$\mathbf{R}_{\langle x,y\rangle} = \left\{ \left. {}^{r_{ts}}/R_{x,y}[t,s] \right| t \in M_{\iota_1}, s \in M_{\iota_2} \right\} \tag{6.38}$$

where R is some binary predicate symbol.

DEFINITION 6.17
Let $\mathcal{T}$ be a theory of FLb. We say that the linguistic statement $\mathcal{A}$, which can be an evaluating predication or a conditional clause is true in $\mathcal{T}$ if it has the intension $\mathbf{A}_{\langle x_1,\dots,x_n\rangle} = \{\ \mathbf{1}/A_{x_1,\dots,x_n}[t_1,\dots,t_n] \mid t_i \in M_{\iota_i}, i = 1,\dots,n\}$.

Our task now is to characterize more concretely the relation in concern, which can be understood as a certain grouping of couples of elements. We can divide it into parts and then characterize the couples of elements forming each part using evaluating predications "$\langle$noun$\rangle_1$ is $\mathcal{A}$ and $\langle$noun$\rangle_2$ is $\mathcal{B}$" where $\langle$noun$\rangle_1$ is a name of the first element of each couple and $\langle$noun$\rangle_2$ the name of the second one. Each considered part of the relation thus can be linguistically described using the conditional clause of the form

$$\mathcal{P} := \textsf{IF } \langle\text{noun}\rangle_1 \textsf{ is } \mathcal{A} \textsf{ and } \langle\text{noun}\rangle_2 \textsf{ is } \mathcal{B} \textsf{ THEN } \mathcal{R}. \tag{6.39}$$

This conditional clause is a linguistic statement, which express imprecisely our knowledge, but which otherwise should be taken as true. This means that the intension of the whole clause $\mathcal{P}$ is

$$\mathbf{P}_{\langle x,y\rangle} := \left\{\ \mathbf{1}/(A_x[t] \wedge B_y[s] \Rightarrow R_{x,y}[t,s])\middle|\, t \in M_{\iota_1}, s \in M_{\iota_2}\right\}$$

provided that $\langle$noun$\rangle_1$ is assigned the variable x and $\langle$noun$\rangle_2$ the variable y (note that this does not contradict to the fact that the knowledge is imprecise). On the basis of this, we may formulate the following theorem.

THEOREM 6.2
Let $\mathcal{LD}^A$ be a simple linguistic description consisting of m rules and $\mathcal{P}_j$, $j = 1,\dots,m$ be the true conditional clauses of the form (6.39). Suppose that $\langle$noun$\rangle_1$ is assigned a variable x and $\langle$noun$\rangle_2$ a variable y and let $\mathcal{A}'$ be a simple evaluating predication with the intension $\mathbf{A}'_{\langle x\rangle}$. Then there is the theory of FLb

$$\mathcal{T} = \{\mathcal{LD}^A, \mathcal{P}_j, \mathcal{A}' \mid j = 1,\dots,m\}$$

such that the following can be proved.

(a) *We may derive a conclusion $\mathcal{B}'_y$ in $\mathcal{T}$ with the intension*

$$\mathbf{B}'_y = \left\{\ b'_s = \bigvee_{t\in M_{\iota_1}} (a'_t \wedge \bigvee_{j=1}^m (a_{j,t} \wedge b_{j,s}))/B'_y[s]\middle|\ s \in M_{\iota_2}\right\}$$

where $B'(y) := (\exists x)(A'(x) \wedge R(x,y))$ and all b'_s for $s \in M_{\iota_2}$ in the multiformula $\mathbf{B}'_{\langle y\rangle}$ are maximal.

(b) *The linguistic statement*

$$\textsf{IF } \widehat{\mathcal{LD}^A} \textsf{ THEN } \mathcal{R} \tag{6.40}$$

is true in $\mathcal{T}$.

PROOF: Because $\mathcal{LD}^A$ is a simple linguistic description and the generating formula $R(x,y)$ of the intension (6.38) is atomic we may construct a fuzzy theory T (6.31) adjoint to $\mathcal{T}$. Then (a) follows from Lemma 6.3.

(b) Using the tautology (P11) and modus ponens we obtain $T \vdash \bigvee_{i=1}^{m}(A_x[t] \wedge B_y[s]) \Rightarrow R_{x,y}[t,s]$ for all $t \in M_{\iota_1}$, $s \in M_{\iota_2}$. This means that the linguistic statement (6.40) has the intension

$$\left\{ \mathbf{1}/(\bigvee_{i=1}^{m}(A_x[t] \wedge B_y[s]) \Rightarrow R_{x,y}[t,s]) \,\middle|\, t \in M_{\iota_1}, s \in M_{\iota_2} \right\}.$$

□

EXAMPLE 6.6

Let us consider breaking of a car whose speed depends on the breaking force. Then $\mathcal{R}$:= "force is in relation with speed". Of course, we know that the higher force, the lower speed of the car and vice-versa. Hence, the statements (6.39) characterizing this relation may take the form

" IF force is big and speed is small THEN force is in relation with speed"
" IF force is small and speed is big THEN force is in relation with speed"

At the same time, we have the linguistic description

{ force is big and speed is small , force is small and speed is big }.

Theorem 6.2 then states that the linguistic statement

IF force is small and speed is big or force is big and speed is small
THEN force is in relation with speed

is true. Furthermore, when knowing that, e.g. "force is very big", the theorem gives use method how to compute the *intension* of the linguistic syntagm characterizing the speed. Finding its linguistic form, however, is a question of the linguistic approximation. □

It follows from our deliberation that we have in FLb basically two inference methods for dealing with linguistic descriptions. The first one deals with the linguistic description $\mathcal{LD}^I$, which consists of linguistically formulated logical implications. The second one is based on the additional assumption and deals with the linguistic description $\mathcal{LD}^A$, which consists of conjunctions of linguistic predications. The question arises, what is the difference between both methods.

We see that the statements (6.39) lead to characterization (6.40) of the relation $\mathcal{R}$ from below — whenever a correspondence between some objects can be characterized using conjunction of evaluating predications then they are in the considered relation. When inspecting the proof of Lemma 6.3, on the basis of which the previous theorem is formulated, we see that reverse of this implication would not give us the proved result. On the other hand, we may characterize $\mathcal{R}$ also in another way, namely using the true conditional clauses of the form

$$\mathcal{P} := \text{ IF } \mathcal{R} \text{ and } \langle\text{noun}\rangle_1 \text{ is } \mathcal{A} \text{ THEN } \langle\text{noun}\rangle_2 \text{ is } \mathcal{B}. \tag{6.41}$$

With respect to Example 6.6, the statement (6.41) may have the form, e.g. " IF force is in relation with speed and force is big THEN speed is small".

THEOREM 6.3
Let $\mathcal{P}_j$, $j = 1, \ldots, m$ be the true conditional clauses of the form (6.41), which determine the theory of FLb

$$\mathcal{T} = \{\mathcal{P}_j \mid j = 1, \ldots, m\}.$$

Then there is a linguistic description $\mathcal{LD}^I$ such that the conditional clause

$$\mathsf{IF}\ \mathcal{R}\ \mathsf{THEN}\ \widehat{\mathcal{LD}^I} \tag{6.42}$$

is true in $\mathcal{T}$.

PROOF: The fuzzy theory T adjoint to $\mathcal{T}$ has the form

$$T = \left\{ \mathbf{1}/(R_{x,y}[t,s] \,\&\, A_{j,x}[t]) \Rightarrow B_{j,y}[s] \big| j = 1, \ldots, m, t \in M_{\iota_1}, s \in M_{\iota_2} \right\}.$$

Then using the tautologies (P10) and (P12) we conclude that

$$T \vdash (R_{x,y}[t,s] \Rightarrow \bigwedge_{j=1}^{m} (A_{j,x}[t]) \Rightarrow B_{j,y}[s])$$

for all $t \in M_{\iota_1}$ and $s \in M_{\iota_2}$. We have thus obtained the multiformula

$$\mathbf{P}_{\langle x,y\rangle} = \left\{ \mathbf{1}/(R_{x,y}[t,s] \Rightarrow \bigwedge_{j=1}^{m} (A_{j,x}[t]) \Rightarrow B_{j,y}[s]) \middle| t \in M_{\iota_1}, s \in M_{\iota_2} \right\},$$

which is the intension of the true conditional clause (6.42) in $\mathcal{T}$. □

It follows from this theorem that the linguistic description $\mathcal{LD}^I$ makes it possible to characterize the relation $\mathcal{R}$ from above — whenever some objects are in the considered relation then it can be characterized using implications between evaluating predications. In general, it follows from various experiments, and also from the theoretical analysis that $\mathcal{LD}^I$ should be used when we need to deduce statements about some facts, while $\mathcal{LD}^A$ is suitable when need to describe a relation or, especially a function. This will be demonstrated in the following subsection.

Fuzzy interpolation of a partially known function. In this subsection, we focus on the E. H. Mamdani's and L. A. Zadeh's idea of finding a suitable functional value of some function characterized by only a finite set of points which, moreover, can be known only imprecisely. The imprecision is characterized by a fuzzy equality. This question has already been elaborated by P. Hájek in [41] from the point of view of basic logic. In this subsection, we will adapt some of his results using the concepts introduced within FLb.

First, we will consider the following classical situation. Let f be a function to be interpolated. Our knowledge of f is limited only to m of its points summarized in the table

$$f: \frac{x \parallel d_1 \mid \cdots \mid d_m}{y \parallel e_1 \mid \cdots \mid e_m} \tag{6.43}$$

The function f in (6.43) can be described using either of two forms

$$H^A(x, y) := \bigvee_{j=1}^{m} ((x = d_j) \wedge (y = e_j)), \tag{6.44}$$

$$H^I(x, y) := \bigwedge_{j=1}^{m} ((x = d_j) \Rightarrow (y = e_j)). \tag{6.45}$$

Apparently, $H^A_{xy}[d_j, f(d_j)] \Leftrightarrow H^I_{xy}[d_j, f(d_j)]$ for all $j = 1, \ldots, m$ due to the uniqueness of the assignment $e_j = f(d_j)$ in (6.43) and thus, in this case both formulas equivalently characterize the function (6.43). Therefore, though there are conjunctions on the right-hand side of (6.44), these together with the implications on the right-hand side of (6.45) may be called IF-THEN rules.

Furthermore, we know that if x is "near to d_j" then y is "near to e_j", $j = 1, \ldots, m$. We are looking for some function g which interpolates f, i.e. it coincides with f in the points d_j, $g(d_j) = f(d_j)$, $j = 1, \ldots, m$ and approximates it for $x \neq d_j$. A question arises, how g can be specified. The interpolation problem will now be formulated in the frame of fuzzy logic when replacing the precise equality by the fuzzy one. We will speak about *fuzzy interpolation* in the sequel.

A two-sorted language J with a fuzzy equality $\doteq$ predicate symbol will be taken into account. Furthermore, let the couples of terms (t_j, s_j), $t_j \in M_{\iota_1}$, $s_j \in M_{\iota_2}$, $j = 1, \ldots, m$ be given, which represent the table (6.43).

We will use fuzzy quantities for the approximate description of the function f and, therefore, we need to specify truth degrees of all the instances $t \doteq t_j$, $s \doteq s_j$ where $t \in M_{\iota_1}$, $s \in M_{\iota_2}$ are closed term of both sorts. This can be done either directly or, e.g. by considering some background (consistent) fuzzy theory BT with fuzzy equality and taking the required truth degrees as the provability degrees in BT, i.e. $BT \vdash_{a_{j,t}} t \doteq t_j$ and $BT \vdash_{b_{j,s}} s \doteq s_j$. Now we can formulate the following lemma.

LEMMA 6.4
Let BT be the above background fuzzy theory and T be a fuzzy theory with fuzzy equality in the language J given by the axioms

$$T = \left\{ {}^{a_{j,t}}/t \doteq t_j,\ {}^{b_{j,s}}/s \doteq s_j \,\middle|\, t \in M_{\iota_1}, s \in M_{\iota_2}, j = 1, \ldots, m \right\}.$$

Furthermore, let $F \notin J$ be a new binary predicate and let us introduce a new axiom so that the following is an extended fuzzy theory

$$T' = T \cup \left\{ \mathbf{1}/(F(x, y) \Leftrightarrow \bigvee_{j=1}^{m} ((x \doteq t_j) \wedge (y \doteq s_j))) \right\}. \tag{6.46}$$

Then

$$T' \vdash_{c_{ts}} F_{x,y}[t, s] \tag{6.47}$$

where $c_{ts} = \bigvee_{j=1}^{m}(a_{j,t} \wedge b_{j,s})$. As a specific case, $T' \vdash F_{x,y}[t_j, s_j]$, $j = 1, \ldots, m$.

PROOF: Using Theorem 4.10(d) on page 131, to each $j = 1, \dots, m$, $t \in M_{\iota_1}$ and $s \in M_{\iota_2}$ we find a proof $w_{j,ts}$ of $(t \doteq t_j) \wedge (s \doteq s_j)$ with the value $\text{Val}(w_{j,ts}) = a_{j,t} \wedge b_{j,s}$. Further, using the tautology (P6) we finally prove that

$$T' \vdash_d \bigvee_{j=1}^{m} ((t \doteq t_j) \wedge (s \doteq s_j))$$

for all $t \in M_{\iota_1}$ and $s \in M_{\iota_2}$ where $d \geq \bigvee_{j=1}^{m}(a_{j,t} \wedge b_{j,s})$.

Since $F \notin J$, we obtain by Theorem 4.15 on page 139 that T' is a conservative extension of T. Let us now construct a model $\mathcal{D} \models T$ such that $\mathcal{D}(t \doteq t_j) = a_{j,t}$ and $\mathcal{D}(s \doteq s_j) = b_{j,s}$.

First, we introduce a fuzzy theory T'' given by special axioms $\text{SAx}(A) = a$ iff $BT \vdash_a A$ and A is a closed atomic formula of $F_{J(BT)}$, and $\text{SAx}(A) = \mathbf{0}$ otherwise. Since SAx is a set of independent evaluated formulas, by Lemma 6.1 there is a model $\mathcal{D} \models T''$ such that $\mathcal{D}(A) = a$ for every above considered closed atomic A. We will now show that the equality axioms (E1)–(E3) for $\doteq$ are true in $\mathcal{D}$ in the degree $\mathbf{1}$.

Axiom (E1): For every term t we have $\mathcal{D}(t \doteq t) = \mathbf{1}$ because $BT \vdash t \doteq t$.

Axiom (E2): Let f be an n-ary functional symbol, $t_1, \dots, t_n \in M_{\iota_1}$ and $s_1, \dots, s_n \in M_{\iota_2}$ closed terms. Since $C := f(t_1, \dots, t_n) \doteq f(s_1, \dots, s_n)$ is a closed atomic formula, $BT \vdash_c C$ for some c. At the same time $BT \vdash_{a_1} t_1 \doteq s_1$, $\dots$, $BT \vdash_{a_n} t_n \doteq s_n$, and since $BT \vdash (t_1 \doteq s_1) \mathbin{\&} \cdots \mathbin{\&} (t_n \doteq s_n) \Rightarrow C$, we obtain $a_1 \otimes \cdots \otimes a_n \leq c$ by Theorem 4.10(a) on page 131. The axiom (E2) is true in the degree $\mathbf{1}$ in $\mathcal{D}$ by this and the definition of the interpretation of open formula w.r.t. to the fact that the support of $\mathcal{D}$ consists of the sets of terms M_{ι_1}, M_{ι_2}.

Analogously we prove that also the axiom (E3) is true in $\mathcal{D}$ in the degree $\mathbf{1}$. From it follows that also $\mathcal{D} \models T$.

Now we extend the structure $\mathcal{D}$ into a structure $\mathcal{D}'$ such that $\mathcal{D}'(A) = \mathcal{D}(A)$ for every $A \in F_{J(T)}$ and put $\mathcal{D}(F_{xy}[t,s]) = \mathcal{D}\left(\bigvee_{j=1}^{m}((t \doteq t_j) \wedge (s \doteq s_j))\right) = c_{ts}$. Then $\mathcal{D}' \models T'$. This, together with the first part of this proof implies that $T' \vdash_{c_{ts}} F_{xy}[t,s]$ by Lemma 4.9(b) on page 124.

The fact $T' \vdash F_{x,y}[t_j, s_j]$ follows from the reflexivity of the fuzzy equality because then $c_{ts} = \mathbf{1}$ in (6.47). □

When working in a fuzzy theory with a fuzzy equality, it is reasonable to work with extensional fuzzy relations introduced by F. Klawonn and R. Kruse in [54]. Stating informally, these are fuzzy relations being closed with respect to all the elements which are (fuzzy) equal to those belonging to them. Note that this concept is trivial in classical logic since we have there only sharp (classical) equality.

DEFINITION 6.18

A formula $A(x_1, \dots, x_n)$ represents an extensional fuzzy relation if

$$T \vdash (A(x_1, \dots, x_n) \mathbin{\&} (x_1 \doteq y_1) \mathbin{\&} \cdots \mathbin{\&} (x_n \doteq y_n)) \Rightarrow A(y_1, \dots, y_n). \tag{6.48}$$

This concept is useful in characterization of the approximation of functions.

We will now consider a function $f : U \longrightarrow V$ where U, V are some sets specified imprecisely using fuzzy quantities in the points $d_1, \ldots, d_m \in U$ of it according to the table

$$f : \frac{x \;\|\; \text{approximately } d_1 \;|\; \cdots \;|\; \text{approximately } d_m}{y \;\|\; \text{approximately } e_1 \;|\; \cdots \;|\; \text{approximately } e_m.} \tag{6.49}$$

We suppose that the intensions of the used fuzzy quantities are characterized on the basis of the above considered background (consistent) fuzzy theory BT in the language J. Furthermore, it seems natural to require that the neighbouring fuzzy quantities are disjoint. The language J is supposed to have at least two sorts ι_1, ι_2. Let M_{ι_1}, M_{ι_2} be sets of closed terms corresponding in the language to the sets U, V, respectively.

THEOREM 6.4
Let $\mathbf{d}_j \in M_{\iota_1}$, $\mathbf{e}_j \in M_{\iota_2}$ be terms (corresponding to d_j, e_j) in the language J and let the fuzzy quantities "approximately d_j", "approximately e_j" have the intensions $\mathbf{Q}[\mathbf{d}_j]_{\langle x\rangle}$ and $\mathbf{Q}[\mathbf{e}_j]_{\langle y\rangle}$, respectively, $j = 1, \ldots, m$. Furthermore, we suppose that the fuzzy quantities "approximately d_j"and "approximately d_k" are disjoint for each $j \neq k$. Then there is a simple linguistic description

$$\mathfrak{LD}^A = \{\text{"approximately } d_j\text{" and "approximately } e_j\text{"} \mid j = 1, \ldots, m\}, \tag{6.50}$$

which determines the theory $\mathfrak{T} = \mathfrak{LD}^A$ of FLb such that the adjoint fuzzy theory T is consistent.

Furthermore, we introduce a new binary predicate symbol $F \notin J(T)$ and extend the fuzzy theory T into

$$T' = T \cup \left\{ \mathbf{1}/(F(x, y) \Leftrightarrow \bigvee_{j=1}^{m} ((x \doteq t_j) \wedge (y \doteq s_j)) \right\}.$$

Then the following holds.

(a) The formula $F(x, y)$ represents an extensional fuzzy relation and fulfils the uniqueness property

$$T' \vdash F(x, y) \,\&\, (x \doteq \mathbf{d}_j) \Rightarrow (y \doteq \mathbf{e}_j) \tag{6.51}$$

for every $j = 1, \ldots, m$.

(b) To every term t we can derive in T' the multiformula

$$\mathbf{C}_{\langle y\rangle, t} = \left\{ \bigvee_{j=1}^{m} (a_{j,t} \wedge b_{j,s}) / F_{xy}[t, s] \right| \left. s \in M_{\iota_2} \right\}, \tag{6.52}$$

which is a fuzzy quantity $\mathbf{C}_{\langle y\rangle, \mathbf{d}_j} = \mathbf{Q}[\mathbf{e}_j]_{\langle y\rangle}$ for each $j = 1, \ldots, m$.

PROOF: The intensions of the fuzzy quantities have the form (6.14). Then we may proceed in the same way as in the proof of Lemma 6.4 and construct the adjoint fuzzy theory T on the basis of the intension of the linguistic description $\mathcal{LD}^A$ and construct a model $\mathcal{D} \models T$, which implies that T is consistent.

The part (b) of this theorem then directly follows from Lemma 6.4. Furthermore, let $t := \mathbf{d}_j$. With respect to the assumption of the disjointness of the neighbouring fuzzy quantities, we have $a_{j,d_j} = \mathbf{1}$ and $a_{k,d_j} = \mathbf{0}$ for all $k \neq j$. Hence, $T' \vdash F_{xy}[\mathbf{d}_j, s] \Leftrightarrow (s \doteq \mathbf{e}_j)$. However, it follows from the construction of T' that the provability degree of the formula $s \doteq \mathbf{e}_j$ in it is the same as in BT for all $s \in M_{\iota_2}$. Consequently (by the equivalence Theorem 4.6 on page 125),

$$\mathbf{C}_{\langle y\rangle,\mathbf{e}_j} = \left\{ b_{j,s}/s \doteq \mathbf{e}_j \,\middle|\, BT \vdash_{b_{j,s}} s \doteq \mathbf{e}_j, s \in M_{\iota_2} \right\},$$

i.e. $\mathbf{C}_{\langle y\rangle,\mathbf{d}_j} = \mathbf{Q}[\mathbf{e}_j]_{\langle y\rangle}$ is a fuzzy quantity.

(a) First, we will show that F is extensional. By the transitivity of $\doteq$ due to Lemma 4.45 on page 159(b)

$$T \vdash ((x' \doteq x) \mathbin{\&} (x \doteq \mathbf{d}_j)) \Rightarrow (x' \doteq \mathbf{d}_j),$$
$$T \vdash ((y' \doteq y) \mathbin{\&} (y \doteq \mathbf{e}_j)) \Rightarrow (y' \doteq \mathbf{e}_j)$$

holds for each $j = 1, \ldots, m$. Then, using modus ponens and the tautologies (P5), (P6) and (P11) we prove that

$$T \vdash \bigvee_{j=1}^{m} ((x \doteq x') \mathbin{\&} (y \doteq y') \mathbin{\&} ((x \doteq \mathbf{d}_j) \wedge (y \doteq \mathbf{e}_j))) \Rightarrow \\ \Rightarrow \bigvee_{j=1}^{m} ((x' \doteq \mathbf{d}_j) \wedge (y' \doteq \mathbf{e}_j)).$$

The form (6.48) then follows from the fact that in the fuzzy predicate calculus, $T \vdash (A \mathbin{\&} (B \vee C)) \Rightarrow ((A \mathbin{\&} B) \vee (A \mathbin{\&} C))$ is (effectively) provable for all formulas A, B, C.

Finally, we will demonstrate (6.51). Let $\mathcal{D}$ be the above model. Then (6.51) holds iff

$$\bigwedge \left\{ \mathcal{D}((x \doteq \mathbf{d}_i) \mathbin{\&} \bigvee_{j=1}^{m} ((x \doteq \mathbf{d}_j) \wedge (y \doteq \mathbf{e}_j))) \,\middle|\, \mathcal{D} \models T \right\} \leq \\ \leq \bigwedge \{\mathcal{D}(y \doteq \mathbf{e}_i) \mid \mathcal{D} \models T\} \tag{6.53}$$

for each $i = 1, \ldots, m$. Using the distributivity of $\otimes$ with respect to supremum and the assumption that $\mathbf{Q}[\mathbf{d}_j]_{\langle x\rangle}$ and $\mathbf{Q}[\mathbf{d}_k]_{\langle x\rangle}$ are disjoint for every $j \neq k$, we obtain that (6.53) holds iff

$$\bigwedge\{\mathcal{D}((x \doteq \mathbf{d_i})^2 \wedge ((x \doteq \mathbf{d_j}) \mathbin{\&} (y \doteq \mathbf{e_j}))) \mid \mathcal{D} \models T\} \leq \bigwedge\{\mathcal{D}(y \doteq \mathbf{e_i}) \mid \mathcal{D} \models T\},$$

which obviously holds true for each $i = 1, \ldots, m$. □

The fuzzy relation represented by the formula $F(x, y)$ from this theorem, i.e. which is extensional and fulfils the uniqueness condition (6.51) will be called the *fuzzy function.* We see from Theorem 6.4 that the linguistic description $\mathcal{LD}^A$ (6.50) determines a fuzzy function. This fuzzy function interpolates the function f in the sense that it coincides with its imprecise characterization given in the table (6.49), and $\mathbf{C}_{\langle y\rangle,t}$ in (6.52) gives us information about the the value of f, provided that its argument represented by the term $t \in M_{\iota_1}$ is given. The value of f is then represented by some term $s_0 \in M_{\iota_2}$. We can seek it using a *defuzzification* procedure. A logical explanation of the defuzzification has not so far been given. However, one of the possible solutions of this task may be the following.

In general, $\mathbf{C}_{\langle y\rangle,t}$ needs not be intension of a syntagm. However, we may seek some syntagm whose intension is approximately equal to it. This task will probably go out of the logical calculus. For example, we should search for a fuzzy quantity $\mathbf{Q}[s_0]_{\langle y\rangle} = \{\ c_{ss_0}/s \doteq s_0\}$ for some term s_0 such that all the biresidua $c_{ss_0} \leftrightarrow \bigvee_{j=1}^{m}(a_{j,t} \wedge b_{j,s})$ are minimal. The term s_0 might then be taken as a result of the defuzzification of $\mathbf{C}_{\langle y\rangle,t}$. In this book, however, we will no more deal with these problems.

7 TOPOI AND CATEGORIES OF FUZZY SETS

Motivation. In Chapters 2 and 3 we have introduced the notion of a topos and showed how classical many sorted logic can be interpreted in a topos. This fact has a very natural interpretation: by interpreting the logic we can define, in fact, a model of some theory. The most frequently used theory is that of classical (sometimes only intuitive) set theory. It is well known that for construction of (also classical) models of set theory we use again model theory (of some different type) and any other model of set theory we build up in scopes of this another set theory. From this point of view, any topos can be considered as a generalization of model theory, mostly for theorems which originally have been intended for interpretation in classical set valued models only. The principal advantages of this generalized model theory are the following: First, the internal logic of this interpretation is not Boolean (in general) and it follows that the differences between results obtained by interpreting theorems in topoi and sets, respectively, are more substantial than formal. Second, by interpreting formulas and theorems in topoi we can obtain objects in these categories with some very special additional properties, which cannot be possible to construct simply in classical set valued models. For example, both these advantages have been used to find some counterexamples of the well known hypotheses in the set theory.

On the other hand, fuzzy set theory can be also thought of as a generalization of set theory constructions, although this interpretation of fuzzy set theory is still unclear and not generally used both by mathematicians and practitioners. Since the only known reasonable generalization of classical model theory (and, hence, of set theory as well) is a topos theory with its internal logic, the natural question is what is the relationship between fuzzy set theory and topos theory. More precisely, does there exist a topos, which can be considered as a generalization of fuzzy sets and, moreover, with internal logic corresponding to fuzzy logic?

Is is worthwhile to emphasize that for construction of these generalizations of set theory , i.e. topoi, categories with properties similar to those of classical category **Set** of sets are used. Hence, to find some topos which could serve as a basis for fuzzy set generalizations, some appropriate category of fuzzy sets seems to be useful. Hence, our principal claim will be to describe some categories of fuzzy sets, which naturally generalize fuzzy sets and which could be used for interpretation of some variant of fuzzy logic.

On the other hand, we have to mention that both theories (i.e. topos theory and fuzzy set theory) as generalizations of classical set theory use generalizations of the concept of characteristic function. Their approaches, however, are different. Fuzzy set approach incorporates propositional logic with a variety of useful connectives for dealing with vagueness and topoi. On the other hand, we have a powerful intuitionistic internal higher order logic, in principal different from fuzzy logic, at least that based on the interval $[0,1]$. Hence, it seems that the principal difference between these two approaches lies in the connectives interpreted in the algebraic structures, which are used in both methods: The *Heyting algebra connectives* associated with topoi based on the unit interval $[0,1]$ can be represented as follows (cf. Lemma 2.9 on page 29 and Definition 2.24 on page 48):

$$a \wedge b = \min(a,b), \quad a \to b = \bigvee\{h \mid h \wedge a \leq b\},$$

$$a \vee b = \max(a,b), \quad \neg a = (a \to 0) = \begin{cases} 0 & \text{if } a \neq 0 \\ 1 & \text{if } a = 0 \end{cases}.$$

The connectives in fuzzy logic based on this unit interval are those of the Łukasiewicz algebra $\mathcal{L}_{\text{Ł}}$ (see Example 2.8 on page 25), i.e.

$$\begin{aligned} a \otimes b &= 0 \vee (a+b-1), & a \to b &= 1 \wedge (1-a+b), \\ a \oplus b &= 1 \wedge (a+b), & \neg a &= 1-a = a \to 0 \\ & & a \leftrightarrow b &= (a \to b) \otimes (b \to a). \end{aligned}$$

To unify both approaches we need to find a categorical structure richer than a topos structure, which could enable us to interpret both systems of these connectives.

In this chapter, we summarize some results concerning this problem and present two principal examples of such categories. The first example will be a topos, the objects of which will be called Ω-sets with a classical intuitionistic internal logic but which does not allow to interpret fuzzy connectives directly (as Łukasiewicz conjunction and implication defined by the Łukasiewicz operations). The other example will be a category, the objects of which will be called Ω-fuzzy sets, which is not a topos, in general, but which allows to interpret either classical logical connectives or fuzzy connectives. We will used the first example also for construction of truth value evaluation of formulas when interpreted in the category.

7.1 Category of Ω-sets as generalization of fuzzy sets

Ω-valued sets. We start this section with an example of a category, which seems to be a natural basis for generalization of fuzzy sets and their internal logic. The category we will be dealing with is connected with Ω-valued sets with identity relation, as stated in the following definition.

DEFINITION 7.1
Let $H = \langle \Omega, \vee, \wedge, \rightarrow \rangle$ be a complete Heyting algebra with the greatest element 1_H and smallest element 0_H. By an Ω-set we understand a couple $\mathbf{A} = (A, \delta)$ such that A is a set and $\delta : A \times A \longrightarrow \Omega$ is a function called a general equality *satisfying the axioms*

$$(\forall x, y \in A)(\delta(x, y) = \delta(y, x)),$$
$$(\forall x, y, z \in A)(\delta(x, y) \wedge \delta(y, z) \leq \delta(x, z)).$$

For any Ω-set $\mathbf{A} = (A, \delta)$ and for any element $a \in A$ we denote by $\mathbf{A}(a)$ the value $\delta(a, a)$. The value $\delta(a, b)$ then represents the measure of identification of elements a and b and $\mathbf{A}(a)$ represents the measure of existence of the element a in the Ω-set $\mathbf{A}$. It is clear that $\delta(a, b) \leq \mathbf{A}(a)$ for all $a, b \in A$.

Among Ω-sets there is a special one, namely the set Ω itself. In fact, by $\widehat{\Omega}$ we denote the Ω-set (Ω, μ) such that $\mu(\alpha, \beta) = (\alpha \leftrightarrow \beta) = (\alpha \rightarrow \beta) \wedge (\beta \rightarrow \alpha)$. It can be easily proved that this definition is correct, i.e. (Ω, μ) is the Ω-set.

Category of Ω-sets. The structure $\Omega - \mathbf{Set}$ of Ω-sets can be transformed into a category by defining morphisms between Ω-sets. Hence, by a morphism between Ω-sets (A, δ) and (B, γ) we understand a map $f : A \times B \longrightarrow \Omega$ satisfying the following conditions.

(i) $(\forall x, z \in A)(\forall y \in B)\ (\delta(x, z) \wedge f(x, y) \leq f(z, y))$,
(ii) $(\forall x \in A)(\forall y, z \in B)\ (\gamma(y, z) \wedge f(x, y) \leq f(x, z))$,
(iii) $(\forall x \in A)(\forall y, z \in B)\ (f(x, y) \wedge f(x, z) \leq \gamma(y, z))$,
(iv) $(\forall x \in A)\ (\delta(x, x) = \bigvee\{f(x, y) : y \in B\})$.

If $f : \mathbf{A} \longrightarrow \mathbf{B}$ and $g : \mathbf{B} \longrightarrow \mathbf{C}$ are two morphisms then their composition is a function $g \circ f : A \times C \longrightarrow \Omega$ such that

$$g \circ f(x, z) = \bigvee_{y \in B} (f(x, y) \vee g(y, z)).$$

It can be proved that a composition of two morphisms is again a morphism. From it follows that we obtain a *category* $\Omega - \mathbf{Set}$ of Ω-sets.

It is clear that the category $\Omega - \mathbf{Set}$ can be considered as some generalization of fuzzy sets. In fact, any Ω-set $\mathbf{A} = (A, \delta)$ determines in some sense an Ω-valued fuzzy set on A with values of the membership function defined by $\mathbf{A}(x)$.

This value can be considered as a truth value of the statement $x \in \mathbf{A}$. These structures (A, δ) are sometimes called *totally fuzzy sets*. Conversely, it could be observed that if a function $\epsilon : A \longrightarrow \Omega$ is defined on the set A then there exists an Ω-set $\mathbf{A} = (A, \delta)$ such that $\mathbf{A}(x) = \epsilon(x)$ for all $x \in A$. In fact, it suffices to define

$$\delta(x, y) = \begin{cases} \epsilon(x) & \text{if } x = y \\ 0_H & \text{otherwise.} \end{cases}$$

Products in the category Ω – Set. We now investigate some basic properties of the category $\Omega - \mathbf{Set}$. It is clear that it has a terminal object. In fact, the Ω-set $\mathbf{To} = (\{*\}, \epsilon)$ such that $\epsilon(*, *) = 1_H$ is the terminal object with the unique morphism $! : \mathbf{A} = (A, \delta) \longrightarrow \mathbf{To}$ such that $!(a, *) = \delta(a, a)$. Furthermore, the category $\Omega - \mathbf{Set}$ has finite limits as stated in the following lemma.

LEMMA 7.1
*The category Ω – **Set** has finite products and pullbacks.*

PROOF: Let $(A, \delta), (B, \rho)$ be two Ω-sets. Let us consider a product $A \times B$ with a function $\sigma : (A \times B)^2 \longrightarrow \Omega$ defined so that $\sigma((a, b), (a', b')) = \delta(a, a') \wedge \rho(b, b')$. It is then clear that $(A \times B, \sigma)$ is Ω-set and we prove that it is a product with projections $\mathrm{pr}_A : (A \times B, \sigma) \longrightarrow (A, \delta)$ and $\mathrm{pr}_B : (A \times B, \sigma) \longrightarrow (B, \rho)$ defined such that

$$(\forall a, a' \in A, b \in B)\ (\mathrm{pr}_A((a, b), a') = \delta(a, a') \wedge \delta(a, a) \wedge \rho(b, b)),$$
$$(\forall a \in A, b, b' \in)\ (\mathrm{pr}_B((a, b), b') = \rho(b, b') \wedge \delta(a, a) \wedge \rho(b, b)).$$

Although it is rather tedious, we present the proof in all details to show methods used in this category. Hence, we have to prove first that $\mathrm{pr}_A, \mathrm{pr}_B$ are morphisms in the category $\Omega - \mathbf{Set}$. From the definition of morphisms we know that we have to prove the following facts (for morphism pr_A):

(1) $\sigma((a, b), (a', b')) \wedge \mathrm{pr}_A((a, b), c) \leq \mathrm{pr}_A((a', b'), c)$,

(2) $\mathrm{pr}_A((a, b), c) \wedge \delta(c, c') \leq \mathrm{pr}_A((a, b), c')$,

(3) $\mathrm{pr}_A((a, b), c) \wedge \mathrm{pr}_A((a, b), c') \leq \delta(c, c')$,

(4) $\sigma((a, b), (a, b)) = \bigvee_{c \in A} \mathrm{pr}_A((a, b), c)$.

The proofs of these facts can be done as follows.
(1) We have

$$\sigma((a, b), (a', b')) \wedge \mathrm{pr}_A((a, b), c) = \delta(a, a') \wedge \rho(b, b') \wedge \delta(a, c) \wedge \delta(a, a) \wedge \rho(b, b) =$$
$$\delta(a, a') \wedge \rho(b, b') \wedge \delta(a, c) \leq \delta(a', c) \wedge \rho(b, b') \leq \delta(a', c) \wedge \delta(a', a') \wedge \rho(b, b') \leq$$
$$\leq \delta(a', c) \wedge \delta(a', a') \wedge \rho(b', b') = \mathrm{pr}_A((a', b'), c).$$

The conditions (2) and (3) follow directly from basic properties of δ, ρ. The condition (4) follows from the fact $\delta(a, c) \leq \delta(a, a)$, for all $a, c \in A$.

Now, let (X,τ) be a Ω-set and let $f : (X,\tau) \longrightarrow (A,\delta)$ and $g : (X,\tau) \longrightarrow (B,\rho)$ be morphisms. Then there exists the unique morphism u such that the following diagram commutes:

$$\begin{array}{ccc} (X,\tau) & \xrightarrow{\ f\ } & (A,\delta) \\ g\downarrow & \overset{u}{\searrow} & \uparrow \mathrm{pr}_A \\ (B,\rho) & \xleftarrow{\ \mathrm{pr}_B\ } & (A\times B,\sigma) \end{array}$$

This map $u : X \times (A \times B)$ can be defined such that $u(x,(a,b)) = f(x,a) \wedge g(x,b)$. In fact, we prove firstly that u is a morphism. The following steps prove this fact.

(1) $\tau(x,x') \wedge u(x,(a,b)) = \tau(x,x') \wedge f(x,a) \wedge g(x,b) \le f(x',a) \wedge g(x',b) = u(x',(a,b))$;

(2) $u(x,(a,b)) \wedge \sigma((a,b),(a',b')) = f(x,a) \wedge g(x,b) \wedge \delta(a,a') \wedge \rho(b,b') \le f(x,a') \wedge g(x,b') = u(x,(a',b'))$;

(3) $u(x,(a,b)) \wedge u(x,(a',b')) = f(x,a) \wedge g(x,b) \wedge f(x,a') \wedge g(x,b') \le \delta(a,a') \wedge \rho(b,b') = \sigma((a,b),(a',b'))$;

(4) $\bigvee_{(a,b)\in A\times B} u(x,(a,b)) = \bigvee_{(a,b)\in A\times B}(f(x,a) \wedge g(x,b)) = (\bigvee_{a\in A} f(x,a)) \wedge (\bigvee_{b\in B} g(x,b)) = \tau(x,x) \wedge \tau(x,x) = \tau(x,x)$.

Further, we have to prove that $\mathrm{pr}_A \circ u = f$ and analogously for pr_B. Let $(x,a) \in X \times A$. We have

$$\begin{aligned} \mathrm{pr}_A \circ u(x,a) &= \bigvee_{(c,d)\in A\times B} (u(x,(c,d)) \wedge \mathrm{pr}_A((c,d),a)) = \\ &= \bigvee_{(c,d)\in A\times B} (f(x,c) \wedge g(x,d) \wedge \delta(c,a) \wedge \delta(c,c) \wedge \rho(d,d)) = \\ &= \bigvee_{(c,d)\in A\times B} (f(x,c) \wedge g(x,d) \wedge \delta(c,a)) = (*). \end{aligned}$$

Since f is a morphism, we have $f(x,c) \wedge \delta(c,a) \le f(x,a) \wedge \delta(a,a)$. Then we obtain

$$(*) = \bigvee_{d\in B} (f(x,a) \wedge \delta(a,a) \wedge g(x,d)) = \bigvee_{d\in B} (f(x,a) \wedge g(x,d)) =$$

$$= f(x,a) \wedge \left(\bigvee_{d\in B} g(x,d) \right) = f(x,a) \wedge \tau(x,x) = f(x,a) \wedge \bigvee_{c\in A} f(x,c) = f(x,a).$$

Hence, we proved that the category $\Omega-$**Set** has finite products. □

Pullbacks in the category $\Omega-$Set. We proceed to show that in this category exist all pullbacks. Since all proofs are rather simple but tedious, we will not present them in all details. For pullbacks we show only how to construct them.

Let us consider the following diagram in the category $\Omega - \mathbf{Set}$.

$$\begin{array}{ccc} & & (A,\delta) \\ & & \downarrow f \\ (B,\rho) & \xrightarrow{g} & (C,\tau). \end{array}$$

The pullback object (P,σ) such that $P = A \times B$ will be constructed and the function $\sigma : P \times P \longrightarrow \Omega$ is defined so that

$$\sigma((a,b),(a',b')) = e(a,b) \wedge e(a',b') \wedge \delta(a,a') \wedge \rho(b,b'),$$

where $e : A \times B \longrightarrow \Omega$ is a function such that

$$e(a,b) = \bigvee_{c \in C} (f(a,c) \wedge g(b,c)).$$

Then the diagram

$$\begin{array}{ccc} (P,\sigma) & \xrightarrow{u} & (A,\delta) \\ v\downarrow & & \downarrow f \\ (B,\rho) & \xrightarrow{g} & (C,\tau) \end{array}$$

is the pullback where the projections $u : P \times A = (A \times B) \times A \longrightarrow \Omega$ and $v : P \times B \longrightarrow \Omega$ are defined so that

$$\begin{aligned} u((a,b),c) &= e(a,b) \wedge \delta(a,c); \quad a,c \in A, b \in B, \\ v((a,b),d) &= e(a,b) \wedge \rho(b,d); \quad a \in A, b,d \in B. \end{aligned}$$

The following theorem then states the importance of the category $\Omega - \mathbf{Set}$.

THEOREM 7.1
The category $\Omega - \mathbf{Set}$ is a topos.

PROOF: As we stated above, we will not present the proof of this theorem in all details. We will show only principal construction of some principal objects of this category. In Lemma 7.1 we have shown that this category has finite limits. In the following we show that this category has also subobject classifier and, more generally, a power object structure for any its object. □

Subobjects in Ω-sets. The subobject classifier is closely connected with monomorphisms and subobjects in the category. Hence, the first natural question arises: What is the relationship between these two structures?

Let $\mathbf{A} = (A,\delta)$ be an Ω-set. Then by a *Ω-subset* of $\mathbf{A}$ we understand a map $s : A \longrightarrow \Omega$ satisfying the conditions

(i) $(\forall x, y \in A)\ (s(x) \wedge \delta(x,y) \leq s(y))$,

(ii) $(\forall x \in A)\ (s(x) \leq \delta(x,x))$.

The fact that s is an Ω-subset of $\mathbf{A}$ will be also denoted by $s \subseteq \mathbf{A}$. It is clear that this construction is a natural generalization of the notion of the extensional fuzzy subset of a fuzzy set $\mathbf{A}$ with the membership function $\mathbf{A} : A \longrightarrow \Omega$. On the other hand, in the category $\Omega - \mathbf{Set}$ a structure of monomorphisms (and epimorphisms, as well) can be described as follows.

LEMMA 7.2
Let $f : \mathbf{A} = (A, \delta) \longrightarrow \mathbf{B} = (B, \rho)$ be a morphism in the category $\Omega - \mathbf{Set}$.

(a) *f is a monomorphism if and only if for all $x, x' \in A$, $y \in B$ we have $f(x, y) \wedge f(x', y) \leq \delta(x, x')$.*
(b) *f is an epimorphism if and only if for all $y \in B$ we have $\mathbf{B}(y) \leq \bigvee_{x \in A} f(x, y)$.*

Now, with any monomorphism $f : \mathbf{A} \longrightarrow \mathbf{B}$ an Ω-subset $\operatorname{Im} f \subseteq \mathbf{B}$ is associated such that

$$(\forall y \in B) \left(\operatorname{Im} f(y) = \bigvee_{x \in A} f(x, y) \right).$$

In fact, we have

$$\begin{aligned} \operatorname{Im} f(y) \wedge \rho(y, y') = \bigvee_{x \in A} f(x, y) \wedge \rho(y, y') \leq \bigvee_{x \in A} (f(x, y) \wedge \rho(y, y')) \leq \\ \leq \bigvee_{x \in A} f(x, y') = \operatorname{Im} f(y'). \end{aligned}$$

The other condition for Ω-subset follows directly from the property $f(x, y) \leq \rho(y, y)$.

Conversely, for any Ω-subset $s \subseteq \mathbf{A} = (A, \delta)$ a monomorphism f_s can be defined such that $f_s : (A, \delta_s) \longrightarrow (A, \delta)$ and

$$\begin{aligned} \delta_s(x, y) = s(x) \wedge s(y) \wedge \delta(x, y), \\ f_s : A \times A \longrightarrow \Omega, \quad f_s(x, y) = \delta_s(x, y). \end{aligned}$$

It is clear that $\mathbf{A}_s = (A, \delta_s)$ is Ω-set and that f_s is a morphism. In fact, for example, we have $\bigvee_{y \in A} f_s(x, y) = s(x) \wedge \bigvee_{y \in A}(s(y) \wedge \delta(x, y)) = s(x) \wedge \delta(x, x) = \delta_s(x, x)$, since $s(y) \wedge \delta(x, y) \leq s(x) \wedge \delta(x, x)$. The other conditions of a morphism can be verified easily. Finally, f_s is a monomorphism, as follows directly from the definition of f_s. This monomorphism f_s is closely connected with the original Ω-subset s. In fact, we have $\operatorname{Im} f_s = s$ as follows from $s(y) \leq s(y) \wedge \delta(y, y) \leq \bigvee_{x \in A}(s(x) \wedge \delta(x, y))$.

From these constructions we obtain the following fact which enables us to describe a structure of Ω-subsets of an Ω-set $\mathbf{A} = (A, \delta)$.

LEMMA 7.3
Let (A, α) be an Ω-set. By $\mathrm{Subset}_\Omega(A, \alpha)$ we denote the set of all Ω-subsets of (A, α) and by $\mathcal{S}(A, \alpha)$ we denote the set of all subobjects (in the category

*$\Omega -$ **Set**) of (A,α) which are of the form $(A,\beta) \xrightarrow{\beta} (A,\alpha)$. Then there is a bijection between these two sets.*

PROOF: Let $\beta : (A,\beta) \to (A,\alpha)$ be an element from $\mathcal{S}(A,\alpha)$. Then we set

$$(\forall a \in A) \left(v(\beta)(a) = \bigvee_{x\in A} \beta(x,a) = \beta(a,a) \right).$$

Then $v(\beta)$ is a monomorphism in the category $\Omega -$ **Set**. In fact, since β is a morphism, it has to satisfy the condition $\alpha(a,b) \wedge \beta(x,a) \le \beta(x,b)$. Then we have

$$v(\beta)(a) \wedge \alpha(a,b) = \bigvee_{x\in A} (\beta(x,a) \wedge \alpha(a,b)) \le \bigvee_{x\in A} \beta(x,b) = \beta(b,b) = v(\beta)(b),$$

and it follows that $v(\beta)$ is Ω-subset in (A,α). Conversely, for Ω-subset S of (A,α) we put $u(S) = (A,\alpha_S) \xrightarrow{\alpha_S} (A,\alpha)$ where

$$(\forall a,b \in A)\ (\alpha_S(a,b) = S(a) \wedge S(b) \wedge \alpha(a,b)).$$

It is clear that $u(S)$ is a subobject. We prove that the maps u and v are mutually inverse. In fact, let $\beta : (A,\beta) \to (A,\alpha) \in \mathcal{S}(A,\alpha)$, then $u\circ v(\beta) = (A,\alpha_{v(\beta)})$ and there exists an isomorphism w in the category $\Omega -$ **Set** such that the following diagram commutes:

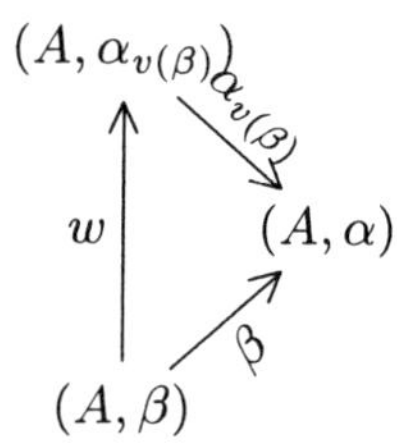

In fact, we can put $w(a,b) = \beta(a,b)$ and it is clear that w is an isomorphism. To show that $\alpha_{v(\beta)} \circ w = \beta$, we have

$$\begin{aligned}(\alpha_{v(\beta)} \circ w)(a,b) &= \bigvee_{x\in A} (\alpha_{v(\beta)}(a,x) \wedge w(x,b)) = \\ &= \bigvee_{x\in A} (\alpha(a,x) \wedge \beta(a,a) \wedge \beta(x,x) \wedge \beta(x,b)) = \\ &= \beta(a,a) \wedge \bigvee_{x\in A} (\alpha(a,x) \wedge \beta(b,x)) \le \beta(a,a) \wedge \beta(a,b) = \beta(a,b).\end{aligned}$$

On the other hand, we have $\beta(a,b) \le \alpha(a,a) \wedge \beta(a,b) \le \bigvee_{x\in A}(\alpha(a,x) \wedge \beta(b,x))$ and the diagram commutes.

Conversely, let $S \in \mathrm{Subset}_\Omega(A,\alpha)$. Then we have $v \circ u(S) = S$ and so, for $a \in A$ we have

$$v \circ u(S)(a) = \alpha_S(a,a) = \alpha(a,a) \wedge S(a) = S(a).$$

□

Subobjects classifier in Ω-sets. To prove that our category is a topos we need only to show that for any Ω-set $\mathbf{A} = (A, \delta)$ there exists another Ω-set $\mathbf{PA}$ being a power set of $\mathbf{A}$ (see 2.17). This power object can be constructed over the set $\mathrm{Sub}_{\Omega-\mathbf{Set}}(\mathbf{A})$ where we put

$$\mathbf{PA} = (\mathrm{Sub}(\mathbf{A}), \backsim),$$

$$(\forall s, t \in \mathrm{Sub}(\mathbf{A})) \ \backsim (s,t) = \bigwedge_{x \in A} (s(x) \leftrightarrow t(x)).$$

This Ω-set $(\mathrm{Sub}(\mathbf{A}), \backsim)$ satisfies the conditions of the power object. For example, for any Ω-set $\mathbf{A}$ there exists a natural isomorphism

$$\phi : \mathrm{Hom}_{\Omega-\mathbf{Set}}(-, \mathbf{PA}) \longrightarrow \mathrm{Sub}(- \times \mathbf{A}).$$

This natural transformation can be defined as follows. At first, for any element $a \in A$ there exists a special Ω-subset $\{a\} \subseteq \mathbf{A}$, called *a singleton of a* such that $\{a\}(x) = \delta(x, a)$, for all $x \in A$. Let $\mathbf{B} = (B, \rho)$ be another Ω-set and let $f \in \mathrm{Hom}_{\Omega-\mathbf{Set}}(\mathbf{B}, \mathbf{PA})$. Then we can set

$$(\forall b \in B, a \in A) \ (\phi_{\mathbf{B}}(f)(b, a) = f(b, \{a\})).$$

Hence, $\phi_{\mathbf{B}}(f) : B \times A \longrightarrow \Omega$ is a Ω-subset and it can be then proved that this defines a natural isomorphism ϕ.

In Section 2.4 we have shown that a subobject classifier is a special case of the power object, namely, the power object $\mathbf{P1}$ of the terminal object. This Ω-set has a known structure, since the following statement holds:

$$\mathbf{P1} = (\mathrm{Sub}(\mathbf{To}), \backsim) = (\Omega, \mu) = \widehat{\Omega}.$$

In fact, $\mathrm{Sub}(\mathbf{To}) = \{s \mid s \text{ is a map } \{*\} \longrightarrow \Omega\} = \Omega$ (where any map $s : \{*\} \longrightarrow \Omega$ is identified with the element $s(*)$, denoted also by s), and for $s, t \in \Omega$ we have $\mu(s,t) = (s \leftrightarrow t) = (s(*) \leftrightarrow t(*)) = \backsim (s,t)$.

The behaviour of the subobject classifier can be also described explicitly. Let us define a morphism $\top : \mathbf{To} \longrightarrow \widehat{\Omega} = \mathbf{P1}$ such that $\top(*, \alpha) = (1_H \leftrightarrow \alpha) = \alpha$ where $\alpha \in \Omega$. For any Ω-set $\mathbf{A} = (A, \delta)$ and any monomorphism $f : \mathbf{X} = (X, \sigma) \longrightarrow \mathbf{A}$ let us consider the corresponding Ω-subset $\mathrm{Im}\, f \subseteq \mathbf{A}$, i.e. $\mathrm{Im}\, f : A \longrightarrow \Omega$. Then the morphism $\chi_f : \mathbf{A} \longrightarrow \widehat{\Omega}$ is defined in such a way that

$$(\forall a \in A, \alpha \in \Omega) \ (\chi_f(a, \alpha) = \mathbf{A}(a) \wedge \mu(\mathrm{Im}(a), \alpha) = \mathbf{A}(a) \wedge (\mathrm{Im}\, f(a) \leftrightarrow \alpha)).$$

Then the diagram

$$\begin{array}{ccc} \mathbf{X} & \overset{f}{\rightarrowtail} & \mathbf{A} \\ {\scriptstyle !}\downarrow & & \downarrow{\scriptstyle \chi_f} \\ \mathbf{To} & \overset{\top}{\longrightarrow} & \widehat{\Omega} = \mathbf{P1} \end{array}$$

is the pullback.

Singletons. We have mentioned the construction of the singleton Ω-set $\{a\}$ of an element $a \in A$ where $\mathbf{A} = (A, \delta)$ is an Ω-set. This singleton of an element is, in fact, a special case of a generally defined *singleton* in $\mathbf{A}$, which is a function $s : A \longrightarrow \Omega$ such that

(i') $(\forall x, y \in A)\ (s(x) \wedge \delta(x, y) \leq s(y))$,

(ii') $(\forall x, y \in A)\ (s(x) \wedge s(y) \leq \delta(x, y))$.

It is clear that any singleton is an Ω-subset in $\mathbf{A}$ since $s(x) \leq \delta(x, x)$ holds. Moreover, any singleton of an element a is also a singleton. Then Ω-set $\mathbf{A}$ is called *complete* if any singleton is a singleton of some element of A, i.e. if for any singleton s there exists the unique $a \in A$ such that $s = \{a\}$. The following proposition shows that to confine on complete Ω-sets is not a lose of generality.

PROPOSITION 7.1
For any Ω-set $\mathbf{A} = (A, \delta)$ there exists a complete Ω-set $\mathbf{A}^*$ and an isomorphism $\mathbf{A} \longrightarrow \mathbf{A}^*$.

PROOF: Let $\mathbf{A}^* = (\mathrm{singl}\,\mathbf{A}, \tau)$, where $\mathrm{singl}\,\mathbf{A}$ is the set of all singletons of $\mathbf{A}$ and τ is defined such that

$$(\forall s, t \in \mathrm{singl}\,\mathbf{A}) \left(\tau(s, t) = \bigvee_{a \in A} (s(a) \wedge t(a)) \right).$$

We show that Ω-set $\mathbf{A}^*$ is complete. In fact, let $S : \mathbf{A}^* \longrightarrow \Omega$ be a singleton. For any element $t \in \mathrm{singl}\,\mathbf{A}$ we define $t_S : A \longrightarrow \Omega$ such that $t_S(x) = t(x) \wedge S(t)$. It is clear that t_S is a singleton. Then we put

$$e_S : A \longrightarrow \Omega, \quad e_S(x) = \bigvee_{t \in \mathrm{singl}\,A} t_S(x).$$

It is again clear that e_S is a singleton in $\mathbf{A}$. For example, we have $e_S(x) \wedge \delta(x, y) = \bigvee_{t \in \mathrm{singl}\,A} (t_S(x) \wedge \delta(x, y)) \leq \bigvee_{s \in \mathrm{singl}\,A} t_S(y) = e_S(y)$. Finally, we prove that $S = \{e_S\}$. Let $s \in \mathrm{singl}\,A$. Then we have

$$\begin{aligned}
\{e_S\}(s) &= \bigvee_{x \in A} (e_S(x) \wedge s(x)) = \bigvee_{x \in A} \left(\left(\bigvee_{t \in \mathrm{singl}\,A} t_S(x) \right) \wedge s(x) \right) = \\
&= \bigvee_{x \in A} \left(\bigvee_{t \in \mathrm{singl}\,A} (t(x) \wedge S(t) \wedge s(x)) \right) = \\
&= \bigvee_{t \in \mathrm{singl}\,A} \left(\bigvee_{x \in A} (t(x) \wedge s(x) \wedge S(t)) \right) = \\
&= \bigvee_{t \in \mathrm{singl}\,A} (\tau(t, s) \wedge S(t)) = S(s) \wedge \tau(s, s) = S(s).
\end{aligned}$$

The uniqueness of $\{e_S\}$ can be proved simply. A morphism $f : \mathbf{A} \longrightarrow \mathbf{A}^*$ can be defined such that for any element $a \in A$ and $s \in \operatorname{singl}\mathbf{A}$, $f(a,s)$ is a similarity between singletons s and $\{a\}$, hence

$$(\forall a \in A, s \in \operatorname{singl}\mathbf{A})\ (f(a,s) = \tau(\{a\},s) = \bigvee_{b\in A}(s(b) \wedge \delta(a,b)) = s(a)).$$

It can be proved easily that f is a morphism which is monomorphism and epimorphism. Hence, both objects are isomorphic. □

Category of complete Ω-sets. Proposition 7.1 allows us to use another description of the category $\Omega - \mathbf{Set}$. By $\mathbf{C}\Omega - \mathbf{Set}$ we denote the category of all *complete* Ω-sets with a little different morphisms. A *strong morphism* between two Ω-sets $\mathbf{A} = (A,\delta)$ and $\mathbf{B} = (B,\rho)$ is a map $f : A \longrightarrow B$ such that

(i) $(\forall x, y \in A)\ (\delta(x,y) \le \rho(f(x), f(y)))$,
(ii) $(\forall x \in A)\ (\mathbf{B}(f(x)) \le \mathbf{A}(x))$.

Then morphisms in a category $\mathbf{C}\Omega - \mathbf{Set}$ will be strong morphisms between complete Ω-sets. It is clear that any strong morphism can define a morphism in the category $\Omega - \mathbf{Set}$. If $f : \mathbf{A} \longrightarrow \mathbf{B}$ is a strong morphism then its graph $[f] : A \times B \longrightarrow \Omega$ defined so that $[f](x,y) = \rho(f(x),y)$ is a morphism in the category $\Omega - \mathbf{Set}$, which can easily be proved.

On the other hand, if $\mathbf{A},\mathbf{B}$ are complete Ω-sets, then for any morphism $f : \mathbf{A} \longrightarrow \mathbf{B}$ we can construct some associated strong morphism as stated in the following lemma.

LEMMA 7.4
Let $\mathbf{A} = (A,\delta)$, $\mathbf{B} = (B,\rho)$ be complete Ω-sets and let $f : \mathbf{A} \longrightarrow \mathbf{B}$ be a morphism in the category $\Omega - \mathbf{Set}$. Then there exists a strong morphism $f^* : \mathbf{A} \longrightarrow \mathbf{B}$ such that $[f^*] = f$.

PROOF: At first, we define a map $\bar{f} : \operatorname{singl}\mathbf{A} \longrightarrow \operatorname{singl}\mathbf{B}$ such that

$$(\forall s \in \operatorname{singl}\mathbf{A}, y \in B)\ (\bar{f}(s)(y) = \bigvee_{x\in A}(f(x,y) \wedge s(x))).$$

It is clear that $\bar{f}(s)$ is a singleton. For example, we have

$$\begin{aligned}\bar{f}(s)(y) \wedge \bar{f}(s)(y') &= \bigvee_{x\in A}(f(x,y)\wedge s(x)) \wedge \bigvee_{x\in A}(f(x,y')\wedge s(x)) \le \\ &\le \bigvee_{x\in A}(f(x,y)\wedge f(x,y')\wedge s(x)) \le \bigvee_{x\in A}(\rho(y,y')\wedge s(x)) = \\ &= \rho(y,y') \wedge \bigvee_{x\in A} s(x) \le \rho(y,y').\end{aligned}$$

Now, since $\mathbf{B}$ is complete, there exists a function $v : \operatorname{singl}\mathbf{B} \longrightarrow B$ such that for any $h \in \operatorname{singl}\mathbf{B}$ we have $h = \{v(h)\}$. Let f^* be the composition of the following maps:

$$\begin{array}{ccc} \operatorname{singl}\mathbf{A} & \xrightarrow{\bar{f}} & \operatorname{singl}\mathbf{B} \\ \{.\}\uparrow & & \downarrow u \\ A & \overset{f^*}{\cdots\cdots\rightarrow} & B \end{array}$$

Hence, for any $x \in A$ we have $\{f^*(x)\} = \bar{f}(\{x\})$. By simple computation we can prove that f^* is a strong morphism. For this, we use the following fact:

$$(\forall x, y \in A)\ \delta(x, y) = \tau_A(\{x\}, \{y\})$$

where τ_A is the standard general equality on singl $\mathbf{A}$ defined in Proposition 7.1. Finally, we have $[f^*] = f$. In fact, for $x \in A, y \in B$ we have

$$[f^*](x,y) = \rho(f^*(x), y) = \tau_B(\{f^*(x)\}, \{y\}) = \tau_B(\bar{f}(\{x\}), \{y\}) =$$
$$= \bigvee_{b \in B} (\bar{f}(\{x\})(b) \wedge \{y\}(b)) = \bigvee_{b \in B} \left(\bigvee_{a \in A} (f(a,b) \wedge \{x\}(a)) \wedge \rho(y,b) \right) =$$
$$= \bigvee_{b \in B} \left(\bigvee_{a \in A} (f(a,b) \wedge \delta(a,x)) \wedge \rho(y,b) \right) = \bigvee_{b \in B} (f(x,b) \wedge \rho(y,b)) = f(x,y).$$

□

From this lemma we obtain the following result which simplifies the category $\Omega -$ **Set**.

THEOREM 7.2
The category $\Omega -$ **Set** *is equivalent to the category* $\mathbf{C}\Omega -$ **Set**.

PROOF: We show only how to construct mutually inverse functors F, G such that

$$\Omega - \mathbf{Set} \underset{G}{\overset{F}{\rightleftarrows}} \mathbf{C}\Omega - \mathbf{Set}.$$

We put

$$(\forall \mathbf{A} = (A, \delta) \in \Omega - \mathbf{Set})\ F(A, \delta) = (\operatorname{singl}\mathbf{A}, \tau_A),$$
$$(\forall \mathbf{B} = (B, \rho) \in \mathbf{C}\Omega - \mathbf{Set})\ G(B, \rho) = (B, \rho).$$

Furthermore, if $f : (A, \delta) \longrightarrow (C, \sigma)$ is a morphism in $\Omega -$ **Set** then by $F(f)$ we denote the strong morphism $\bar{f}$ defined in the proof of Lemma 7.4. Analogously, for a strong morphism $g : (B, \rho) \longrightarrow (D, \psi)$ in a category $\mathbf{C}\Omega -$ **Set** by $G(g)$ we denote a morphism $[g]$ which was also introduced in this lemma. Then it is not difficult to prove that F and G are functors, which are mutually inverse. □

Properties of the category $\mathbf{C}\Omega -$ Set. It seems that structures of objects and morphisms in the category of complete Ω-sets are simpler than in the original category of Ω-sets. We will illustrate this situation on some categorical constructions in this category $\mathbf{C}\Omega - \mathbf{Set}$. From Theorem 7.2 it follows that all important objects in this category can be obtained by applying functor F to corresponding objects in the original category $\Omega - \mathbf{Set}$. For example, the terminal object $\mathbf{To}_{\mathbf{C}\Omega-\mathbf{Set}}$ of the category of complete Ω-sets is equal to $F(\mathbf{To}_{\Omega-\mathbf{Set}})$, where $\mathbf{To}_{\Omega-\mathbf{Set}} = (\{*\}, \epsilon)$ is the terminal object in the category of Ω-sets. But from the construction of the functor F presented in Theorem 7.4 it follows that $F(\mathbf{To}_{\Omega-\mathbf{Set}}) = (\mathrm{singl}\{*\}, \tau_{\{*\}})$. Since any map $s : \{*\} \longrightarrow \Omega$ is a singleton and only these maps are singletons, we can identify the set singl$\{*\}$ with the set Ω. Analogously it can be proved that according to this identification, $\tau_{\{*\}}$ corresponds to the map $\rho : \Omega \times \Omega \longrightarrow \Omega$ such that $\rho(\alpha, \beta) = \alpha \wedge \beta$. Hence, the terminal object in $\mathbf{C}\Omega - \mathbf{Set}$ is (Ω, ρ).

Analogously, we can translate constructions of all the important objects and constructions in the category $\mathbf{C}\Omega - \mathbf{Set}$. We will not present these translations here in details and present only results of these constructions for the most important objects of this category.

THEOREM 7.3
Let $\mathbf{A} = (A, \alpha)$, $\mathbf{B} = (B, \beta)$, $\mathbf{C} = (C, \gamma)$ *be objects in the category* $\mathbf{C}\Omega - \mathbf{Set}$.

(a) *The product of* (A, α) *and* (B, β) *is a complete* Ω*-set* $(A \cdot B, \delta) = (\{(a, b) \in A \times B \mid \alpha(a, a) = \beta(b, b)\}, \delta)$ *where* $\delta((a, b), (a', b')) = \alpha(a, a') \wedge \beta(b, b')$ *with projections from the category* $\mathbf{Set}$.

(b) *Let* $\mathbf{A} \xrightarrow{f} \mathbf{C} \xleftarrow{g} \mathbf{B}$ *be morphisms in* $\mathbf{C}\Omega - \mathbf{Set}$. *Let* $\mathbf{D} = (D, \tau)$ *be such that* $\mathbf{D} = \{(a, b) \in A \cdot B \mid f(x) = g(y)\}$ *and* $\tau = \delta | D$. *Then the following diagram*

$$\begin{array}{ccc} \mathbf{D} & \xrightarrow{f'} & \mathbf{B} \\ {\scriptstyle g'}\downarrow & & \downarrow{\scriptstyle g} \\ \mathbf{A} & \xrightarrow{f} & \mathbf{C} \end{array}$$

is the pullback with f', g' *classical set projections.*

(c) *A morphism* $\mathbf{A} \xrightarrow{f} \mathbf{C}$ *in* $\mathbf{C}\Omega - \mathbf{Set}$ *is a monomorphism if and only if* f *is a monic map.*

(d) *Let* $\Omega^* = \{(\alpha, \beta) \in \Omega \times \Omega \mid \alpha \leq \beta\}$ *and let* $\mu^* : \Omega^* \times \Omega^* \longrightarrow \Omega$ *be defined such that*

$$\mu^*((\alpha, \beta), (\alpha', \beta')) = (\alpha \Leftrightarrow \alpha') \wedge \beta \wedge \beta'.$$

Then (Ω^*, μ^*) *with a morphism* $\top : \mathbf{To}_{\mathbf{C}\Omega-\mathbf{Set}} \longrightarrow \Omega^*$ *such that* $\top(\alpha) = (\alpha, \alpha)$ *is the subobjects classifier in the category* $\mathbf{C}\Omega - \mathbf{Set}$.

If $\mathbf{A} = (A, \delta) \rightarrowtail \mathbf{B} = (B, \rho)$ is a monomorphism (hence, A can be considered as a subset in B) then the characteristic morphism $\chi_{\mathbf{A}} : \mathbf{B} \longrightarrow \Omega^*$ is defined such that

$$(\forall b \in B)\ (\chi_{\mathbf{A}}(b) = \left(\bigvee_{a \in A} \rho(a, b), \rho(b, b) \right) \in \Omega^*).$$

Example of Ω-sets. In this example we describe special elements of topos of Ω-sets for $\Omega = ([0,1], \wedge,\vee,\Rightarrow)$ with classical Heyting connectives. We will be dealing with classical fuzzy sets $\mathbf{A} = (A, \alpha)$, where $\alpha : A \longrightarrow [0,1]$. For any such fuzzy set $\mathbf{A}$ we can construct the *corresponding totally fuzzy set* $\mathbf{A}^t = (A, \delta_{\mathbf{A}})$ such that

$$(\forall x, y \in A)\ \delta_{\mathbf{A}^t}(x,y) = \begin{cases} \alpha(x) & \text{iff } x = y \\ 0 & \text{otherwise.} \end{cases}$$

LEMMA 7.5
Let (A,δ) and (B,β) be fuzzy sets and let $f : A \times B \longrightarrow [0,1]$ be a map. Then the following statements are equivalent.

(a) *$f : \mathbf{A}^t \longrightarrow \mathbf{B}^t$ is a morphism in the category $[0,1]-\mathbf{Set}$.*

(b) *There exists a function $g : A \longrightarrow B$ such that*

  (i) $(\forall x \in A, y \in B)\ (f(x,y) \le \delta_{\mathbf{A}^t}(g(x),y))$,

  (ii) $(\forall x \in A)\ (f(x, g(x)) = \alpha(x))$.

PROOF: (a)$\Rightarrow$(b). Since $f(a,b) \wedge f(a,b') \le \delta_{\mathbf{B}^t}(b,b')$ for all $a \in A, b, b' \in B$, it follows that $f(a,b) \wedge f(a,b') = 0$ for all $b \ne b'$. Hence, for all $x \in A$ there exists the unique $g(x) \in B$ such that

$$(\forall x \in A, y \in B)\ f(x,y) = \begin{cases} \ge 0 & \text{iff} \quad g(x) = y, \\ 0 & \text{otherwise.} \end{cases}$$

Thus, we obtained a function $g : A \longrightarrow B$. For this function we have $\beta(g(x)) = \delta_{\mathbf{B}^t}(g(x), g(x)) \ge f(x, g(x)) = \bigvee_{y \in B} f(x,y) = \alpha(x)$ and $f(x,y) \le \delta_{\mathbf{B}^t}(g(x), y)$.

(b)$\Rightarrow$(a). Let a function $g : A \longrightarrow B$ satisfy the conditions (i) and (ii). It is clear that any function $f : A \times B \longrightarrow [0,1]$ satisfies the conditions (a) and (b) from the definition of morphisms in the category Ω-**Set**. Now, for $a \in A$ and $b \in B$ we have $f(a,b) \le \delta_{\mathbf{B}^t}(g(a), b) \le \delta_{\mathbf{B}^t}(b,b)$ and for $b \ne b' \in B$ we have $f(a,b) \wedge f(a,b') \le \delta_{\mathbf{B}^t}(g(a),b) \wedge \delta_{\mathbf{B}^t}(g(a),b') = 0 = \delta_{\mathbf{B}^t}(b,b')$ and the condition (c) from this definition holds. Finally, we have $\delta_{\mathbf{A}^t}(a,a) = \alpha(a) = f(a, g(a)) = \bigvee_{b \in B} f(a,b)$ and the condition (d) is also satisfied. □

Morphisms in the category $[0,1]-\mathbf{Set}$ can be constructed directly from the maps $A \longrightarrow B$. If $\mathbf{A}$ and $\mathbf{B}$ are two fuzzy sets then for any map $f : A \longrightarrow B$ we can define a map $[f] : A \times B \longrightarrow [0,1]$ such that $[f](a,b) = \delta_{\mathbf{B}^t}(f(a), b)$. The following lemma then characterizes those maps f for which $[f]$ is a morphism between the corresponding totally fuzzy sets.

LEMMA 7.6
Let $\mathbf{A} = (A,\alpha), \mathbf{B} = (B,\beta)$ be fuzzy sets and let $f : A \longrightarrow B$ be a map. Then the following statements are equivalent.

(a) *$[f] : \mathbf{A}^t \longrightarrow \mathbf{B}^t$ is a morphism in the category $[0,1]-\mathbf{Set}$.*

(b) *$f : \mathbf{A}^t \longrightarrow \mathbf{B}^t$ is a strong morphism in the category $[0,1]-\mathbf{Set}$.*

(c) $(\forall x \in A)\ (\beta(f(x)) = \alpha(x))$.

PROOF: (a)⇒(c). Suppose that $[f]$ is a morphism. Then for all $a \in A$ we have $\alpha(a) = \bigvee_{b\in B}(a, b) = \bigvee_{b\in B}(f(a), b) = \delta_{\mathbf{B}^t}(f(a), f(a)) = \beta(f(a))$.

(c)⇒(b). Both conditions for strong morphism follow directly from (c).

(b)⇒(a). This is trivial. □

For the fuzzy set $\mathbf{A} = (A, \alpha)$ singletons and $[0,1]$-subsets can be characterized in a simple way. In fact, a map $s : A \longrightarrow [0,1]$ is a singleton in $\mathbf{A}^t$ iff there exists an element $a_s \in A$ such that $s(a_s) \leq \alpha(a_s)$ and $s(x) = 0$ for $x \in A, x \neq a_s$. Let us mention that, in general, $\mathbf{A}^t$ is not a complete $[0,1]$-set. Analogously, a map $s : A \longrightarrow [0,1]$ is a $[0,1]$-subset in $\mathbf{A}^t$ iff for all $x \in A$, $s(x) \leq \alpha(x)$ holds. Hence, these $[0,1]$-subsets correspond classical subsets in fuzzy set theory.

Topologies in Ω-sets. As we know from Section 2.5, the internal logic in a topos $\Omega -$ **Set** has the Heyting connectives but, in general, not the fuzzy ones. It means that any logical formula containing these connectives can be interpreted in this category. On the other hand, although the category $\Omega -$ **Set** seems to be very similar to fuzzy sets and representing a natural generalization of fuzzy sets, it is in some sense different from fuzzy sets at least from the point of view of the possibility to interpret non-Heyting connectives (e.g. the Łukasiewicz ones).

On the other hand, a well-behaving structure of Ω-sets enables us to introduce a lot of constructions known from classical sets without any artificial, "ad hoc" approach. The principal advantage of the structure $\Omega -$ **Set** is that it enables the interpretation of any reasonable formula by which we can define a classical **Set** structure. An example of such structure can be a topology. As we know from the Section 3.4, a topology can be defined in any topos by interpreting appropriate formulas, expressing principal properties of classical **Set** topology. Since $\Omega -$ **Set** is a topos, by interpreting these formulas in this category we obtain a topology in fuzzy sets (or, more precisely, in Ω-valued sets). We present more details of this approach which could serve as a typical method how to define various classical structures in fuzzy sets.

For simplicity we will be dealing with complete Ω-sets only, i.e. we work in the category $\mathbf{C}\Omega -$ **Set**. As we saw in the previous parts, the subobject classifier in this category is

$$\top : (\Omega, \rho) = \mathbf{To} \longrightarrow (\Omega^*, \mu^*)$$

where $\Omega^* = \{(\alpha, \beta) \in \Omega^2 : \alpha \leq \beta\}$, $\mu^*((\alpha,\beta),(\alpha',\beta')) = (\alpha \Leftrightarrow \alpha') \wedge \beta \wedge \beta'$, $\rho(\alpha,\beta) = \alpha \wedge \beta$ and $\top(\alpha) = (\alpha, \alpha)$.

Let $j : \Omega^* \longrightarrow \Omega^*$ be a topology as introduced in Section 3.4 and let $(J, \epsilon) \hookrightarrow (\Omega^*, \mu^*)$ be a corresponding Ω-subset, i.e. J is the following pullback in the category $\mathbf{C}\Omega -$ **Set**

$$\begin{array}{ccc} (J,\epsilon) & \hookrightarrow & (\Omega^*,\mu^*) \\ {\scriptstyle !}\downarrow & & \downarrow{\scriptstyle j} \\ \mathbf{To} & \xrightarrow{\top} & (\Omega^*,\mu^*). \end{array}$$

Then according to Theorem 7.3 we can describe explicitly elements of the subobject J as follows:

$$\begin{aligned} J =&\{(\alpha,(\beta,\gamma)) \mid \beta \le \gamma, \rho(\alpha,\alpha) = \mu^*((\beta,\gamma),(\beta,\gamma)), j(\beta,\gamma) = \top(\alpha)\} = \\ =&\{(\alpha,(\beta,\gamma)) \mid \beta \le \gamma, \alpha = \gamma, j(\beta,\gamma) = (\alpha,\alpha)\} = \\ =&\{(\beta,\gamma) \mid \beta \le \gamma, j(\beta,\gamma) = (\gamma,\gamma)\} \hookrightarrow \Omega^*, \end{aligned}$$

and $\epsilon = \mu^*|(J \times J)$. Moreover, we can describe properties which the morphism j has to satisfy. The property $j \circ \top = \top$ can be rewritten as follows:

$$(\forall \alpha \in \Omega)\ (j(\alpha,\alpha) = (\alpha,\alpha)),$$

and analogously for the property $j \circ j = j$ we have $j(j(\alpha,\beta)) = j(\alpha,\beta)$. To interpret the property $j \circ \wedge = \wedge \circ (j \times j)$, we need firstly to interpret $\vee$. From the definition of this morphism $\wedge : \Omega^* \times \Omega^* \longrightarrow \Omega^*$ it follows that it is the characteristic morphism of the monomorphism $\top \times \top : \mathbf{To} \longrightarrow (\Omega^* \times \Omega^*, \sigma)$ (see Definition 2.30 on page 57). At first, we have

$$\Omega^* \times \Omega^* = \{((\alpha,\beta),(\alpha',\beta)) \mid \alpha,\alpha' \le \beta\},$$

as it can be seen by simple computation. Moreover, we have

$$\begin{aligned} \sigma(((\alpha,\beta),(\alpha',\beta)),((\rho,\delta),(\rho',\delta))) &= \mu^*((\alpha,\beta),(\rho,\delta)) \wedge \mu^*((\alpha',\beta),(\rho',\delta)) = \\ &= (\alpha \Leftrightarrow \rho) \wedge \beta \wedge \delta \wedge (\alpha' \Leftrightarrow \rho'). \end{aligned}$$

Then from the Theorem 7.3 we obtain

$$\begin{aligned} &(\forall ((\alpha,\beta),(\alpha',\beta)) \in \Omega^* \times \Omega^*)\ (\alpha,\beta) \wedge (\alpha',\beta) = \chi_{\top\times\top}((\alpha,\beta),(\alpha',\beta)) = \\ &= \left(\bigvee_{\omega\in\Omega} \sigma(((\omega,\omega),(\omega,\omega)),(\alpha,\beta),(\alpha',\beta))), \sigma(((\alpha,\beta),(\alpha',\beta)),((\alpha,\beta),(\alpha',\beta)))\right) \\ &= \left(\bigvee_{\omega\in\Omega} ((\omega \Leftrightarrow \alpha) \wedge (\omega \Leftrightarrow \alpha') \wedge \omega \wedge \beta), \beta\right) = (\alpha \wedge \alpha', \beta). \end{aligned}$$

Then the condition $j \circ \wedge = \wedge \circ (j \times j)$ can be equivalently expressed as follows:

$$(\forall \alpha,\alpha',\beta \in \Omega, \alpha,\alpha' \le \beta)\ (j(\alpha \wedge \alpha',\beta) = j(\alpha,\beta) \wedge j(\alpha',\beta)).$$

We can show how the closure $\bar{S}$ is defined for any subobject S in Ω-set A. Since we are dealing with (Ω-valued) fuzzy sets, we will consider Ω-sets defined by these fuzzy sets, i.e. if (A,α) is a fuzzy set (i.e. $\alpha : A \longrightarrow \Omega$ is a map) then the corresponding Ω-set is (A,δ_α) where

$$(\forall a,b \in A)\ \delta_\alpha(a,b) = \begin{cases} 0_H & \text{iff } a \ne b, \\ \alpha(a) & \text{iff } a = b. \end{cases}$$

For Ω-sets defined in such a way, some constructions are simpler. For example, if $(A,\alpha) \hookrightarrow (B,\beta)$ is a subobject (i.e. $\alpha = \beta|A$), for the characteristic morphism $\chi_A : (B,\delta_\beta) \longrightarrow (\Omega^*,\mu^*)$ we have for all $b \in B$

$$\chi_A(b) = \left(\bigvee_{a\in A} \delta_\beta(a,b), \delta_\beta(b,b)\right) = \begin{cases} (\beta(b),\beta(b)) & \text{iff } b \in A, \\ (0_H,\beta(b)) & \text{iff } b \notin A. \end{cases}$$

Now, let $(S, \beta) \hookrightarrow (A, \tau)$ be a subobject (here $\beta = \tau|S$). Then the closure of S in the topology j is the pullback

$$\begin{array}{ccc} (\bar{S}, \delta_\tau) & \hookrightarrow & (A, \delta_\tau) \\ \downarrow & & \downarrow \chi_S \\ (J, \epsilon) & \hookrightarrow & (\Omega^*, \mu^*). \end{array}$$

Hence, we have

$$\begin{aligned} \bar{S} =& \{((\alpha, \beta), a) \in J \times A \mid \epsilon((\alpha, \beta)(\alpha, \beta)) = \delta_\tau(a, a), \chi_S(a) = (\alpha, \beta)\} = \\ =& \{a \in A \mid (0_H, \tau(a)) \in J \text{ for } a \notin S \text{ or } (\tau(a), \tau(a)) \in J \text{ for } a \in S\}. \end{aligned}$$

7.2 Category of Ω-fuzzy sets

The category $\Omega - \mathbf{Set}$ studied in the previous section has on one hand a nice structure (it is a topos with the best possibility to interpret some logic, based on the Heyting connectives), but on the other hand, it offers quite poor possibilities to interpret other logical connectives, not directly based on the Heyting connective. Unfortunately, the Łukasiewicz connectives are of this nature, which is a principal reason that the category $\Omega - \mathbf{Set}$ cannot be taken as the best generalization of fuzzy sets. It seems that the principal reason for the disadvantages of the category $\Omega - \mathbf{Set}$ lies directly in the Heyting algebra Ω. In principle, we are able to interpret only those logical formulas, which are constructed over the connectives, which can be interpreted in Ω. In case that such connective (contained in a formula) could not be interpreted naturally in Ω, it would be almost impossible to interpret the formula in a reasonable way. Unfortunately, Heyting algebra structure enables the interpretation of classical connectives but not the interpretation of the Łukasiewicz fuzzy conjunction and implication, since these are based on the connectives, which are not present in this algebra. Hence, a method improving this situation could be based on some modification of the underlying lattice Ω. This method results in a possibility to interpret these extended logics and to create a category which could be considered as a generalization of fuzzy sets in a more convenient way. On the other hand, for possibility to interpret Łukasiewicz connectives in these categories we have to pay some penalty — these new categories, in general, are not topoi, although most of their properties are very similar to those of topoi.

In this section we present example of such method, due to U. Höhle [45, 46, 48, 49, 50] which is based on a modification of a Heyting algebra Ω.

7.2.1 Ω-fuzzy sets over MV-algebras

In Chapter 2, we have introduced the notion of MV-algebra. Recall that according to Theorem 2.2(b) on page 33, the residuated lattice $\Omega = \langle L, \vee, \wedge, \otimes, \rightarrow, \mathbf{1}, \mathbf{0}\rangle$ is an MV-algebra iff $(a \rightarrow b) \rightarrow b = a \vee b$ holds for every $a, b \in L$. It is clear that this structure is a generalization both of Łukasiewicz algebra $\mathcal{L}_\mathrm{L}$ as well as Heyting algebra. Moreover, a complete MV-algebra satisfies the infinite distributive laws (cf. Lemma 2.12 on page 34)

$$a \otimes (\bigvee_{i\in I} b_i) = \bigvee_{i\in I} (a \otimes b_i),$$
$$a \wedge (\bigvee_{i\in I} b_i) = \bigvee_{i\in I} (a \wedge b_i).$$

It follows directly from the definition that the lattice structure of any MV-algebra is a complete Heyting algebra (if we put $\otimes = \wedge$) and conversely, we can understand these algebras as complete Heyting algebras (but only with respect to $\wedge, \vee$) with some defined additional monoidal structure $\otimes$ (adjoint to another operation $\rightarrow$). To tell the truth, this represents the principal advantage of these structures: MV-algebras have all advantages of Heyting algebras (at least, well defined internal logic) and, moreover, have some possibility to interpret additional logical connectives (such as the Łukasiewicz ones) as these monoidal structures.

Ω-fuzzy sets over MV-algebras. We now introduce the notion of an Ω-fuzzy set based on a MV-algebra Ω. This definition is a modification of the Ω-set introduced in the previous section, which utilizes the existence of additional operation $\otimes$.

DEFINITION 7.2
Let $\Omega = \langle L, \vee, \wedge, \otimes, \rightarrow\rangle$ be a complete MV-algebra. Then the pair (A, α) is Ω-fuzzy set if A is a set and $\alpha : A \times A \longrightarrow \Omega$ is a map such that

(i) $\alpha(x, y) \leq \alpha(x, x) \wedge \alpha(y, y)$,

(ii) $\alpha(x, y) = \alpha(y, x)$,

(iii) $\alpha(x, y) \otimes (\alpha(y, y) \rightarrow \alpha(y, x)) \leq \alpha(x, z)$.

Moreover, a Ω-fuzzy set (A, α) is called *separated* if it satisfies the axiom

$$\alpha(x, x) \vee \alpha(y, y) \leq \alpha(x, y) \quad \text{implies} \quad x = y.$$

The category $\Omega-$ **FSet** of Ω-fuzzy sets then consists of Ω-fuzzy sets as objects and morphisms between the objects $(A, \alpha), (B, \beta)$, which are maps $f : A \longrightarrow B$ such that

(a) $(\forall x, y \in A)(\beta(f(x), f(y)) \geq \alpha(x, y))$,

(b) $(\forall x \in A)(\alpha(x, x) = \beta(f(x), f(x)))$.

The composition of morphisms is the usual composition of maps. By $\mathcal{L}$ we understand a complete Heyting algebra $\langle L, \vee, \wedge\rangle$ of a MV-algebra Ω. It is then natural to investigate the relationship between the categories $\Omega -$ **FSet** and $L -$ **Set**. Recall that the category $L -$ **Set** is equivalent to the category **C**$L -$ **Set** of complete L-sets with strong morphisms (see Theorem 7.2) and we can identify both these categories. In this case **C**$L -$ **Set** is a complete subcategory in the category $\Omega -$ **FSet**. In fact, for any L-set (A, α) we have

$\alpha(x,y) \otimes (\alpha(y,y) \to \alpha(y,z)) \leq \alpha(x,y) \wedge \alpha(y,z) \leq \alpha(x,z)$. From it follows that (A,α) is also Ω-fuzzy set from $\Omega - \mathbf{FSet}$.

With any Ω-fuzzy set (A,α) we can associate a separated Ω-fuzzy set (A^*,α^*) such that $A^* = A|\equiv$ where $\equiv$ is an equivalence relation on A defined by

$$x \equiv y \quad \text{iff} \quad \alpha(x,x) \vee \alpha(y,y) \leq \alpha(x,y).$$

The map α^* is defined by α and representatives of the equivalence classes. Since separated Ω-fuzzy sets have some appropriate properties, we will be dealing only them. For simplicity, by $\Omega - \mathbf{FSet}$ we will understand a category, *the objects of which are separated Ω-fuzzy sets only.*

Limits in the category $\Omega - \mathbf{FSet}$. We want to investigate principal properties of the category $\Omega - \mathbf{FSet}$ from the point of view of possibilities to interpret a logic in this category. As seen in Section 3.4, one of the best categories from this point of view is a topos. Hence, we will investigate how close is this category to the topos structure. The first significant property of a topos is the existence of all limits. The following proposition states that this is also a property of our category.

PROPOSITION 7.2
*Let $\Omega = \langle L, \vee, \wedge, \otimes, \to \rangle$ be an MV-algebra. Then the category Ω – **FSet** is complete.*

PROOF: It is clear that the initial object of this category is $\mathbf{Io} = (\emptyset, \emptyset)$ and the terminal object is $\mathbf{To} = (L, \wedge)$. We will show how the products in this category can be constructed.

Let $\{(A_i,\alpha_i) \mid i \in I\}$ be a set of (separated) Ω-fuzzy sets and put

$$A = \left\{ (a_i)_i \in \prod_{i \in I} A_i \,\middle|\, (\forall i,j \in I)(\alpha_i(a_i,a_i) = \alpha_j(a_j,a_j)) \right\},$$

$$\alpha((a_i)_i,(b_i)_i)) = \bigwedge_{i \in I} \alpha_i(a_i,b_i).$$

Clearly, (A,α) is the Ω-fuzzy set and the restriction of classical set-projections $A \longrightarrow A_i$ are morphisms in this category. Moreover, this object is the product. Analogously, it can be proved (by using analogous construction in the category **Set**) that $\Omega - \mathbf{FSet}$ has equalizers. Hence, this category has limits. □

Epi-mono decomposition in $\Omega - \mathbf{FSet}$. In Section 2.5 we have introduced a notion of epi-mono decomposition of morphisms, which seems to be important for the interpretation of formulas of the types $(\exists x)A$ and $(\forall x)A$. We want to show that our category $\Omega - \mathbf{FSet}$ has an analogous property. At first we need to characterize morphisms which are mono or epi. We will need the following lemma, which will be useful for identification of "subobject classifier" in our category.

LEMMA 7.7
Let $\Omega = \langle L, \vee, \wedge, \otimes, \rightarrow\rangle$ be an MV-algebra and let $\gamma_\Omega : (L\times L)\times(L\times L) \longrightarrow L$ be defined such that

$$\gamma((\omega_1,\tau_1),(\omega_2,\tau_2)) = (\omega_1 \otimes (\tau_1 \rightarrow \tau_2)) \wedge (\omega_2 \otimes (\tau_2 \rightarrow \tau_1)).$$

Then $(L \times L, \gamma_\Omega)$ is Ω-fuzzy set.

PROOF: We show only how the property (iii) from the definition of Ω-fuzzy sets can be proved - the other properties can be proved simply. Hence, we have

$$\begin{aligned}\gamma_\Omega((\omega_1,\tau_1),(\omega_2,\tau_2)) &\otimes (\omega_2 \rightarrow \gamma((\omega_2,\tau_2),(\omega_3,\tau_3))) \leq \\ &\leq \omega_2 \otimes (\omega_2 \rightarrow (\omega_1 \otimes (\tau_1 \rightarrow \tau_2))) \otimes (\omega_2 \rightarrow (\omega_2 \rightarrow (\tau_2 \rightarrow \tau_3))) \leq \\ &\leq \omega_1 \otimes (\tau_1 \rightarrow \tau_2) \otimes (\tau_2 \rightarrow \tau_3) \leq \omega_1 \otimes (\tau_1 \rightarrow \tau_3),\end{aligned}$$

$$\begin{aligned}\gamma_\Omega((\omega_1,\tau_1),(\omega_2,\tau_2)) &\otimes (\omega_2 \rightarrow \gamma((\omega_2,\tau_2),(\omega_3,\tau_3))) \leq \\ &\leq \omega_2 \otimes (\tau_2 \rightarrow \tau_1) \otimes (\omega_2 \rightarrow (\omega_3(\tau_3 \rightarrow \tau_2))) \leq \\ &\leq (\tau_2 \rightarrow \tau_1) \otimes \omega_3(\tau_3 \rightarrow \tau_2) \leq \omega_3 \otimes (\tau_3 \rightarrow \tau_1).\end{aligned}$$

□

The following lemma solves (at least partly) the problem of epimorphisms.

LEMMA 7.8
*Let $\Omega = \langle L, \vee, \wedge, \otimes, \rightarrow\rangle$ be an MV-algebra and let (B,β) be a Ω-fuzzy set such that for any $b \in B$ there exists $\omega \in L$ such that $\beta(b,b) = \omega \otimes \omega$ (i.e. $\beta(b,b)$ is a square). Furthermore, let $(A,\alpha) \xrightarrow{f} (B,\beta)$ be a morphism in the category $\Omega -$ **FSet**. Then the following statements are equivalent.*

(a) *f is an epimorphism.*
(b) $(\forall b \in B)\ (\beta(b,b) = \bigvee_{a\in A}((\alpha(a,a) \rightarrow \beta(f(a),b)) \otimes \beta(f(a),b)))$.

PROOF: (a)⇒(b). Let us consider an Ω-fuzzy set (C,γ_Ω) such that $C = L\times L$ and $\gamma_\Omega : (L\times L)\times(L\times L) \longrightarrow L$ is defined in Lemma 7.7. Since this fuzzy set is not, in general, separated, let $(\widehat{C},\widehat{\gamma_\Omega})$ be a separated Ω-fuzzy set associated with (C,γ_Ω) and let $a \mapsto \bar{a}$ be the canonical map $C \longrightarrow \widehat{C}$. We define two maps $(B,\beta) \overset{g_1}{\underset{g_2}{\rightrightarrows}} (\widehat{C},\widehat{\gamma_\Omega})$ such that

$$g_1(b) = \bigvee\{\lambda \in L \mid \lambda\otimes\lambda \leq \bigvee_{a\in A}((\alpha(a,a) \rightarrow \beta(f(a),b)) \otimes \beta(f(a),b)\},$$

$$g_2(b) = \bigvee\{\lambda \in L \mid \lambda\otimes\lambda \leq \beta(b,b)\}.$$

It could be then proved that any g_i defines a morphism $\widehat{g}_i : (B,\beta) \longrightarrow (\widehat{C},\widehat{\gamma_\Omega})$ such that

$$\widehat{g}_i(b) = \overline{(\beta(b,b),g_i(b))} \in \widehat{C}.$$

Moreover, since $\widehat{g}_1 \circ f = \widehat{g}_2 \circ f$, we have $\widehat{g}_1 = \widehat{g}_2$ and it follows that $\beta(b,b) = \beta(b,b) \otimes (g_2(b) \to g_1(b))$ for all $b \in B$; or equivalently $g_2(b) \otimes g_2(b) = g_1(b) \otimes g_1(b)$. Since $g_i(b)$ is the square root of $g_i(b) \otimes g_i(b)$, from properties of square roots it follows that

$$\beta(b,b) = g_2(b) \otimes g_2(b) = g_1(b) \otimes g_1(b) \leq \\ \leq \bigvee_{a\in A} ((\alpha(a,a) \to \beta(f(a),b)) \otimes \beta(f(a),b)),$$

and the assertion (b) is proved.

(b)$\Rightarrow$(a). Let us consider the following diagram where $r_1 \circ f = r_2 \circ f$:

$$(A,\alpha) \xrightarrow{f} (B,\beta) \underset{r_2}{\overset{r_1}{\rightrightarrows}} (D,\tau).$$

Using the condition (b) we obtain

$$\tau(r_1(b), r_2(b)) \geq \bigvee_{a\in A} \tau(r_1(b), r_1 \circ f(a)) \otimes (\alpha(a,a) \to \tau(r_2 \circ f(a), r_2(b))) \geq \\ \geq \bigvee_{a\in A} \beta(b, f(a)) \otimes (\alpha(a,a) \to \beta(f(a),b)) \geq \beta(b,b).$$

Since (D,τ) is separated, we obtain that r_1 and r_2 coincide. $\square$

A solution of the problem for monomorphism is more simple as stated in the following lemma.

LEMMA 7.9
*A morphism f in the category $\Omega-$***FSet** *is a monomorphism iff f is an injective map.*

The proof of this lemma can be done analogously as in the category **Set** and it will be omitted.

To show that the category $\Omega-$ **FSet** has an analogy of the epi-mono decomposition property we introduce the following two sets of morphisms of this category. Let $\mathcal{E}$ be the set of all morphisms in $\Omega-$ **FSet** which satisfy the condition (b) of Lemma 7.8. Furthermore, let $\mathcal{M}$ be the set of morphisms $(A,\alpha) \xrightarrow{f} (B,\beta)$ from our category, which satisfy the following two conditions:

(i) $\alpha(a_1,a_2) = \beta(f(a_1), f(a_2))$, for all $a_1, a_2 \in A$,

(ii) $f(A) = \{b \in B \mid \bigvee_{a\in A}((\alpha(a,a) \to \beta(f(a),b)) \otimes \beta(f(a),b)) = \beta(b,b)\}$.

It is clear that any morphism from $\mathcal{E}$ is an epimorphism and any morphism from $\mathcal{M}$ is a monomorphism. In fact, if $f(a_1) = f(a_2)$ then we have

$$\alpha(a_1,a_1) \vee \alpha(a_2,a_2) = \beta(f(a_1), f(a_1)) \vee \beta(f(a_2), f(a_2)) = \\ = \beta(f(a_1), f(a_2)) = \alpha(a_1,a_2),$$

and since (A, α) is separated, we obtain $a_1 = a_2$. Then the following proposition states the "epi-mono" decomposition property in the category $\Omega - \mathbf{FSet}$.

PROPOSITION 7.3
Any morphism $f : (A, \alpha) \longrightarrow (B, \beta)$ in the category $\Omega - \mathbf{FSet}$ can be factored as $f = g \circ h$ where $h \in \mathcal{E}$ and $g \in \mathcal{M}$.

PROOF: Let the set D be defined such that

$$D = \left\{ b \in B \,\middle|\, \bigvee_{a \in A} ((\alpha(a,a) \to \beta(f(a), b)) \otimes \beta(f(a), b)) = \beta(b,b) \right\},$$

and let $\tau = \beta|(D \times D)$. Then (D, τ) is an Ω-fuzzy set, by g we denote the inclusion $D \hookrightarrow B$ and h is determined by f, i.e. $h(a) = f(a)$ for all $a \in A$. Then it is clear that $h \in \mathcal{E}$ and that g satisfies the condition (i). We show that g satisfies the condition (ii). Hence, let $b \in B$ be such that $\beta(b,b) = \bigvee_{d \in D}((\tau(d,d) \to \beta(g(d), b)) \otimes \beta(g(d), b))$. Then we obtain

$$\begin{aligned}
\beta(b,b) &= \bigvee_{d \in D} ((\beta(d,d) \otimes (\beta(d,d) \to \beta(d,b)) \otimes (\beta(d,d) \to \beta(d,b)) = \\
&= \bigvee_{a \in A} \left(\bigvee_{d \in D} \alpha(a,a) \otimes (\alpha(a,a) \to \beta(f(a),d)) \otimes (\alpha(a,a) \to \beta(f(a),b)) \otimes \right. \\
&\qquad \left. \otimes (\beta(d,d) \to \beta(d,b)) \otimes (\beta(d,d) \to \beta(d,b)) \right) \leq \\
&\leq \bigvee_{a \in A} \alpha(a,a) \otimes (\alpha(a,a) \to \beta(f(a),b)) \otimes (\alpha(a,a) \to \beta(f(a),b)).
\end{aligned}$$

Hence $b \in D$. □

Subobject classifier in $\Omega - \mathbf{FSet}$. A significant property of any topos is the existence of subobject classifier which enables us to interpret any formula as a morphism into this subobject. In the category $\Omega - \mathbf{FSet}$, in general, such object does not exist. On the other hand, there exists an object which has very similar property and which can classify "almost all" subobjects. It seems that this object is very similar to the subobject classifier from the category $\mathbf{C}\Omega - \mathbf{Set}$ as was introduced in the Theorem 7.3. In Lemma 7.7 we proved that for an MV-algebra $\Omega = \langle L, \vee, \wedge, \otimes, \to \rangle$, $(L \times L, \gamma_\Omega)$ is the Ω-fuzzy set. This fuzzy set can be used for construction of this modified subobject classifier. In fact, let

$$\Omega^* = \{(\alpha, \beta) \in L \times L \mid \alpha \geq \beta\},$$

and let μ be the restriction of γ_Ω onto Ω^*. Then it is clear that (Ω^*, μ) is the Ω-fuzzy set. Now we modify a little a notion of the Ω-subset. Recall that for an Ω-set $\mathbf{A} = (A, \delta)$ (where Ω is considered as Heyting algebra) we introduced the notion of an Ω-subset as a map $s : A \longrightarrow \Omega$ such that

(i) $(\forall x, y \in A)\ (s(x) \wedge \delta(x,y) \leq s(y))$,

(ii) $(\forall x \in A)\ (s(x) \leq \delta(x,x))$.

For the category $\Omega-\mathbf{FSet}$ this definition of Ω-subset will be modified as follows.

We say that a map $s : A \longrightarrow L$ is a Ω-subset of Ω-fuzzy set $\mathbf{A} = (A, \alpha)$ (again abbreviated as $s \subseteq \mathbf{A}$), if

(i') $(\forall x, y \in A)\ (s(x) \otimes (\alpha(x,x) \to \alpha(x,y)) \leq s(y))$,

(ii') $(\forall x \in A)\ (s(x) \leq \alpha(x,x))$.

Hence, if $\Omega = \langle L, \vee, \wedge, \to, \otimes \rangle$ is a complete MV-algebra then $\langle L, \vee, \wedge \rangle$ is a complete Heyting algebra for $\otimes_L = \wedge$ and the natural question is whether any L-set (A, α) is also Ω-fuzzy set. The answer is affirmative since we have

$$\begin{aligned}\alpha(x,y) \otimes (\alpha(y,y) \to \alpha(y,z)) \leq & \alpha(y,y) \otimes (\alpha(y,y) \to \alpha(y,z)) = \\ = & \alpha(y,y) \wedge \alpha(y,z) = \alpha(y,z), \\ \alpha(x,y) \otimes (\alpha(y,y) \to \alpha(y,z)) \leq & \alpha(x,y) \wedge (\alpha(y,y) \to \alpha(y,z)) \leq \alpha(x,y),\end{aligned}$$

and it follows that

$$\alpha(x,y) \otimes (\alpha(y,y) \to \alpha(y,z)) \leq \alpha(x,y) \wedge \alpha(y,z) \leq \alpha(x,z).$$

Analogously, any Ω-subset of (A, α) (which is considered as L-set) is a Ω-subset of (A, α) considered as Ω-fuzzy set, as well. In fact, we have

$$\begin{aligned}s(x) \otimes (\alpha(x,x) \to \alpha(x,y)) \leq & \alpha(x,x) \otimes (\alpha(x,x) \to \alpha(x,y)) = \\ = & \alpha(x,x) \wedge \alpha(x,y) = \alpha(x,y), \\ s(x) \otimes (\alpha(x,x) \to \alpha(x,y)) \leq & s(x),\end{aligned}$$

and it follows that

$$s(x) \otimes (\alpha(x,x) \to \alpha(x,y)) \leq s(x) \wedge \alpha(x,y) \leq s(y).$$

As observed, the terminal object in our category is $\mathbf{To} = (L, \wedge)$ and it is clear that the unique morphism ! from Ω-fuzzy set (A, α) to $\mathbf{To}$ is such that $!(a) = \alpha(a,a)$. Moreover, let $\top : \mathbf{To} \longrightarrow (\Omega^*, \mu)$ be defined such that $\top(\omega) = (\omega, \omega)$. Then we have $\mu(\top(x), \top(y)) = (x \otimes (x \to y)) \wedge (y \otimes (y \to x)) = x \wedge y$ and it follows that $\top$ is a morphism.

The relationship between Ω-fuzzy set Ω^* and Ω-subsets is presented in the following theorem.

THEOREM 7.4

Let $\Omega = \langle L, \vee, \wedge, \otimes, \to \rangle$ be a complete MV-algebra. Let $\mathbf{A} = (A, \alpha)$ be Ω-fuzzy set and let $\mathcal{S} = \{s \mid s \subseteq \mathbf{A}\}$ be the set of all Ω-subsets of $\mathbf{A}$, i.e. maps $s : A \longrightarrow L$ satisfying the conditions (i') and (ii').

(a) There exists a bijection $\chi : \mathcal{S} \longrightarrow \mathrm{Hom}_{\Omega-\mathbf{FSet}}(\mathbf{A}, \Omega^*)$.

(b) *For any $s \in \mathcal{S}$ let $S_s = \{a \in A \mid s(a) = \alpha(a,a)\}$. Then the following diagram is the pullback in the category $\Omega - \mathbf{FSet}$.*

$$\begin{array}{ccc} (S_s,\alpha) & \hookrightarrow & (A,\alpha) \\ !\downarrow & & \downarrow\chi(s) \\ \mathbf{To} & \xrightarrow{\top} & (\Omega^*,\mu) \end{array}$$

PROOF: (a) For $s \in \mathcal{S}$ we set

$$(\forall a \in A)\ (\chi(s)(a) = (\alpha(a,a), s(a)) \in \Omega^*).$$

Then $\chi(s)$ is a morphism. In fact, since s satisfies the condition (i'), we have $\alpha(a,a) \to \alpha(a,b) \le s(a) \to s(b)$ and it follows that

$$\begin{aligned} \mu(\chi(s)(a),\chi(s)(b)) &= (\alpha(a,a) \otimes (s(a) \to s(b))) \wedge (\alpha(b,b) \otimes (s(b) \to s(a))) \ge \\ &\ge (\alpha(a,a) \otimes (\alpha(a,a) \to \alpha(a,b))) \wedge (\alpha(b,b) \otimes (\alpha(b,b) \to \alpha(a,b))) = \\ &= \alpha(a,a) \wedge \alpha(b,b) \wedge \alpha(a,b) = \alpha(a,b). \end{aligned}$$

Conversely, for a morphism $f : (A,\alpha) \longrightarrow \Omega^*$ we define a map $s = \chi^{-1}(f) : A \longrightarrow L$ such that $s = \mathrm{pr}_2 \circ f$, where $\mathrm{pr}_2 : \Omega^* \longrightarrow L$ is the second projection. Then $s \subseteq \mathbf{A}$. In fact, let $a \in A$ and let $f(a) = (f_1, f_2)$. Since f is a morphism, we have $\alpha(a,a) = \mu(f(a), f(a)) = f_1$ and it follows that $s(a) = f_2 \le f_1 = \alpha(a,a)$. Further, for $a, b \in A$ we have $\alpha(a,b) \le \mu(f(a), f(b)) = (\alpha(a,a) \otimes (s(a) \to s(b)) \wedge (\alpha(b,b) \otimes (s(b) \to s(a))$. It follows from this inequality that

$$\begin{aligned} s(a) \otimes (\alpha(a,a) \to \alpha(a,b)) &= \alpha(a,b) \otimes (\alpha(a,a) \to s(a)) \le \\ &\le \alpha(a,a) \otimes (s(a) \to s(b)) \otimes (\alpha(a,a) \to s(a)) = \\ &= (\alpha(a,a) \wedge s(a)) \otimes (s(a) \to s(b)) = \\ &= s(a) \otimes (s(a) \to s(b)) = s(a) \wedge s(b) \le s(b). \end{aligned}$$

We see that χ and χ^{-1} are mutually converse maps.

(b) Let $s \in \mathcal{S}$. It is clear that the diagram commutes. Further, let (X,τ) be Ω-fuzzy set with morphisms $u : (X,\tau) \longrightarrow (A,\alpha)$ and $v : (X,\tau) \longrightarrow \mathbf{To}$ such that $\chi(s) \circ u = \top \circ v$. Since $v =!$, for each $x \in X$ it follows easily that $\alpha(u(x), u(x)) = \tau(x,x) = s(u(x))$. Hence, $u(x) \in S_s$ and $u : (X,\tau) \longrightarrow S_s, \alpha)$ is the morphism. It is then clear that the diagram is the pullback. □

Using Theorem 7.4 we can interpret formulas in a way analogous to the approach introduced in Section 3.4. Recall that according to this approach, a formula A with a set of free variables X such that $FV(A) \subseteq X$ is interpreted as a morphism $\|A\| : M(X) \longrightarrow \Omega$ where $M(X) = \prod_{x \in X} M(\iota_x)$ is the product (in the appropriate category, mostly in topos) of interpretations of the types ι_x corresponding to the variables $x \in X$ and Ω is the subobject classifier. According to Theorem 3.19 on page 89, such morphism $\|A\|$ is the characteristic morphism of some subobject $[A] \hookrightarrow M(X)$. Moreover, for definition of morphisms $\|A\|$ some morphisms $\Omega^n \longrightarrow \Omega$ are used, which represent the interpretation of logical connectives, such as $\wedge, \vee$, etc. This approach shows a way how to introduce the interpretation of formulas in our category $\Omega - \mathbf{FSet}$, which, in general, is not topos.

In the previous theorem, we have shown that for any Ω-subset $s \subseteq \mathbf{A} = (A, \alpha)$ (i.e. a map $s : A \longrightarrow L$ satisfying (i'),(ii')) there exists a subset $S_s \subseteq A$ and its characteristic morphism $\chi(s) : \mathbf{A} \longrightarrow (\Omega^*, \mu)$. In the following proposition we will show that characteristic morphism can be constructed for the so called complete subsets in A.

THEOREM 7.5
Let $\Omega = \langle L, \vee, \wedge, \otimes, \rightarrow \rangle$ be a complete MV-algebra, $\mathbf{A} = (A, \alpha)$ be an Ω-fuzzy set and let $S \subseteq A$. Then there exists an Ω-subset $s_S \subseteq \mathbf{A}$ such that the diagram

$$\begin{array}{ccc} (S, \alpha) & \hookrightarrow & (A, \alpha) \\ {!}\downarrow & & \downarrow\chi(s_S) \\ \mathbf{To} & \xrightarrow{\top} & (\Omega^*, \mu). \end{array}$$

commutes. Moreover, this diagram is the pullback iff $S = \{a \in A \mid s_S(a) = \alpha(a, a)\}$. In this case, S is called a complete subset *and $\chi(s_S)$ is called the* characteristic morphism of S.

PROOF: For any $a \in A$ we define

$$s_S(a) = \bigvee_{x \in S} \alpha(a, x).$$

Then we prove that s_S is an Ω-subset, $s_S \subseteq \mathbf{A}$. Clearly, $s_S(a) \leq \alpha(a, a)$ for all $a \in A$. Further, let $a, b \in A$ and let us consider the following possible cases.

(1) $a, b \in S$. Then we have

$$\begin{aligned} &s_S(a) = \alpha(a, a), s_S(b) = \alpha(b, b), \\ &\alpha(a, a) \otimes (\alpha(a, a) \rightarrow \alpha(a, b)) = \alpha(a, b) \leq \alpha(b, b). \end{aligned}$$

(2) $a \in S, b \notin S$. Then we have

$$\begin{aligned} s_S(a) \otimes (\alpha(a, a) \rightarrow \alpha(a, b)) \leq & \bigvee_{x \in S} (s_S(x) \otimes (\alpha(x, x) \rightarrow \alpha(x, b))) = \\ = & \bigvee_{x \in S} \alpha(x, b) = s_S(b). \end{aligned}$$

(3) $a \notin S$, $b \in S$. Then we have

$$\begin{aligned} \left(\bigvee_{x \in S} \alpha(a, x)\right) \otimes (\alpha(a, a) \rightarrow \alpha(a, b)) \leq \alpha(a, a) \otimes (\alpha(a, a) \rightarrow \alpha(a, b)) = \\ = \alpha(a, b) \leq \alpha(b, b) = s_S(b). \end{aligned}$$

(4) $a, b \notin S$. Then from properties of α it follows

$$\begin{aligned}(\bigvee_{x\in S} \alpha(a,x)) \otimes (\alpha(a,a) \to \alpha(a,b)) &= \bigvee_{x\in S} (\alpha(a,x) \otimes (\alpha(a,a) \to \alpha(a,b))) \leq \\ &\leq \bigvee_{x\in S} \alpha(x,b) = s_S(b).\end{aligned}$$

Hence, $s_S \subseteq \mathbf{A}$. Since for $a \in S$ we have $s_S(a) = \alpha(a,a)$, the diagram commutes. Now, let S be complete and let (B,β) be an Ω-fuzzy set with a morphism u such that the diagram

$$\begin{array}{ccc} (B,\beta) & \xrightarrow{u} & (A,\alpha) \\ !\downarrow & & \downarrow\chi(s_S) \\ \mathbf{To} & \xrightarrow{\top} & (\Omega^*,\mu) \end{array}$$

commutes. Then for all $x \in B$ we have $\alpha(u(x),u(x)) = \beta(u(x),u(x)) = s_S(u(x))$ and it follows that $u(x) \in S$. Hence $u : (B,\beta) \longrightarrow (S,\alpha)$ and the diagram is the pullback.

Conversely, let the diagram be pullback and let us suppose that there exists $a \in A \setminus S$ such that $s_S(a) = \alpha(a,a)$. Then the diagram

$$\begin{array}{ccc} (S\cup\{a\},\alpha) & \hookrightarrow & (A,\alpha) \\ !\downarrow & & \downarrow\chi(s_S) \\ \mathbf{To} & \xrightarrow{\top} & (\Omega^*,\mu), \end{array}$$

commutes and there must exist a morphism $r : (S\cup\{a\},\alpha) \longrightarrow (S,\alpha)$ such that the diagram

$$\begin{array}{ccc} (S,\alpha) & \hookrightarrow & (A,\alpha) \\ r\uparrow & \nearrow & \\ (S\cup\{a\},\alpha) & & \end{array}$$

commutes. But in this case we have $a = r(a) \in S$ — a contradiction. □

Truth-valued morphisms in Ω – FSet. Theorem 7.5 can be used for construction of the truth-valued morphisms in the category Ω – **FSet**, which are analogous to the truth-valued morphisms constructed in any topos (see Definition 2.30). We show only some examples of this construction, namely the construction of the truth-valued morphisms $\cap_{\Omega^*}, \otimes_{\Omega^*} : \Omega^* \times \Omega^* \longrightarrow \Omega^*$ which could be used for interpretation of the conjunction and Łukasiewicz conjunction of formulas. The rest of truth-valued morphisms can be constructed analogously with some help of constructions presented in Definition 2.30 on page 57.

First, we have to do some auxiliary computations. For definition of binary logical truth-valued morphisms we need the product $(\Omega^* \times \Omega^*, \sigma)$ of Ω-fuzzy set (Ω^*, μ). From the construction of the product in this category (see Proposition 7.2) it follows that

$$\Omega^* \times \Omega^* = \{((\alpha_1, \beta_1), (\alpha_2, \beta_2)) \mid (\alpha_i, \beta_i) \in \Omega^*, \mu((\alpha_1, \beta_1), (\alpha_1, \beta_1)) = \mu((\alpha_2, \beta_2), (\alpha_2, \beta_2))\} = \{((\alpha, \beta_1), (\alpha, \beta_2)) \mid \alpha \geq \beta_1, \beta_2\},$$

and

$$\sigma(((\alpha, \beta_1), (\alpha, \beta_2)), ((\rho, \tau_1), (\rho, \tau_2))) = \mu((\alpha, \beta_1), (\rho, \tau_1)) \wedge \mu((\alpha, \beta_2), (\rho, \tau_2)).$$

Let us consider the monomorphism $\langle \top, \top \rangle : \mathbf{To} \longrightarrow \Omega^* \times \Omega^*$. Then we have

$$(\forall \omega \in L) \ (\langle \top, \top \rangle(\omega) = ((\omega, \omega), (\omega, \omega))),$$

and for any $((\alpha, \beta_1), (\alpha, \beta_2)) \in \Omega^* \times \Omega^*$ and $\omega \in L$ we have

$$\sigma(((\omega, \omega), (\omega, \omega)), ((\alpha, \beta_1), (\alpha, \beta_2))) = \omega \wedge \beta_1 \wedge \beta_2,$$

as it can be simply proved. Now, the monomorphism $\langle \top, \top \rangle$ defines a subset $S_\cap \subseteq \Omega^* \times \Omega^*$ such that

$$S_\cap = \{((\omega, \omega), (\omega, \omega)) \mid \omega \in L\} \subseteq \Omega^* \times \Omega^*.$$

This subset $S_\cap$ is a complete subset. In fact, according to the proof of Theorem 7.5, the Ω-subset $s_{S_\cap} : (\Omega^* \times \Omega^*, \sigma) \longrightarrow L$ is defined as follows:

$$s_{S_\cap}((\alpha, \beta_1), (\alpha, \beta_2)) = \bigvee_{\omega \in L} \sigma(((\omega, \omega), (\omega, \omega)), ((\alpha, \beta_1), (\alpha, \beta_2))) = \\ = \bigvee_{\omega \in L} (\omega \wedge \beta_1 \wedge \beta_2) = \beta_1 \wedge \beta_2.$$

Hence, to prove that $S_\cap$ is complete we must prove that

$$S_\cap = \{\mathbf{a} \in \Omega^* \times \Omega^* \mid s_{S_\cap}(\mathbf{a}) = \sigma(\mathbf{a}, \mathbf{a})\}.$$

The inclusion $\subseteq$ is trivial.

Let $\mathbf{a} = ((\alpha, \beta_1), (\alpha, \beta_2))$ be such that $s_{S_\cap}(\mathbf{a}) = \sigma(\mathbf{a}, \mathbf{a})$. Then we have $\beta_1 \wedge \beta_2 = s_{S_\cap}(\mathbf{a}) = \sigma(\mathbf{a}, \mathbf{a}) = \alpha$ and since $\alpha \geq \beta_1, \beta_2$ we have $\alpha = \beta_1 = \beta_2$. Hence $\mathbf{a} \in S_\cap$ and $S_\cap$ is complete. According to Theorem 7.5, for the set $S_\cap$ there exists the characteristic morphism $\Omega^* \times \Omega^* \longrightarrow \Omega^*$ which we denote by $\cap_{\Omega^*}$.

We show explicitly how this morphism $\cap_{\Omega^*}$ works. Since $\cap_{\Omega^*} = \chi(s_{S_\cap})$, according to Theorem 7.4, for any $\mathbf{a} = ((\alpha, \beta_1), (\alpha, \beta_2))$ we have

$$\cap_{\Omega^*}((\alpha, \beta_1), (\alpha, \beta_2)) = \chi(s_{S_\cap})(\mathbf{a}) = (\sigma(\mathbf{a}, \mathbf{a}), s_{S_\cap}(\mathbf{a})) = \\ = (\mu((\alpha, \beta_1), (\alpha, \beta_1)) \wedge \mu((\alpha, \beta_2), (\alpha, \beta_2)), \beta_1 \wedge \beta_2) = (\alpha, \beta_1 \wedge \beta_2).$$

This result can be used as a method for defining other truth-valued morphisms. For example, the monoidal product (representing the Łukasiewicz conjunction) $\otimes_{\Omega^*} : \Omega^* \times \Omega^* \longrightarrow \Omega^*$ can be defined analogously as $\cap_{\Omega^*}$ as follows:

$$\otimes_{\Omega^*}((\alpha, \beta), (\alpha, \beta_1)) = (\alpha, \beta_1 \otimes \beta_2).$$

Then $\otimes_{\Omega^*}$ is the morphism in $\Omega - \mathbf{FSet}$. In fact, for $\mathbf{a} = ((\alpha,\beta_1),(\alpha,\beta_2)) \in \Omega^* \times \Omega^*$ we have

$$\mu(\otimes_{\Omega^*}(\mathbf{a}), \otimes_{\Omega^*}(\mathbf{a})) = \alpha = \sigma(\mathbf{a},\mathbf{a}).$$

Furthermore, for $\mathbf{a} = ((\alpha,\beta_1),(\alpha,\beta_2))$ and $\mathbf{b} = ((\rho,\tau_1),(\rho,\tau_2))$ we have

$$\begin{aligned}\sigma(\mathbf{a},\mathbf{b}) =&(\alpha \otimes (\beta_1 \to \tau_1)) \wedge (\alpha \otimes (\beta_2 \to \tau_2)) \wedge (\rho \otimes (\tau_1 \to \beta_1)) \wedge (\rho \otimes (\tau_2 \to \beta_2))\\ =&\alpha \otimes ((\beta_1 \to \tau_1) \wedge (\beta_2 \to \tau_2)) \wedge \rho \otimes ((\tau_1 \to \beta_1) \wedge (\tau_2 \to \beta_2)) \leq\\ \leq&\alpha \otimes (\beta_1 \otimes \beta_2 \to \tau_1 \wedge \tau_2) \wedge \rho \otimes (\tau_1 \otimes \tau_2 \to \beta_1 \wedge \beta_2) \leq\\ \leq&\alpha \otimes (\beta_1 \otimes \beta_2 \to \tau_1 \otimes \tau_2) \wedge \rho \otimes (\tau_1 \otimes \tau_2 \to \beta_1 \otimes \beta_2) =\\ =&\mu(\otimes_{\Omega^*}(\mathbf{a}), \otimes_{\Omega^*}(\mathbf{b})).\end{aligned}$$

It is then clear that a complete subset $S_\otimes$ corresponding to the characteristic morphism $\otimes_{\Omega^*}$ is

$$S_\otimes = \{((\beta_1 \otimes \beta_2, \beta_1 \otimes \beta_2),(\beta_1 \otimes \beta_2, \beta_1 \otimes \beta_2)) \in \Omega^* \times \Omega^* \mid \beta_1, \beta_2 \in L\}.$$

Hence $S_\cap \subseteq S_\otimes$. This morphism $\otimes_{\Omega^*}$ can be used for interpretation of the Łukasiewicz conjunction $\otimes$. It means that Łukasiewicz logic can be also interpreted in this category.

On the other hand, the internal logic in the category $\Omega - \mathbf{FSet}$ is not well known and we are not able to prove an analogy of Theorem 2.19 on page 58. Most difficulties which concern properties of the subobject structures in this category have their origin in the problems with subobjects classification in this category. As seen in the previous theorems of this section, the classification of Ω-subsets is possible. Moreover, there are some other classification possibilities in this category but we will not be dealing with them in this book.

7.3 Interpretation of formulas in the category $\mathbf{C}\Omega - \mathbf{Set}$

In the previous section, we have mentioned the interpretation of logic in the category, the objects of which are similar to fuzzy sets. In this section, we want to derive this notion in more details and in a different category — in the category of complete Ω-sets. In Section 4.3, we have exposed first-order fuzzy logic and its semantics, which can be generalized to many sorted fuzzy logic, as well. Recall that a structure for many sorted FLn with the language J with truth-valued lattice L and with the set Υ of sorts is

$$\mathcal{D} = \langle\{M(\iota) \mid \iota \in \Upsilon\}, \{P_\mathcal{D} \mid P \in \mathcal{P}\}, \{f_\mathcal{D} \mid f \in \mathcal{F}\}\rangle$$

where for $P \in \mathcal{P}$ such that $P \subseteq \iota_1 \times \cdots \times \iota_n$ we have $P_\mathcal{D} : M(\iota_1) \times \cdots \times M(\iota_n) \to L$ and for $f \in \mathcal{F}$ with $f : \iota_1 \times \cdots \times \iota_n \to \iota$ we have $f_\mathcal{D} : M(\iota_1) \times \cdots \times M(\iota_n) \longrightarrow M(\iota)$. This structure enables us to introduce a truth-valued interpretation of formulas with values in the lattice $\mathcal{L}$, i.e. a function $\mathcal{D} : F_J \longrightarrow L$ which is defined by induction on the complexity of the corresponding formula. This valuation is then a basis for the definition of the semantics of formulas from F_J.

On the other hand, in Section 3.4 we have introduced a model theory in the topos $\mathcal{E}$, which enables us to compute a truth-value $||A||$ of any formula

A with a set of free variables X containing $FV(A)$ as a morphism $||A|| : \prod_{x\in X} M(\iota_x) \longrightarrow \Omega$ where Ω is the subobject classifier in this topos. Using this interpretation, we have received another type of semantics of formulas from F_J. In this section we show that, at least for the Heyting algebra $\mathcal{L}$, both these approaches are identical if we take an appropriate category of fuzzy sets for the topos $\mathcal{E}$.

Truth valuations of formulas in Heyting algebras. Let $\Omega = \langle L, \vee, \wedge\rangle$ be a complete Heyting algebra and let us consider a many sorted language J with a set Υ of sorts, a set $\mathcal{F}$ of functional symbols and a set $\mathcal{P}$ of predicate symbols. With any sort ι we associate a complete Ω-set (A_ι, α_ι) and for any functional symbol $f : \iota_1 \times \cdots \times \iota_n \longrightarrow \iota$ let $f_{\mathcal{D}} : A_{\iota_1} \times \cdots \times A_{\iota_n} \longrightarrow A_\iota$ be a morphism in the category $\mathbf{C}\Omega - \mathbf{Set}$ (where $\times$ is the product in this category) and for any $P \subseteq \iota_1 \times \cdots \times \iota_n$ let $P_{\mathcal{D}} : (A_{\iota_1} \times \cdots \times A_{\iota_n}, \alpha) \longrightarrow L$ be an Ω-subset in this category, i.e. it must satisfy the conditions

(i) $(\forall \mathbf{a}, \mathbf{b})\ P_{\mathcal{D}}(\mathbf{a}) \wedge \alpha(\mathbf{a}, \mathbf{b}) \le P_{\mathcal{D}}(\mathbf{b})$,
(ii) $(\forall \mathbf{a})\ P_{\mathcal{D}}(\mathbf{a}) \le \alpha(\mathbf{a}, \mathbf{a})$.

In this way we obtain a structure $\mathcal{D}$ for the language J. For any variable x, we will denote by ι_x the type of x. For any set of variables X we put $(\mathcal{D}(X), \alpha_0) = \prod_{x\in X}(A_{\iota_x}, \alpha_{\iota_x})$, the product of complete Ω-sets A_{ι_x} in the category $\mathbf{C}\Omega - \mathbf{Set}$. Recall that according to Theorem 7.3,

$$\mathcal{D}(X) = \{((a_x)_{x\in X}) \in \prod_{x\in X} A_{\iota_x} \mid (\forall x, y \in X)\ \alpha_{\iota_x}(a_x, a_x) = \alpha_{\iota_y}(a_y, a_y)\},$$

$$\alpha_0((a_x)_{x\in X}, (b_x)_{x\in X}) = \bigwedge_{x\in X} \alpha_{\iota_x}(a_x, b_x).$$

Instead of $\alpha_0(\mathbf{a}, \mathbf{a})$, we will sometimes use the abbreviation $\mathcal{D}_X(\mathbf{a})$, which indicates explicitly a set X of variables. Let us mention that for any element $\mathbf{a} \in \mathcal{D}(X)$, we have $\alpha_0(\mathbf{a}, \mathbf{a}) = \bigwedge_{x\in X} \alpha_{\iota_x}(a_x, a_x) = \alpha_{\iota_y}(a_y, a_y)$ for all $y \in X$. If t is a term of the sort ι then for any set X of variables containing all the variables of t we can simply define a map

$$t_{\mathcal{D},X} : \mathcal{D}(X) \longrightarrow A_\iota.$$

This map is defined inductively on the complexity of the term t. For example, if $t = f(t_1, \ldots, t_n)$, then we suppose that $t_{i,\mathcal{D},X} : \mathcal{D}(X) \longrightarrow (A_{\iota_i}, \alpha_{\iota_i})$ is defined for any $i = 1, \ldots, n$ and for $\mathbf{a} \in \mathcal{D}(X)$ we put

$$t_{\mathcal{D},X}(\mathbf{a}) = f_{\mathcal{D}}(t_{1,\mathcal{D},X}(\mathbf{a}), \ldots, t_{n,\mathcal{D},X}(\mathbf{a})).$$

This definition is correct since it can be proved that

$$(t_{1,\mathcal{D},X}(\mathbf{a}), \ldots, t_{n,\mathcal{D},X}(\mathbf{a})) \in (A_1, \alpha_1) \times \cdots \times (A_n \alpha_n),$$

as follows from the statement

$$(\forall \mathbf{a} \in \mathcal{D}(X))\ \alpha_\iota(t_{\mathcal{D},X}(\mathbf{a}), t_{\mathcal{D},X}(\mathbf{a})) = \bigwedge_{x\in X} \alpha_{\iota_x}(a_x, a_x).$$

The proof of this fact can be done by induction.

It is not surprising that in this way we obtain a morphism in the category $\mathbf{C\Omega} - \mathbf{Set}$, as stated in the following lemma.

LEMMA 7.10
Let t be a term of a sort ι. Then $t_{\mathcal{D},X} : (\mathcal{D}(X), \alpha_0) \longrightarrow (A_\iota, \alpha_\iota)$ is a morphism in the category $\mathbf{C\Omega} - \mathbf{Set}$.

PROOF: Let us suppose that $t = f(t_1, \dots, t_n)$ and let the statement of lemma hold for all terms t_i. Then for any $\mathbf{a}, \mathbf{b} \in \mathcal{D}(X)$ we have

$$\alpha_0(\mathbf{a}, \mathbf{b}) \leq \alpha_i(t_{i,\mathcal{D},X}(\mathbf{a}), t_{i,\mathcal{D},X}(\mathbf{b})), \quad i = 1, \dots, n,$$

and it follows that

$$\begin{aligned}\alpha_0(\mathbf{a}, \mathbf{b}) &\leq \bigwedge_{i=1}^{n} \alpha_i(t_{i,\mathcal{D},X}(\mathbf{a}), t_{i,\mathcal{D},X}(\mathbf{b})) = \\ &= \alpha((t_{1,\mathcal{D},X}(\mathbf{a}), \dots, t_{n,\mathcal{D},X}(\mathbf{a})), (t_{1,\mathcal{D},X}(\mathbf{b}), \dots, t_{n,\mathcal{D},X}(\mathbf{b}))) \leq \\ &\leq \alpha_\iota(t_{\mathcal{D},X}(\mathbf{a}), t_{\mathcal{D},X}(\mathbf{b})).\end{aligned}$$

Hence, $t_{\mathcal{D},X}$ is a morphism. □

Now, for any formula A and for any set X such that $FV(A) \subseteq X$ we can define a valuation map

$$\mathcal{D}_X(A) : \mathcal{D}(X) \longrightarrow L$$

by induction depending on the complexity of A. Moreover, any valuation should satisfy the condition

$$(\forall \mathbf{a} = ((a_x)_{x \in X}) \in \mathcal{D}(X)) \quad \mathcal{D}_X(A)(\mathbf{a}) \leq \bigwedge_{x \in X} \alpha_{\iota_x}(a_x, a_x) \quad (= \alpha_0(\mathbf{a}, \mathbf{a})).$$

The valuation map is defined as follows. For any formula A and any set X of free variables such that $FV(A) \subseteq A$ we set for all $\mathbf{a} \in \mathcal{D}(X)$,

$$\mathcal{D}_X(A)(\mathbf{a}) = \alpha_0(\mathbf{a}, \mathbf{a}) \wedge \mathcal{D}_X[A](\mathbf{a}),$$

where the valuation $\mathcal{D}_X[A] : \mathcal{D}(X) \longrightarrow L$ is defined as follows.

(i) $A := P(t_1, \dots, t_n)$. Let us assume that $t_{i,\mathcal{D},X}$ are already defined for $i = 1, \dots, n$. Then for $\mathbf{a} \in \mathcal{D}(X)$ we set

$$\mathcal{D}_X[P(t_1, \dots, t_n)](\mathbf{a}) = P_{\mathcal{D}}(t_{1,\mathcal{D},X}(\mathbf{a}), \dots, t_{n,\mathcal{D},X}(\mathbf{a})).$$

(ii) $A := (t_1 = t_2)$, where t_i are of a sort ι. Let us assume that $t_{i,\mathcal{D},X}$ are defined for $i = 1, 2$. Then we set

$$\mathcal{D}_X[t_1 = t_2](\mathbf{a}) = \alpha_\iota(t_{1,\mathcal{D},X}(\mathbf{a}), t_{2,\mathcal{D},X}(\mathbf{a})).$$

(iii) Let $A := \bigwedge \Psi$ where Ψ is a set of formulas. Then we set

$$\mathcal{D}_X\left[\bigwedge \Psi\right](\mathbf{a}) = \bigwedge_{B\in\Psi} \mathcal{D}_X[B](\mathbf{a}).$$

The valuation for $A := \bigvee \Psi$ can be defined analogously.

(iv) Let $A := B \Rightarrow C$. Then we set

$$\mathcal{D}_X[B \Rightarrow C](\mathbf{a}) = (\mathcal{D}_X[B](\mathbf{a}) \to \mathcal{D}_X[C](\mathbf{a})).$$

(v) Let $A := (\exists x)B$ where x is of the sort ι. Then we put

$$\mathcal{D}_X[(\exists x)B](\mathbf{a}) = \bigvee_{a\in A_\iota} (\alpha_\iota(a,a) \to \mathcal{D}_X[B_x[a]](\mathbf{a})),$$

where $B_x[a]$ is the result of the substitution of the term a for x at each occurrence of x in the formula B.

(vi) Let $A := (\forall x)B$ where x is of a sort ι. Then we set

$$\mathcal{D}_X[(\forall x)B](\mathbf{a}) = \bigwedge_{a\in A_\iota} (\alpha(a,a) \to \mathcal{D}_X[B_x[a]](\mathbf{a})).$$

LEMMA 7.11
Let A be a formula and let X be a set of variables containing such that $FV(A) \subseteq A$. Then $\mathcal{D}_X(A) : \mathcal{D}(X) \to L$ is a Ω-subset in the category $\mathbf{C}\Omega - \mathbf{Set}$.

PROOF: The proof will be done by induction on the complexity of A. The condition $\mathcal{D}_X(A)(\mathbf{a}) \le \alpha_0(\mathbf{a},\mathbf{a})$ follows from the definition of valuation.

(i) Let $A := P(t_1,\dots,t_n)$ and let $(S,\alpha) = (A_1,\alpha_1)\times\dots\times(A_n,\alpha_n)$. Then according to Lemma 7.10 and the properties of the Ω-subset $P_\mathcal{D}$ we have

$$\begin{aligned}
&\alpha_0(\mathbf{a},\mathbf{b}) \wedge \mathcal{D}_X(A)(\mathbf{a}) = \bigwedge_{x\in X} \alpha_{\iota_x}(a_X, b_x) \wedge P_\mathcal{D}(t_{1,\mathcal{D},X}(\mathbf{a}),\dots,t_{n,\mathcal{D},X}(\mathbf{a})) = \\
&= \alpha_0(\mathbf{a},\mathbf{b}) \wedge \bigwedge_{i=1}^{n} \alpha_i(t_{i,\mathcal{D},X}(\mathbf{a}), t_{i,\mathcal{D},X}(\mathbf{b})) \wedge P_\mathcal{D}(t_{1,\mathcal{D},X}(\mathbf{a}),\dots,t_{n,\mathcal{D},X}(\mathbf{a})) \le \\
&\le \alpha_0(\mathbf{a},\mathbf{b}) \wedge (P_\mathcal{D}(t_{1,\mathcal{D},X}(\mathbf{b}),\dots,t_{n,\mathcal{D},X}(\mathbf{b})) \le \\
&\le \alpha_0(\mathbf{b},\mathbf{b}) \wedge P_\mathcal{D}(t_{1,\mathcal{D},X}(\mathbf{b}),\dots,t_{n,\mathcal{D},X}(\mathbf{b})) = \mathcal{D}_X(A)(\mathbf{b}).
\end{aligned}$$

for any $\mathbf{a},\mathbf{b} \in \mathcal{D}(X)$.

(ii) Let $A := (t_1 = t_2)$ where t_i are terms of the sort ι. Then we have

$$\begin{aligned}
\alpha_0(\mathbf{a},\mathbf{b}) \wedge \mathcal{D}_X(A)(\mathbf{a}) &= \alpha_0(\mathbf{a},\mathbf{b}) \wedge \alpha_\iota(t_{1,\mathcal{D},X}(\mathbf{a}), t_{2,\mathcal{D},X}(\mathbf{a})) \le \\
&\le \alpha_\iota(t_{1,\mathcal{D},X}(\mathbf{a}), t_{1,\mathcal{D},X}(\mathbf{b})) \wedge \alpha_\iota(t_{2,\mathcal{D},X}(\mathbf{a}), t_{2,\mathcal{D},X}(\mathbf{b})) \wedge \\
&\quad \wedge \alpha_\iota(t_{1,\mathcal{D},X}(\mathbf{a}), t_{1,\mathcal{D},X}(\mathbf{a})) \wedge \alpha_0(\mathbf{b},\mathbf{b}) \le \\
&\le \alpha_\iota(t_{1,\mathcal{D},X}(\mathbf{b}), t_{2,\mathcal{D},X}(\mathbf{b})) \wedge \alpha_0(\mathbf{b},\mathbf{b})) = \\
&= \mathcal{D}_X(A)(\mathbf{b}).
\end{aligned}$$

(iii) Let $A := \bigwedge \Phi$. Then we have

$$\begin{aligned}\alpha_0(\mathbf{a},\mathbf{b}) \wedge \mathcal{D}_X(A)(\mathbf{a}) =& \alpha_0(\mathbf{a},\mathbf{b}) \wedge \bigwedge_{B\in\Phi} \mathcal{D}_X(B)(\mathbf{a}) = \\ &= \bigwedge_{B\in\Phi} (\mathcal{D}_X(B)(\mathbf{a}) \wedge \alpha_0(\mathbf{a},\mathbf{b})) \wedge \alpha_0(\mathbf{b},\mathbf{b}) \leq \\ &\leq \bigwedge_{B\in\Phi} \mathcal{D}_X(B)(\mathbf{b}) \wedge \alpha_0(\mathbf{b},\mathbf{b}) = \mathcal{D}_X(A)(\mathbf{b}).\end{aligned}$$

Analogously we can proceed for $A := \bigvee \psi$.

(iv) Let $A := B \to C$. We must prove that

$$\alpha_0(\mathbf{a},\mathbf{b}) \wedge (\mathcal{D}_X(B)(\mathbf{a}) \to \mathcal{D}_X(C)(\mathbf{a})) \leq (\mathcal{D}_X(B)(\mathbf{b}) \to \mathcal{D}_X(C)(\mathbf{b})).$$

By the adjunction of $\to$ and $\wedge$ in the Heyting algebra we obtain

$$\begin{aligned}&\alpha_0(\mathbf{a},\mathbf{b}) \wedge \mathcal{D}_X(B)(\mathbf{b}) \wedge (\mathcal{D}_X(B)(\mathbf{a}) \to \mathcal{D}_X(C)(\mathbf{a})) \leq \\ &\leq \alpha_0(\mathbf{a},\mathbf{b}) \wedge \mathcal{D}_X(B)(\mathbf{a}) \wedge (\mathcal{D}_X(B)(\mathbf{a}) \to \mathcal{D}_X(C)(\mathbf{a})) = \\ &= \alpha_0(\mathbf{a},\mathbf{b}) \wedge \mathcal{D}_X(B)(\mathbf{a}) \wedge \mathcal{D}_X(C)(\mathbf{a}) \leq \alpha_0(\mathbf{a},\mathbf{b}) \wedge \mathcal{D}_X(C)(\mathbf{a}) \\ &\leq \mathcal{D}_X(C)(\mathbf{b})\end{aligned}$$

and from this the required property follows.

(v) Let $A = (\forall x)B$, where x is of a sort interpreted as (E, τ). Then we have

$$\begin{aligned}&\alpha_0(\mathbf{a},\mathbf{b}) \wedge (\tau(e,e) \to \mathcal{D}_X(B_x[e])(\mathbf{a})) \wedge \tau(e,e) = \\ &= \tau(e,e) \wedge \alpha_0(\mathbf{a},\mathbf{b}) \wedge \mathcal{D}_X(B_x[e])(\mathbf{a}) \leq \mathcal{D}_X(B_x[e])(\mathbf{b}),\end{aligned}$$

as follows from the induction assumption. Hence, we have obtained

$$(\tau(e,e) \to \mathcal{D}_X(B_x[e])(\mathbf{a})) \wedge \alpha_0(\mathbf{a},\mathbf{b}) \leq \tau(e,e) \to \mathcal{D}_X(B_x[e])(\mathbf{b}).$$

Then

$$\begin{aligned}\alpha_0(\mathbf{a},\mathbf{b}) \wedge \mathcal{D}_X(A)(\mathbf{a}) &= \alpha_0(\mathbf{a},\mathbf{b}) \wedge \bigwedge_{e\in E} (\tau(e,e) \to \mathcal{D}_X(B_x[e])(\mathbf{a})) = \\ &= \bigwedge_{e\in E} \alpha_0(\mathbf{a},\mathbf{b}) \wedge (\tau(e,e) \to \mathcal{D}_X(B_x[e])(\mathbf{a})) \leq \\ &\leq \alpha_0(\mathbf{b},\mathbf{b}) \wedge \bigwedge_{e\in E} (\tau(e,e) \to \mathcal{D}_X(B_x[e])(\mathbf{b})) = \\ &= \mathcal{D}_X(A)(\mathbf{b}).\end{aligned}$$

Analogously we can proceed for $A = (\exists x)B$. □

PROPOSITION 7.4
Let $X \subseteq Y$ be sets of free variables such that $FV(A) \subseteq X$ and let $p_{YX} : \mathcal{D}(Y) \longrightarrow \mathcal{D}(X)$ be the canonical projection. Then we have

(a) $(\forall \mathbf{a} \in \mathcal{D}(Y))\ \mathcal{D}_Y[A](\mathbf{a}) = \mathcal{D}_X[A](p_{YX}(\mathbf{a}))$,

(b) $(\forall \mathbf{a} \in \mathcal{D}(Y))\ \mathcal{D}_Y(A)(\mathbf{a}) = \mathcal{D}_X(A)(p_{YX}(\mathbf{a}))$.

PROOF: (a) It can be proved that for any term t we have $t_{\mathcal{D},Y} = t_{\mathcal{D},X} \circ p_{YX}$. Then the assertion can be proved using an induction on the complexity of A and this property of terms.

(b) It follows from (a) and from the equality $\mathcal{D}_Y(\mathbf{a}) = \mathcal{D}_X(p_{YX}(\mathbf{a}))$. In fact, if β_0 is associated with $\mathcal{D}(Y)$, then for every $\mathbf{a} \in \mathcal{D}(Y)$ we have

$$\beta_0(\mathbf{a}, \mathbf{a}) = \bigwedge_{y \in Y} \alpha_{\iota_y}(a_y, a_y) = \alpha_{\iota_x}(a_x, a_x) = \alpha_0(p_{YX}(\mathbf{a}), p_{YX}(\mathbf{a})).$$

□

For any (Gentzen) sequent $\Phi \Rightarrow \Psi$ we say that the structure $\mathcal{D}$ is a model of this sequent (with respect to a set of variables X), in symbols $\mathcal{D} \models_X \Phi \Rightarrow \Psi$, if

$$(\forall \mathbf{a} \in \mathcal{D}(X))\ (\mathcal{D}_X\left(\bigwedge_{A \in \Phi} A\right)(\mathbf{a}) \leq \mathcal{D}_X\left(\bigvee_{B \in \Psi} B\right)(\mathbf{a})).$$

Although the X set of variables is mentioned explicitly, this definition does not depend on X. In fact, the following lemma holds.

LEMMA 7.12
Let $\Phi \Rightarrow \Psi$ be a Gentzen sequent and let X, Y be sets of variables such that $FV(\Phi) \cup FV(\Psi) \subseteq X \subseteq Y$. Then

$$\mathcal{D} \models_X \Phi \Rightarrow \Psi \quad \textit{iff} \quad \mathcal{D} \models_Y \Phi \Rightarrow \Psi.$$

PROOF: Let $\mathbf{b} \in \mathcal{D}(Y)$. Then according to Proposition 7.4, we have

$$\mathcal{D}_Y\left(\bigwedge \Phi\right)(\mathbf{b}) = \mathcal{D}_X\left(\bigwedge \Phi\right)(p_{YX}(\mathbf{b})) \leq \mathcal{D}_X\left(\bigvee \Psi\right)(p_{YX}(\mathbf{b})) = \\ = \mathcal{D}_Y\left(\bigvee \Psi\right)(\mathbf{b}).$$

The converse inclusion follows analogously from the fact that for any $\mathbf{a} \in \mathcal{D}(X)$ there exists $\mathbf{b} \in \mathcal{D}(Y)$ such that $p_{YX}(\mathbf{b}) = \mathbf{a}$. □

From this lemma it follows that the relation $\mathcal{D} \models_X \Phi \Rightarrow \Psi$ is independent on a set X of variables such that $FV(\Phi) \cup FV(\Psi) \subseteq X$ and we will write simply $\mathcal{D} \models \Phi \Rightarrow \Psi$.

Two methods for evaluation of formulas. We have derived two methods how to evaluate validity of the Gentzen sequent $\Phi \Rightarrow \Psi$.

(i) The first method is based on the results of Section 3.4 (i.e. it is based on the interpretation of formulas in $\mathcal{E}$-structure $\mathcal{M}$ where $\mathcal{E}$ is a topos) and the sequent $\Phi \Rightarrow \Psi$ is valid iff $[\bigwedge \Phi]^{\mathcal{M}}(X) \leq [\bigvee \Psi]^{\mathcal{M}}(X)$ as subobjects of the object $M(X)$ in the category $\mathcal{E}$,

(ii) The second method is based on the truth valuations of formulas in the Heyting algebras based on the structure $\mathcal{D}$ for interpretation of the language J in the category $\mathbf{C}\Omega - \mathbf{Set}$, i.e. the sequent $\Phi \Rightarrow \Psi$ is valid iff $\mathcal{D}_X(\bigwedge \Phi)(\mathbf{a}) \leq \mathcal{D}_X(\bigvee \Psi)(\mathbf{a})$ for all $\mathbf{a} \in \mathcal{D}(X)$ as elements of the corresponding Heyting algebras.

It is clear that the second method is closely connected with the truth valuation of formulas, as introduced in Chapter 4. Our goal now is to show that both these approaches are, in principle, the same, at least for logic based on the Heyting algebras (i.e. intuitionistic logic).

We will start with a structure $\mathcal{D}$ for interpretation of the language J in the category $\mathbf{C}\Omega - \mathbf{Set}$ where $\Omega = \langle L, \vee, \wedge \rangle$ is a complete Heyting algebra, i.e.

$$\mathcal{D} = \langle \{(A_\iota, \alpha_\iota) \mid \iota \in \Upsilon\}, \{P_\mathcal{D} \mid P \in \mathcal{P}\}, \{f_\mathcal{D} \mid f \in \mathcal{F}\} \rangle,$$

where, if $P \in \mathcal{P}$ is of the type $\iota_1 \times \cdots \times \iota_n$ then

$$P_\mathcal{D} : \prod_{k=1}^{n} (A_{\iota_k}, \alpha_{\iota_k}) \longrightarrow L$$

is Ω-subset of this product, which is considered in the category $\mathbf{C}\Omega - \mathbf{Set}$ and which will be sometimes denoted by (Π_P, α_P) and where $f_\mathcal{D}$ is a corresponding morphism in our category.

On the basis of the structure $\mathcal{D}$, we can define a $\mathbf{C}\Omega - \mathbf{Set}$-structure $\mathcal{M}$ in the topos $\mathbf{C}\Omega - \mathbf{Set}$, as introduced in Definition 3.35 on page 84. The structure $\mathcal{M}$ will be defined as follows:

$$\mathcal{M} = \langle \{(A_\iota, \alpha_\iota) \mid \iota \in \Upsilon\}, \{P_\mathcal{M} \mid P \in \mathcal{P}\}, \{f_\mathcal{M} \mid f \in \mathcal{F}\} \rangle$$

where $P_\mathcal{M} \rightarrowtail (\Pi_P, \alpha_P)$ has to be a subobject in the category $\mathbf{C}\Omega - \mathbf{Set}$ (contrary to the Ω-subset $P_\mathcal{D}$). Hence, the only problem is how this subobject $P_\mathcal{M}$ can be defined from the Ω-subset $P_\mathcal{D}$.

In general, for any Ω-subset $S : (A, \alpha) \longrightarrow L$ we can define a subobject $(A_S, \alpha) \hookrightarrow (A, \alpha)$, which is called *associated with* S, such that

$$A_S = \{a \in A \mid S(a) = \alpha(a, a)\}.$$

Then $P_\mathcal{M} = (S_P, \alpha_P)$ will be the subobject of (Π_P, α_P) associated with $P_\mathcal{D}$.

It should be observed that the construction of the subobject associated with Ω-subset is, in some sense, similar to the bijection presented in Lemma 7.3 on page 265. In fact, according to this bijection, Ω-subset S of (A, α) is associated with a subobject $(A, \alpha_S) \overset{\alpha_S}{\rightarrowtail} (A, \alpha)$ in the category $\Omega - \mathbf{Set}$ where $\alpha_S(a, b) = S(a) \wedge S(b) \wedge \alpha(a, b)$. Unfortunately, the Ω-set (A, α_S) is not, in general, *complete* and, hence, we cannot use this subobject for our purposes. On the other hand, let us consider the following commutative diagram:

$$\begin{array}{ccc}
(A, \alpha_S) & \xrightarrow{\alpha_S} & (A, \alpha) \\
\{.\}\big\downarrow & \overset{\widehat{\alpha_S}}{\nearrow} & \big\uparrow u \\
(\mathrm{singl}(A, \alpha_S), \tau) & \overset{\overline{\alpha_S}}{\rightarrowtail} & \mathrm{singl}(A, \alpha)
\end{array}$$

where the morphism $\overline{\alpha_S}$ was defined in the proof of Lemma 7.4 on page 269 and the morphism u maps the singleton t onto an element $u(t) \in A$ such that $t = \{u(t)\}$. Then we have the following result.

LEMMA 7.13
Let (A, α) be a complete Ω-set and let S be Ω-subset of (A, α). Then the image of the complete Ω-set $(\text{singl}(A, \alpha_S), \tau)$ in the morphism $\widehat{\alpha_S}$ in the category $\mathbf{C\Omega - Set}$ *is equal to (A_S, α).*

PROOF: From the proof of Lemma 7.4 it follows that

$$(\forall s \in (\text{singl}(A, \alpha_S), \tau))(\forall a \in A)\ \overline{\alpha_S}(s)(a) = \bigvee_{x \in A} (\alpha_S(x, a) \wedge s(a)).$$

It can be proved that for the image $\operatorname{Im} \widehat{\alpha_S}$ we have

$$\operatorname{Im} \widehat{\alpha_S} = \{a \in A \mid \alpha(a, a) = \bigvee_{s \in \text{singl}(A, \alpha_S)} \alpha(\widehat{\alpha_S}(s), a)\}.$$

For any singleton $s \in \text{singl}(A, \alpha_S)$ we denote $a_s = \widehat{\alpha_S}(s) \in A$. From it follows that $\alpha(a_s, x) = \overline{\alpha_S}(s)(x)$ for all $x \in A$. Then we have

$$\alpha(a_s, x) = \bigvee_{y \in A} (\alpha_S(y, x) \wedge s(y)) = \alpha(x, x) \wedge S(x) \wedge s(x) = S(x) \wedge s(x),$$

as follows from the properties of S and s. Let $a \in \operatorname{Im} \widehat{\alpha_S}$. Then we have

$$\alpha(a, a) = \bigvee_{s \in \text{singl}(A, \alpha_S)} \alpha(a_s, a) = \bigvee_{s \in \text{singl}(A, \alpha_S)} (S(a) \wedge s(a)) =$$
$$= S(a) \wedge \bigvee_{s \in \text{singl}(A, \alpha_S)} s(a).$$

Furthermore, for $s = \{a\} \in \text{singl}(A, \alpha_S)$ we have

$$\bigvee_{s \in \text{singl}(A, \alpha_S)} s(a) = \{a\}(a) = \alpha_S(a, a) \leq \alpha(a, a),$$

which follows that $\alpha(a, a) = S(a)$. The converse inclusion is obvious. □

The relation between both methods. For any formula A with the set of free variables contained in X we can now define two objects:

(i) The truth-value function $\mathcal{D}_X(A) : \mathcal{D}(X) \longrightarrow L$, which is the Ω-subset (see Lemma 7.13).

(ii) the subobject $[A]^{\mathcal{M}}(X) \rightarrowtail (\mathcal{D}(X), \alpha_0)$ in the topos $\mathbf{C\Omega - Set}$.

As mentioned above, a subobject $(\mathcal{D}(X)_{\mathcal{D}_X(A)}, \alpha_0)$ can be associated with the Ω-subset $\mathcal{D}_X(A)$. We will denote the former by $([A]^{\mathcal{D}}(X), \alpha_0)$. The following theorem shows the principal relationship between both these subobjects.

THEOREM 7.6
For any formula A and any set X of variables such that $FV(A) \subseteq A$ we have

$$[A]^{\mathcal{M}}(X) = ([A]^{\mathcal{D}}(X), \alpha_0).$$

PROOF: The proof will be done by induction on the complexity of A. We will demonstrate few cases only. The rest can be proved analogously.

(1) $A := P(t_1, \ldots, t_n)$ where $P \subseteq \iota_1 \times \cdots \times \iota_n$. According to the results of Section 3.4, $[A]^{\mathcal{M}}(X)$ is the following pullback in the category $\mathbf{C\Omega - Set}$.

$$\begin{array}{ccc} [A]^{\mathcal{M}}(X) & \rightarrowtail & (\mathcal{D}(X), \alpha_0) \\ \downarrow & & \downarrow \langle t_{1,\mathcal{D},X}, \ldots, t_{n,\mathcal{D},X} \rangle \\ P_{\mathcal{M}} = (S_P, \alpha_P) & \hookrightarrow & (\Pi_P, \alpha_P). \end{array}$$

For simplicity we will write t_i instead of $t_{i,\mathcal{D},X}$, if possible. According to Theorem 7.3 on page 271 we have $[A]^{\mathcal{M}}(X) = (C, \alpha_0)$ where

$$C = \{\mathbf{a} \in \mathcal{D}(X) \mid \mathbf{b} = (t_1(\mathbf{a}), \ldots, t_n(\mathbf{a})) \in S_P, \alpha_0(\mathbf{a}, \mathbf{a}) = \alpha_P(\mathbf{b}, \mathbf{b})\}.$$

Then we have $C = [A]^{\mathcal{D}}(X)$ as can be easily proved.

(2) $A := (t_1 = t_2)$, where t_i are terms of the type ι, which is interpreted as (E, τ). Then $t_{i,\mathcal{D},X} : (\mathcal{D}(X), \alpha_0) \longrightarrow (E, \tau)$ is a morphism and it is clear that it is the same as $[t_i]^{\mathcal{M}}(X)$, constructed according to the results from Section 3.4. Then we have

$$\begin{aligned} (\forall \mathbf{a} \in \mathcal{D}(X)) \; (\mathcal{D}_X(A)(\mathbf{a}) &= \tau(t_{1,\mathcal{D},X}(\mathbf{a}), t_{2,\mathcal{D},X}(\mathbf{a}))), \\ [A]^{\mathcal{M}}(X) &= \{\mathbf{a} \in \mathcal{D}(X) \mid t_{1,\mathcal{D},X}(\mathbf{a}) = t_{2,\mathcal{D},X}(\mathbf{a})\}. \end{aligned}$$

Since $t_{i,\mathcal{D},X}$ are morphisms (according to Lemma 7.10), for $\mathbf{a} \in [A]^{\mathcal{M}}(X)$ we have

$$\alpha_0(\mathbf{a}, \mathbf{a}) = \tau(t_1(\mathbf{a}), t_1(\mathbf{a})) = \tau(t_1(\mathbf{a}), t_2(\mathbf{a})) = \mathcal{D}_X(A)(\mathbf{a}).$$

Conversely, for $\mathbf{a} \in [A]^{\mathcal{D}}(X)$ we have

$$\tau(t_1(\mathbf{a}), t_2(\mathbf{a})) = \tau(t_1(\mathbf{a}), t_1(\mathbf{a})) = \tau(t_2(\mathbf{a}), t_2(\mathbf{a})).$$

The result can be then obtained from the following simple lemma.

LEMMA 7.14
Let (A, α) be a complete Ω-set and let $a, b \in A$ be such that $\alpha(a, a) = \alpha(b, b) = \alpha(a, b)$. Then $a = b$.

The proof of this lemma follows from the fact that the singletons $\{a\}$ and $\{b\}$ are identical.

(3) $A := \bigwedge \Phi$. Then we have

$$[A]^{\mathcal{D}}(X) = \left\{ \mathbf{a} \in \mathcal{D}(X) \,\middle|\, \alpha_0(\mathbf{a}, \mathbf{a}) = \bigwedge_{B \in \Phi} \mathcal{D}_X(B)(\mathbf{a}) \right\},$$

$$[A]^{\mathcal{M}}(X) = \bigcap_{B \in \Phi} [B]^{\mathcal{M}}(X).$$

According to the induction assumption, we have $[B]^{\mathcal{D}}(X) = [B]^{\mathcal{M}}(X)$ from which the result immediately follows.

(4) $A := (\exists y)B$, where y is interpreted as (E, τ). Recall that the image of a subobject in the category $\mathbf{C\Omega} - \mathbf{Set}$ presented by the diagram

$$\begin{array}{ccc} (C,\delta) & \rightarrowtail\!\!\!\!\rightarrow & (A,\alpha) \\ \exists_f\downarrow & & \downarrow f \\ (D,\omega) & \rightarrowtail & (B,\beta), \end{array}$$

is defined in such a way that

$$D = \left\{ b \in B \;\middle|\; \beta(b,b) = \bigvee_{c \in C} \beta(f(c), b) \right\}.$$

Furthermore, we need the following simple lemma, the proof of which could be done induction on the complexity of formulas.

LEMMA 7.15
Let B be a formula, X a set of variables such that $FV(B) \subseteq B$ and let y be a variable such that $y \notin X$. Let $(\mathbf{a}, e) \in (\mathcal{D}(X), \alpha_0) \times (E, \tau)$. Then we have

$$\mathcal{D}_{X \cup \{y\}}(B)(\mathbf{a}, e) = \mathcal{D}_X(B_y[e])(\mathbf{a}).$$

Let us now consider the following diagram describing the interpretation of the formula A:

$$\begin{array}{ccc} [B]^{\mathcal{M}}(X \cup \{y\}) & \rightarrowtail & (\mathcal{D}(X \cup \{y\}, \alpha_y) \\ \exists_{p_y}\downarrow & & \downarrow p_y \\ [A]^{\mathcal{M}}(X) & \rightarrowtail & (\mathcal{D}(X), \alpha_0) \end{array}$$

where p_y is the projection. Then, according to the induction assumption and the previous lemma, we have

$$\begin{aligned} [B]^{\mathcal{D}}(X \cup \{y\}) &= \{(\mathbf{x}, e) \in \mathcal{D}(X) \times E \mid \alpha_0(\mathbf{x},\mathbf{x}) = \tau(e,e) = \\ &= \mathcal{D}_{X \cup \{y\}}(B)(\mathbf{x}, e) = \mathcal{D}_X(B_y[e])(\mathbf{x})\}. \end{aligned}$$

Let $\mathbf{a} \in [A]^{\mathcal{M}}(X)$. Then

$$\alpha_0(\mathbf{a},\mathbf{a}) = \bigvee_{(\mathbf{x},e) \in [B]^{\mathcal{D}}(X \cup \{y\})} \alpha_0(p_y(\mathbf{x}, e), \mathbf{a}).$$

For the element $\mathbf{b} = (\mathbf{x}, e) \in [B]^{\mathcal{D}}(X \cup \{y\})$, we have

$$\begin{aligned} \tau(e,e) &= \alpha_0(\mathbf{x},\mathbf{x}) = \mathcal{D}_X(B_y[e])(\mathbf{x}), \\ \mathcal{D}_X(B_y[e])(\mathbf{a}) &\geq \mathcal{D}_X(B_y[e])(\mathbf{x}), \end{aligned}$$

which follows that

$$\begin{aligned} \tau(e,e) \to \mathcal{D}_X(B_y[e])(\mathbf{a}) &\geq \tau(e,e) \to \mathcal{D}_X(B_y[e])(\mathbf{x}) \wedge \alpha_0(\mathbf{x},\mathbf{a}) = \\ &= (\tau(e,e) \to \mathcal{D}_X(B_y[e])(\mathbf{x})) \wedge (\tau(e,e) \to \alpha_0(\mathbf{x},\mathbf{a})) = \\ &= \tau(e,e) \to \alpha_0(\mathbf{x},\mathbf{a}) == \alpha_0(\mathbf{x},\mathbf{x}) \to \alpha_0(\mathbf{x},\mathbf{a}) \geq \\ &\geq \alpha_0(\mathbf{x},\mathbf{a}). \end{aligned}$$

Hence, we obtain

$$\bigvee_{r\in E}(\tau(r,r)\to \mathcal{D}_X(B_y[r])(\mathbf{a})) \geq \bigvee_{(\mathbf{x},r)\in[B](X\cup\{y\})}(\tau(r,r)\to \mathcal{D}_X(B_y[r])(\mathbf{a})) \geq$$

$$\geq \bigvee_{(\mathbf{x},r)\in[B](X\cup\{y\})} \alpha_0(\mathbf{x},\mathbf{a}) = \alpha_0(\mathbf{a},\mathbf{a}).$$

Therefore,

$$\mathcal{D}_X(A)(\mathbf{a}) = \alpha_0(\mathbf{a},\mathbf{a}) \wedge \bigvee_{r\in E}(\tau(r,r)\to \mathcal{D}_X(B_y[r])(\mathbf{a})) = \alpha_0(\mathbf{a},\mathbf{a}),$$

and $\mathbf{a}\in [A]^{\mathcal{D}}(X)$. The converse inclusion can be proved analogously.

The other constructions can be proved similarly. □

On the basis of the above theorem, the following theorem, which uses notation presented above, can be immediately derived.

THEOREM 7.7
Let $\Phi \Rightarrow \Psi$ be a Gentzen sequent. Then the following statements are equivalent.

(a) $\mathcal{M} \models \Phi \Rightarrow \Psi$,

(b) $\mathcal{D} \models \Phi \Rightarrow \Psi$.

Truth degrees of formulas interpreted in the category $\mathbf{C}\Omega - \mathbf{Set}$. In Section 4.1, we have introduced the notion of the truth degree of a formula A in a fuzzy theory T, $T \models_a A$. This notion has been defined only for the classical interpretation of logic with the language J within the category **Set**. In this subsection, we show that using a truth valuation of formulas in a the Heyting algebra Ω we can define the notion of the degree of truth value of a formula A in a fuzzy theory T, even in case when formulas are interpreted in the topos $\mathbf{C}\Omega - \mathbf{Set}$. Since for interpretation of formulas in this topos we have formally the same tools as for interpretation in sets, the definition of this relation can formally proceed analogously as in classical case.

Let us consider a many sorted fuzzy logic with the language J and let

$$\mathcal{D} = \langle\{(A_\iota,\alpha_\iota) \mid \iota\in\Upsilon\}, \{P_{\mathcal{D}} \mid P\in\mathcal{P}\}, \{f_{\mathcal{D}} \mid f\in\mathcal{F}\}\rangle,$$

be a structure for interpretation of J in the topos $\mathbf{C}\Omega-\mathbf{Set}$ where $\Omega = \langle L,\vee,\wedge\rangle$ is a complete Heyting algebra. Let F_J be the set of all the formulas of our logic. For any set X of variables, we denote by $F_{J,X}$ the set of all the formulas from F_J with the free variables contained in X. The structure $\mathcal{D}$ then defines a truth valuation $\|\cdot\|_{\mathcal{D},X} : F_{J,X} \longrightarrow L$ of formulas from $F_{J,X}$ such that

$$\|A\|_{\mathcal{D},X} = \bigvee_{\mathbf{a}\in\mathcal{D}(X)} \mathcal{D}_X(A)(\mathbf{a}).$$

for all $A\in F_{J,X}$. Using this valuation we can introduce the following definition.

DEFINITION 7.3

(i) *A function* $T : F_J \longrightarrow L$ *is called a fuzzy theory in* F_J.

(ii) *We say that a truth valuation* $\mathcal{D}_X$ *is a model of a fuzzy theory* T, *in symbols* $\mathcal{D}_X \models T$, *if*

$$\|A\|_{\mathcal{D},X} \geq T(A).$$

for all $A \in F_{J,X}$).

(iii) *Let* $A \in F_{J,X}$ *be a formula. Then we say that* A *is true in the degree* $a \in L$ *in the fuzzy theory* T, *in symbols* $T \models_a A$, *if*

$$a = \bigwedge\{\|A\|_{\mathcal{D},FV(A)} \mid \textit{for every structure } \mathcal{D} \quad \textit{such that} \quad \mathcal{D}_{FV(A)} \models T\}.$$

By using this definition a lot of the results presented in Section 4.1 for the category **Set** can be obtained also for the category **CΩ** − **Set** of fuzzy sets. For example, the following lemma holds (recall that for a topos $\mathcal{E}$ and a Gentzen sequent ϕ, $\models_{\mathcal{E}} \phi$ if $\mathcal{D} \models \phi$ for any $\mathcal{E}$-structure $\mathcal{D}$).

LEMMA 7.16

Let A *and* B *be formulas and let the following conditions hold:*

(a) $\models_{\mathbf{C\Omega-Set}} \Rightarrow (A \Rightarrow B)$,

(b) $\models_a A$

(c) $\models_b B$.

Then we have $a \leq b$.

PROOF: Let $\mathcal{D}$ be a structure for the language J in the topos **CΩ** − **Set** and let $\mathcal{M}$ be a **CΩ** − **Set**-structure associated with $\mathcal{D}$. Since $\mathcal{M} \models \Rightarrow (A \Rightarrow B)$, it follows from Theorem 7.7 that $\mathcal{D} \models \Rightarrow (A \Rightarrow B)$. Hence, for all $\mathbf{a} \in \mathcal{D}(X)$ we have

$$\alpha_0(\mathbf{a},\mathbf{a}) = \mathcal{D}_X(\emptyset) \leq \mathcal{D}_X(A \Rightarrow B)(\mathbf{a}) = \alpha_0(\mathbf{a},\mathbf{a}) \wedge (\mathcal{D}_X(A)(\mathbf{a}) \to \mathcal{D}_X(B)(\mathbf{a})),$$

which follows that

$$\mathcal{D}_X(A)(\mathbf{a}) = \mathcal{D}_X(A)(\mathbf{a}) \wedge \alpha_0(\mathbf{a},\mathbf{a}) \leq \mathcal{D}_X(A)(\mathbf{a}) \wedge (\mathcal{D}_X(A)(\mathbf{a}) \to \mathcal{D}_X(B)(\mathbf{a})) \leq$$
$$\leq \mathcal{D}_X(B)(\mathbf{a}).$$

Hence, $a \leq b$. □

8 FEW HISTORICAL AND CONCLUDING REMARKS

In this short chapter, we will briefly remind few moments from the history of fuzzy (and many-valued) logic and outline some of its problems for the future research. Since there are a lot of authors participating on its development, we are not able to cite all of them. Our selection is thus incomplete, influenced by our interest, the topic of this book and, of course, limited by our knowledge. Therefore, we apologize to all whose work has not been explicitly mentioned. Let us hope that an exhausting history of fuzzy and many-valued logic will be written in the future.

Beginning of many-valued logic. Probably the earliest, hidden mentions of many-valued logic can be found in [74]. However, the real beginning of its development can be assigned to J. Łukasiewicz who in [65] proposed his well known three-valued system. Shortly after him, also Post in [107] proposed a system of many-valued logic but unlike Łukasiewicz, he never continued his work. J. Łukasiewicz returned to the work on many-valued logic in thirties when he extended three-valued logic to finite and infinite-valued systems. He can be regarded as the founder of many-valued logic and influenced most workers following him. However, the development continued slowly.

Axiomatization of the Łukasiewicz three-valued system was proposed by M. Wajsberg in [131]. Several other important works appeared in fifties. Let us mention, e.g., the works by A. Rose, J. B. Rosser and A. R. Turquette [114, 115] where the latter became the standard book on many-valued logic for years. C. C. Chang has published two seminal papers [12, 13], in which he proved completeness of Łukasiewicz axioms and demonstrated that important theoretical results can be obtained by means of the deep algebraic study of structures of truth values. Besides other, he introduced the concept of MV-algebra. Since then, the work has speeded up. The book [113] summarizes many deep results and became the new standard of many-valued logic after the mentioned J. B. Rose and A. R. Turquette's one.

Beginning of fuzzy logic. The history of fuzzy logic roots to middle of sixties when the seminal paper [136] of L. A. Zadeh was published in 1965. He introduced the idea of the set with unsharp boundaries — the *fuzzy set*. Because every classical set can be unambiguously grasped using the characteristic function, the fuzzy set can be considered as its generalization, i.e., as a function from a universal set to some scale, which has originally been proposed to be the interval $[0,1]$. One year later, J. A. Goguen published the paper [31], in which he proposed to generalize this set to a lattice L. The connection with the many valued logic was first demonstrated also by the same author in the paper [32], which was an inspiration also for the subsequent development of the fuzzy logic in narrow sense presented in this book. Fuzzy and many-valued logics begun to be developed parallel and influenced each other.

Though not many papers have been published until the first half of the seventies, most of them became seminal for the future development. Let us mention, e.g., the papers by L. A. Zadeh [137, 138, 141, 139].

Probably most influential in that time, except for the mentioned L. A. Zadeh's papers, were the papers written by E. H. Mamdani and S. Assilian [69, 70] where the concept of fuzzy control appeared first time. They initiated the rapid development of fuzzy logic and later on, in the late eighties, caused the phenomenon called "fuzzy boom".

Other direction, which is sometimes also included into fuzzy logic, is the possibility theory, initiated again by L. A. Zadeh [142]. This theory became the basis of the *possibilistic logic*. The main contribution to the possibilistic logic has been given by D. Dubois and H. Prade (e.g. [23]). Let us stress that this logic deals with uncertainty rather than with vagueness. As uncertainty is not truth functional, connectives in possibilistic logic are not truth functional, as well. However, note that the early works on the possibility theory, the truth functionality has been kept (cf. [142]).

Formalization of fuzzy logic. The end of seventies and eighties can be characterized by the algebraic development of various aspects of fuzzy logic. There were also some attempts to formulate the resolution principle (e.g. [63, 120]). At the same time, various generalizations of classical logic and extensions of fuzzy one have been studied. Let us mention, e.g. the linguistic logic, i.e. the logic dealing with imprecisely expressed truth values [141], various models of approximate reasoning (e.g. [2]), or the linear logic (cf. [29]).

The breakthrough to fuzzy logic in narrow sense has been done by J. Pavelka [103]. His paper, devoted to propositional fuzzy logic, contains the definition of its evaluated syntax and semantics and it is finished by the proof of the completeness theorem. In addition to this, it also contains the metatheorem (Theorem 4.2 on page 105) stating that fuzzy logic with the implication connective, whose set of truth values forms the interval $[0, 1]$, can be syntactico-semantically complete only if the corresponding implication operation is continuous. Unfortunately, Pavelka's papers had been left almost unnoticed. In the end of the eighties, this branch of fuzzy logic has been extended to the first-order by V. Novák and the generalization of the Gödel completeness theorem has been proved in the paper [88]. The same author published in 1989 the book on fuzzy set theory [87], in which the unifying point of view based on the mentioned results in fuzzy logic has been taken.

Since the middle of eighties, we can recognize the rapidly growing interest in fuzzy logic. This was caused especially by the famous applications in Japan (washing machine, subway, camcoder, etc.). So far, several tens of books have been published and their number is still growing.

Fuzzy logic as a strict formal theory. Recent years can be characterized by the strong endeavour to establish fuzzy logic as a strict formal theory based on the deep algebraic results, especially in the theory of MV-algebras. This was especially thanks to the works of the Italian school led by A. diNola, D. Mundici, R. Cignolli and others, as well as to the Spanish school led by F. Esteva and L. Godo. The basic book summarizing most important results in the theory of MV-algebras is [84].

Besides the algebraic development, parallel deep mathematical research has been done in the theory of t-norms. Among many researchers in this area, let us mention E. P. Klement, R. Mesiar, E. Pap, J. Fodor or B. De Baets. The prepared monograph [56] will surely become one of the standard books on t-norms.

An outstanding role in the strict formalization of fuzzy logic and a thorough elaboration of its metatheory is played by the works of P. Hájek who summarized his point of view and many significant results in the monograph [41]. A lot of his metatheory is based on the highly abstract but very significant algebraic results obtained by U. Höhle (see the citations), as well as on the mentioned results in the theory of t-norms. The Höhle's endeavour is directed both to the algebra as well as to the category theory. The latter work is a part of a wider program, which is the formal development of fuzzy set theory.

Categorical aspects of fuzzy set theory. Fuzzy set theory and its logical aspects follow the abstract set theory development and its generalization very closely. One of the most important generalization of set theory has been done by A. Grothendieck and F. Giraud, and later F. W. Lawvere and M. Tierney who introduced the notions of topos and elementary topos and showed importance of these special categories for classical mathematical constructions generalizations well known in set theory. Topos theory approach enables us to give a categorical description of what is needed to do the constructions well known in ordinary mathematics, investigate the consequences of the categorical description and find interesting examples of categories with the needed properties. Hence, it is not surprising that many authors tried to follow the same categorical approach in ordinary fuzzy set theory.

Unfortunately, the situation in fuzzy set theory is more complicated and not so clearly connected with topos theory. In one of the first paper on categorical properties of fuzzy sets, M. Eytan [26] made the claim that a certain category whose objects are fuzzy sets with values in a complete Heyting algebra is a topos. Unfortunately, A. M. Pitts [105] proved that this is not true. His paper started still not finished period of searching appropriate category which can be thought as generalization of fuzzy set theory.

Several authors presented examples of categories of fuzzy sets, which seem to be (at least from some point of view) suitable for abstract fuzzy set theory development. Let us mention for example [31, 44, 109, 110]. Although these categories have many properties similar to those of fuzzy sets, none of them has generally been accepted as a "generalized fuzzy set", i.e. none of them could have the same position in fuzzy set theory as topoi have in ordinary set theory. The principal problem seems to be the internal logic of these categories, i.e. the logic based on the order structure of subobjects of these categories: The internal logic of any topos is based on the structure of Heyting algebra defined on these subobjects, while the internal logic of any category serving as generalization of fuzzy set theory has to be based probably on the more general structure, for example the MV-algebra. Recently, several attempts appeared mostly in the works by U. Höhle ([45, 46, 48, 49, 50]), which are based on even more general structures, called GL-monoids. However, the research of the abstract categorical background of fuzzy set theory is not yet finished since, as mentioned, there is no generally accepted type of the category playing the same role as topoi in ordinary set theory. Hence, the future research in category theory must be focused on searching of such category.

Some up-to-date problems of fuzzy logic. In Sections 4.1 and 4.2., we have proved (the above cited Theorem 4.2 and Lemma 4.3 on page 112) that we cannot have a complete fuzzy logic with implication, which is not continuous. This led us to the assumption that the algebra of truth values we work in forms the structure of the Łukasiewicz algebra $\mathcal{L}_{\text{Ł}}$. Such a conclusion is in accordance with the deliberation given in Section 4.2, namely that the vagueness phenomenon is continuous and thus, continuity of all the connectives is required. On the other hand, F. Esteva et al. in [25] posed a question, whether there exists a way to get completeness when including one or both of the residuations $\rightarrow_P$, $\rightarrow_G$ (cf. Examples 2.6 and 2.1 on page 17) as the interpretations of the additional implications $\Rightarrow_P$, $\Rightarrow_G$. A clue seems to accept an infinitary deduction rule. For $\Rightarrow_P$, such a deduction rule may look as follows:

$$\frac{\{\, ^{1}/A \Rightarrow_P \mathbf{a} \mid a \in L \setminus \{0\}\}}{^{1}/A \Rightarrow_P \bot}.$$

This rule overcomes the discontinuity of $\rightarrow_P$ in the point $(0, 0)$.

Even worse situation can be encountered with the implication $\Rightarrow_G$ since the corresponding residuation $\rightarrow_G$ is discontinuous in an infinite number of points.

A possible solution have been given in [41] where a more complex infinitary deduction rule

$$\frac{\{\,\mathbf{1}/A \Rightarrow_G \mathbf{a},\ \mathbf{1}/\mathbf{b} \Rightarrow_G B \mid b < c < a, b, a \in \mathbb{Q} \cap [0,1], c \in [0,1)\}}{\mathbf{1}/A \Rightarrow_P B}.$$

has been introduced. Using it, the proof of the completeness theorem for the logic (called TT∀) which contains all three conjunctions ($\&, \wedge, \bullet$; the latter is interpreted by the ordinary product) and all three corresponding implications $\Rightarrow, \Rightarrow_G, \Rightarrow_P$ has been given.

Another problem, mentioned in Chapter 6, is introduction of the linguistic quantifiers and the appropriate inference schemes. The development of the mathematical background is still not finished, though progressed (cf. [41, 79]). However, its elaboration based on the linguistic theory has not yet been done.

A problem considered by L. A. Zadeh as a part of the fuzzy logic agenda (cf. page 9) is defuzzification. We have touched it in Section 5.4 from the functional point of view. Its elaboration inside the formalism of fuzzy logic is quite difficult since it requires the weighted average operation, which has no formal counterpart in the language. From our point of view, the defuzzification means finding a typical term "representing" the multiformula $\mathbf{A}_{\langle x \rangle}$. Till now, this is done directly only in models since the multiformulas $\mathbf{A}_{\langle x \rangle}$ represent fuzzy sets and thus, we may use the membership degrees to find the required element. Of course, additional assumptions on the membership function must be imposed, e.g. measurability, linear ordering of the support, etc. In the formalism of fuzzy logic, the required element should be represented by some term. However, it is unclear, what typically logical conditions should be imposed to find such a term. A very general idea is to find a crisp formula $\overline{A}$ (i.e. a formula such that $T \vdash \overline{A}(x) \vee \neg \overline{A}(x)$) such that $T \vdash \overline{A}_x[t_0]$. This task is not solved yet.

We hope to make the reader convinced that fuzzy logic has already found its place in mathematics and that, despite scepticism of mathematicians accompanying it for many years, it offers a lot of interesting and well established problems to be solved in the future.

References

[1] Anderson, C. A. (1984). "General intensional logic," in Gabbay, D. and F. Guenthner (eds.), *Handbook of Philosophical Logic, Vol. II*, D. Reidel: Dordrecht, 355–385.

[2] Appelbaum, L. and Ruspini E. H. (1985). "ARIES an approximate reasoning inference engine," in: Gupta M. M. et al. (eds.), *Approximate Reasoning in Expert Systems*, Amsterdam: North Holland, 745–765.

[3] Barr, M. and Wells, C. (1984). *Toposes, Triples and Theories*. Berlin: Springer Verlag.

[4] Bellman, R. E. and Zadeh L. A. (1975). "Local and fuzzy logics." In: Dunn, J. M. and Epstein, G. eds. *Modern Uses of Multiple-Valued Logic*. Boston: D. Reidel, 105–165.

[5] Bělohlávek, R. and V. Novák: Learning Rule Base of the Linguistic Expert Systems. Int. J. of Soft Computing (submitted).

[6] Biacino, L. and Gerla, G. (1986). "Weak Decidability and Weak Recursive Enumerability for L-subsets," Preprint No. 45, Università Degli Studi di Napoli, Napoli.

[7] Biacino, L. and Gerla, G. (1988). "Recursively Enumerable L-sets," Zeit. Math. Logic. Grundl. Math.

[8] Barwise, J. (ed.) (1977). *Handbook of Mathematical Logic*, Amsterdam: North-Holland.

[9] Birkhoff, G. (1967) *Lattice Theory*. Amer. Math. Soc. Providence: R. I.

[10] Black , M. (1937). "Vagueness: An Exercise in Logical Analysis," Philosophy of Science **4**, 427–455. Reprinted in Int. J. of General Systems **17**(1990), 107–128.

[11] Butnariu, D. and Klement, E. P. (1993). *Triangular Norm-based Measures and Games with Fuzzy Coalitions*. Dordrecht: Kluwer.

[12] Chang, C. C. (1958). "Algebraic analysis of many valued logics," Trans. AMS, **93**, 74–80.

[13] Chang, C. C. (1959). "A new proof of the completeness of the Łukasiewicz axioms," Trans. AMS, **88**, 467–490.

[14] Chang, C. C. and Keisler, H. J. (1966). *Continuous Model Theory*, Princeton: Princeton University Press.

[15] Chang, C. C. and Keisler, H. J. (1973). *Model Theory*, Amsterdam: North–Holland.

[16] Cohen, P. M. (1965), *Universal algebra*, New York: Harper & Row.

[17] van Dalen, D. (1994). *Logic and Structure*, Berlin: Springer.

[18] Dilworth, R. P. (1938). "Abstract residuation over lattices," Trans. Amer. Math. Soc. **44**, 262–268.

[19] Dilworth, R. P. and Ward, M. (1939) "Residuated lattics," Trans. Amer. Math. Soc. **45**, 335–354.

[20] diNola, A. (1993). "MV-Algebras in the Treatment of Uncertainty." In: P. Löwen, M. Roubens, eds. *Proc. of the Int. IFSA Congress.* Doredrecht: Kluwer, 123–131.

[21] di Nola, A. and Gerla, G. (1986). "Fuzzy models of first order languages." Zeitschr. f. math. Logik und Grundlagen d. Math. **32**, 331–340.

[22] Dubois D. and Prade, H. (1980). *Fuzzy Sets and Systems: Theory and Applications.* New York:Academic Press.

[23] Dubois, D. and Prade, H. (1990). "Fuzzy sets in approximate reasoning — Part 1: Inference with possibility distributions," Fuzzy Sets and Systems **40**, 143–202.

[24] Dubois, D., Lang, J. and Prade, H. (1991). "Fuzzy sets in approximate reasoning — Part 2: Logical approaches," Fuzzy Sets and Systems **40**, 203–244.

[25] Esteva, F., Godo, L., Hájek, P. and Navara, M. (1999). "Residuated fuzzy logics with an involutive negation," Archive for Mathematical Logic. (to appear).

[26] Eytan, M. (1981). "Fuzzy sets: a topos-logical point of view," Fuzzy Sets and Systems **5**,47-67.

[27] Gallin, D. (1975). *Intensional and Higher-order Modal Logic.* Amsterdam: North Holland.

[28] Gerla, G. (1994). "Comparing fuzzy and crisp deduction systems," Fuzzy Sets and Systems **67**, 317–328.

[29] Girard J. Y. (1987). "Linear logic," Theor. Comp. Sci., **50**, 1–102.

[30] Glushankof, D. (1992). "Prime deductive systems and injective objects in the algebras of Lukasiewicz infinite-valued calculi," Algebra Universalis **29**, 354–377.

[31] Goguen, J. A. (1967). "L-fuzzy sets," J. of Math. Anal. Applic. **18**, 145–174.

[32] Goguen, J. A. (1968-69). "The logic of inexact concepts," Synthese **19**, 325–373.

[33] Goldblatt, R. (1979). *TOPOI, The categorical analysis of logic.* Amsterdam: North-Holland.

[34] Gottwald, S. (1989). *Mehrwertige Logik.* Berlin: Akademie-Verlag.

[35] Gottwald, S. (1993).*Fuzzy Sets and Fuzzy Logic.* Wiesbaden: Vieweg.

[36] Gottwald, S. and Novák, V. (1997). "On the Consistency of Fuzzy Theories," *Proc. of VIIth IFSA World Congress, Prague.* Prague: Academia, 168–171.

[37] Grätzer, G. (1978). *General Lattice Theory.* Berlin: Akademie-Verlag.

[38] Hájek P. (1995). "Fuzzy logic and arithmetical hierarchy," Fuzzy Sets and Systems **73**, 359–363.

[39] Hájek, P. (1996). "Fuzzy logic as logic." In: G. Coletti et al., ed.: *Mathematical Models of Handling Partial Knowledge in Artificial Intelligence.* Oxford: Pergamon Press.

[40] Hájek P. (1997). "Fuzzy logic and arithmetical hierarchy II," Studia Logica **58**, 129–141.

[41] Hájek, P. (1998). *Metamathematics of fuzzy logic.* Dordrecht: Kluwer.

[42] Hájek P., Esteva, F. and Godo, L/ (1995). "Fuzzy logic and probability," in P. Besnard and S. Hanks, eds., Proc. 11th Conf. on Uncertainty and Artificial Intelligence Montreal, 237–244.

[43] Hájek, P, Paris, J. and Shepherdson, J. (1998). The liar paradox and fuzzy logic. (to appear)

[44] Higgs, D. (1973). *A category approach to Boolean/valued set theory.* Manuscript:University of Waterloo.

[45] Höhle, U. (1991). "Monoidal closed categories, weak topoi and generalized logics," Fuzzy Sets and Systems **42**,15-35.

[46] Höhle, U. (1992). "M-Valued Sets and Sheaves over Integral Commutative CL-Monoids," in Rodabaugh, S. E, Klement, E. P., Höhle, U., eds., *Applications of Category to Fuzzy Subsets.* Dordrecht: Kluwer, 33–72.

[47] Höhle, U. (1995). "Monoidal logic." In: Kruse, R., ed. *Foundations in Fuzzy Systems.* Wiesbaden: Vieweg.

[48] Höhle, U. (1995). "Commutative residuated l-monoids," in [51], 53–106.

[49] Höhle, U. (1995). "Presheaves over GL-monoids," in [51], 127–157.

[50] Höhle, U. (1999). "Classification of Subsheaves over GL-algebras," Proc. of Logic Colloquium 98 Prague. Berlin: Springer Verlag.

[51] Höhle, U. and Klement,E. P. (eds.) (1995), *Non-Classical Logics and Their Applications to Fuzzy Subsets. A Handbook of the Mathematical Foundations of Fuzzy Set Theory.* Dordrecht: Kluwer.

[52] Johnstone, P.T. (1977). *Topos Theory.* London: Academic Press.

[53] Kelley, J. L. (1957). *General Topology.* Princeton, New Jersey: Van Nostrand.

[54] Klawonn, F. and Kruse, R. (1993), "Equality Relations as a Basis for Fuzzy Control," Fuzzy Sets and Systems **54**, 147–156.

[55] Klawonn, F. and Novák, V. (1996). "The Relation between Inference and Interpolation in the Framework of Fuzzy Systems," Fuzzy Sets and Systems **81**, 331–354.

[56] Klement, E. P., Mesiar, R. and Pap, E. *Triangular Norms.* Monograph in preparation.

[57] Klir, G.J. and Yuan, B. (1995). *Fuzzy Sets and Fuzzy Logic: Theory and Applications.* New York: Prentice-Hall.

[58] Kruse, R., Gebhardt, J. and Klawonn, F. (1994). *Foundations of Fuzzy Systems*, Chichester: Wiley.

[59] Kulka, J. and Novák, V. (1984). "Have fuzzy operations a psychological correspondence?" Studia psychologica **26**, 131–141.

[60] Kuz'min, V. B. (1985). "About semantical structure of linguistic hedges:an experimental hypothesis," BUSEFAL **24**, 118 - 125, Université Paul Sabatier, Toulouse.

[61] Lakoff, G. (1973). "Hedges: A study in meaning criteria and logic of fuzzy concepts," J. Philos. Logic **2**, 458–508.

[62] Lawvere, F.W., ed. (1971). *Toposes, Algebraic Geometry and Logic.* Berlin: Springer Verlag.

[63] Lee R. C. T. (1972). "Fuzzy logic and the resolution principle,"J. Assoc. Comput. Mach., **19**, 109–119.

[64] Lehmke, S. (1995). *On Resolution-Based Theorem Proving in Propositional Fuzzy Logic with 'Bold' Connectives.* Diploma thesis. University of Dortmund.

[65] Łukasiewicz J. (1920). "O logice tròjwartościowej (On three-valued logic)," Ruch Filozoficzny, **5**, 170–171.

[66] Łukasiewicz, J. (1970). *Selected works* (col. by L. Borkowski). Amsterdam.

[67] Mac Lane, S. and Moerdijk, I. (1992). *Sheaves in Geometry and Logic, A First Introduction to Topos Theory.* Berlin: Springer Verlag.

[68] Makkai, M. and Reyes, E.G. (1977). *First Order Categorical Logic.* Berlin: Springer Verlag.

[69] Mamdani, E.H. (1974). "Applications of fuzzy algorithms for simple dynamic plant," Proc. IEE **121**, 1585–1588.

[70] Mamdani E.H. and Assilian, S. (1975). "An experiment in linguistic synthesis with a fuzzy logic controller," Int. J. of Man-Machine Studies, **7**, 1–13.

[71] Marks-II, R. J. (1994). *Fuzzy logic technology and applications.* IEEE Technological Activities Board.

[72] Mareš, M. (1994). *Computation over fuzzy quantities.* Boca Raton: CRC Press.

[73] McLarty, C. (1995). *Elementary Categories, Elementary Toposes.* Oxford: Oxford Science Publications, Clarendon Press.

[74] McColl H. (1897). "Symbolic reasoning. II," Mind, N.S., **6**, 493–510.

[75] McNaughton, R. (1951). "A theorem about infinite-valued sentential logic," J. Symb. Log., **16**, 1–13.

[76] Mendelson, E. (1979). *Introduction to Mathematical Logic.* New York: D. Van Nostrand.

[77] Menger, K. (1942). "Statistical Metrics", Proc. Nat. Acad. Sci. **28**, 535–537.

[78] Mesiar, R. and V. Novák (1977). "On Fitting Operations," Proc. of VIIth IFSA World Congress, Prague: Academia, 286–290.

[79] Mesiar R. and and H. Thiele (1999). "On T-quantifiers and S-quantifiers," in: Novák V. and I. Perfilieva (eds.), *Discovering the World with Fuzzy Logic.* Heidelberg: Springer (to appear).

[80] Mundici, D. (1986). "Interpretation of AF C^*-Algebras in Łukasiewicz Sentential Calculus," J. of Functional Analysis **65**, 15–63.

[81] Mundici, D. (1993). "Ulam games, Łukasiewicz logic, and AF C^*-algebras." Fund. Inf. **18**, 151–161.

[82] Mundici, D. (1993). "An Elementary Proof of Chang's Completeness Theorem for the Infinite-valued Calculus of Łukasiewicz." Studia Logica **58**, 79–97.

[83] Mundici, D. (1994). "A constructive proof of McNaughton's theorem in infinite-valued logic." J. of Symbolic Logic **59**, 596–602.

[84] Mundici, D., Cignoli, R. and D'Ottaviano, I.M.L. (1995). *Algebras of Lukasiewicz Logics*, (in Portuguese), Collection CLE, vol. 12, UNICAMP, Campinas, San Paulo, Brazil, (1994) 256 pages. Second edition, 1995. Expanded Edition in English, entitled "Algebraic Foundations of many-valued Reasoning", in preparation.

[85] Negoita C. V. and Ralescu, D. A. (1975). *Application of Fuzzy Sets to System Analysis.* Stuttgart: Birkhauser.

[86] Novák, V. (1989). "Sorites-like First–order Fuzzy Theories.," Proc. of IIIrd IFSA Congress, Seattle.

[87] Novák, V. (1989). *Fuzzy Sets and Their Applications.* Bristol: Adam Hilger.

[88] Novák, V. (1990). "On the syntactico–semantical completeness of first–order fuzzy logic. Part I, II." Kybernetika **26**, 47–66; 134–154.

[89] Novák, V. (1992). *The Alternative Mathematical Model of Linguistic Semantics and Pragmatics.* New York: Plenum.

[90] Novák, V. (1992). "Fuzzy Logic As a Basis of Approximate Reasoning," in: Zadeh, L. A., Kacprzyk, J. (eds.), *Fuzzy Logic for the Management of Uncertainty.* New York: J. Wiley & Sons.

[91] Novák, V. (1992). "On the Logical Basis of Approximate Reasoning," in: Novák, V., Ramík, J., Mareš, M., Černý, M., Nekola, J. (eds.), *Fuzzy Approach to Reasoning and Decision–Making.* Dordrecht: Kluwer; Praha: Academia, 17–27.

[92] Novák, V. (1994). "Fuzzy Control from the Point of View of Fuzzy Logic," Fuzzy Sets and Systems **66**, 159–173.

[93] Novák, V. (1995). "A New Proof of Completeness of Fuzzy Logic and Some Conclusions for Approximate Reasoning," Proc. Int. Conference FUZZ-IEEE/IFES'95, Yokohama, 1461–1468.

[94] Novák, V. (1995), "Ultraproduct Theorem and Recursive properties of Fuzzy Logic,"in [51], 341–370.

[95] Novák, V. (1995). "Linguistically Oriented Fuzzy Logic Controller and Its Design," Int. J. of Approximate Reasoning 1995, **12**, 263–277.

[96] Novák, V. (1995). "Towards Formalized Integrated Theory of Fuzzy Logic," in: Bien, Z, and K. Min (eds.), *Fuzzy Logic and Its Applications to*

Engineering, Information Sciences, and Intelligent Systems, Dordrecht: Kluwer, 353–363.

[97] Novák, V. (1996). "On the Hilbert-Ackermann Theorem in Fuzzy Logic," Acta Mathematica et Informatica Universitatis Ostraviensis **4**, 57–74.

[98] Novák, V: *Open Theories, Consistency and Related Results in Fuzzy Logic.* Int. J. of Approximate Reasoning **18**, 191–200.

[99] Novák, V. and Perfilieva, I. (1997). "On Logical and Algebraic Foundations of Approximate Reasoning," Proc. FUZZ-IEEE'97, Barcelona, 693–698.

[100] Novák, V. and Perfilieva, I. (1997). "On Model Theory in Fuzzy Logic in Broader Sense," Proc. Int. Conference EUFIT'97, Verlag Mainz, Aachen, 143–147.

[101] Novák, V. and Perfilieva, I. (1999). "Evaluating Linguistic Expressions and Functional Fuzzy Theories in Fuzzy Logic," in: Zadeh, L. A., Kacprzyk, J. (eds.), *Computing with Words in Systems Analysis.* Heidelberg: Springer Verlag (to appear).

[102] Parikh, R. (1971). "Existence and Feasibility in Arithmetics," J. of Symb. Logic., **36**, 494–508.

[103] Pavelka, J. (1979). "On fuzzy logic I, II, III," Zeit. Math. Logic. Grundl. Math. **25**, 45–52; 119–134; 447–464.

[104] Perfilieva I. and Tonis, A. (1995). "Functional System in Fuzzy Logic Formal Theory," BUSEFAL, **64**, 42–50.

[105] Pitts, A.M. (1982) "Fuzzy sets do not form a topos," Fuzzy Sets and Systems **8**,101-104.

[106] Ponasse, D. (1988). "Categorical studies of fuzzy sets," Fuzzy Sets and Systems **28**,235-244.

[107] Post E. L. (1921). "Introduction to a general theory of elementary propositions," American J. of Math., **43**, 163–185.

[108] Powell, M. J. D. (1980). *Approximation theory and methods.* Cambridge: Cambridge university press.

[109] Pultr,A. (1976). "Fuzzy mappings and fuzzy sets," Comment. Mat. Univ. Carolin., **17**(3),

[110] Pultr, A. (1976). "Closed categories of L-fuzzy sets," Vortrage zur Automaten und Algorithmentheorie, TU Dresden.

[111] Pultr, A. (1984). "Fuzziness and Fuzzy Equality," in: Skala, H.J., Termini, S. and Trillas, E. (eds.) *Aspects of Vaguenes.* Dordecht:D.Reidel.

[112] Rasiowa, H. and Sikorski, R. (1963). *The Mathematics of Metamathematics.* Warszawa: PWN.

[113] Rescher N. (1969). *Many-Valued Logic.* New York: McGraw Hill.

[114] Rose, A. and Rosser, J. B. (1958). "Fragments of many-valued statement calculi," Trans. A. M. S. **87**, 1–53.

[115] Rosser, J. B. and Turquette, A. R. (1952). *Many-Valued Logics.* Amsterdam 1952.

[116] Rosenfeld, A. (1971). "Fuzzy groups," J. Math. Anal. Applic. **35**, 512–517.

[117] Russel, B. (1923). "Vagueness," Australian J. Phi. **1**, 84–92.

[118] Scarpellini, B. (1962). "Die Nichaxiomatisierbarkeit des unendlichwertigen Prädikatenkalküls von Łukasiewicz," J. of Symbolic Logic **27**, 159–170.

[119] Sgall, P., Hajičová, E. and Panevová, J. (1986). *The Meaning of the Sentence in Its Syntactic and Pragmatic Aspects.* D. Reidel: Dordrecht.

[120] Shen Z. L., L. Ding and Mukaidono M. (1988). Fuzzy resolution principle. *Proc. 8th Int. Symp. on Multiple-valued Logic*, Palma de Mallorca, Spain, 210–214.

[121] Shoenfield, J. R. (1967). *Mathematical Logic.* New York: Addison–Wesley.

[122] B. Schweizer and Sklar, A. (1983). *Probabilistic metric spaces.* New York: North Holland.

[123] Stout, L.N. (1991). "A survey of fuzzy set and topos theory," Fuzzy Sets and Systems **42**,3-14.

[124] A. Tarski (1956). *Logic, semantics, metamathematics.* Oxford: Oxford University Press.

[125] Takeuti, G. and Titani, S. (1992). "Fuzzy logic and fuzzy set theory," Arch. Math. Logic, **32**, 1–32.

[126] Trillas, E. and Valverde, L. (1985). "On the Implication of Indistinguishability in the Setting of Fuzzy Logic," in: Kacprzyk, J. and Yager, R. R., eds., *Management Decision Support Systems Using Fuzzy Sets and Possibility Theory*, Köln: Verlag TÜV Rheinland, 198–212.

[127] Turunen, E. (1992). "Algebraic Structures in Fuzzy Logic," Fuzzy Sets and Systems **52**, 181–188.

[128] Turunen, E. (1995). "Well-defined Fuzzy Logic. Fuzzy Sentential Logic," Math. Logic Quaterly **41**, 236–248.

[129] Vopěnka, P. (1979). *Mathematics In the Alternative Set Theory.* Leipzig: Teubner.

[130] Vopěnka P. (1990). *Fundamentals of the Mathematics In the Alternative Set Theory.* Bratislava: Alfa (in Slovak).

[131] Wajsberg M. (1931). "Aksjomatyzacja trójwartosciowego rachunku zdań (Axiomatization of three-valued calculus)," Société des Sciences et Letters Varsovie, cl. III, **24**, 126–148.

[132] Wyler, O. (1995). "Fuzzy logic and categories of fuzzy sets," [51], 235–268.

[133] Ying, M. (1991), "Deduction theorem for many-valued inference," Z. Math. Logic Grundlagen Math. **37**, 533–537.

[134] Ying, M. (1992), "The fundamental theorem of ultraproduct in Pavelka's logic," Z. Math. Logic Grundlagen Math. **38**, 197–201.

[135] Ying, M. (1992), "Compactness, the Loöwenheim-Skolem property and the direct product of lattices of truth values," Z. Math. Logic Grundlagen Math. **38**, 521–524.

[136] Zadeh, L. A. (1965). "Fuzzy Sets," Inf. Control. **8**, 338–353.

[137] Zadeh, L.A. (1971). "Similarity relations and fuzzy orederings," Inf. Sci. **3**, 177–200.

[138] Zadeh, L.A. (1973). "Quantitative Fuzzy Semantics," Inf.Sci. **3**, 159–176.

[139] Zadeh, L.A. (1973). "Outline of a New Approach to the Analysis of Complex Systems and Decision Processes," IEEE Trans. Syst. Man and Cybern, **1**, 28–44.

[140] Zadeh, L.A. (1975). "Fuzzy logic and approximate reasoning," Synthese **30**, 407–428.

[141] Zadeh, L.A. (1975). "The concept of a linguistic variable and its application to approximate reasoning I, II, III," Inf. Sci., **8**, 199–257, 301–357; **9**, 43–80.

[142] Zadeh, L.A. (1978). "Fuzzy Sets as a Basis for a Theory of Possibility," Fuzzy Sets And Systems," **1**, 3–28.

[143] Zadeh, L.A. (1983) "A computational approach to fuzzy quantifiers in natural languages," Comp. Math. with Applic. **9**, 149-184.

Index

A

B

C

D

E

F

M

N

O

P

Q

R

S

T

U

V

W